Plant Breeding for Water-Limited Environments

Abraham Blum

Plant Breeding for Water-Limited Environments

Springer

Abraham Blum
Tel Aviv
Israel
ablum@plantstress.com

ISBN 978-1-4419-7490-7 e-ISBN 978-1-4419-7491-4
DOI 10.1007/978-1-4419-7491-4
Springer New York Dordrecht Heidelberg London

Library of Congress Control Number: 2010937416

Printed on acid-free paper

Springer is part of Springer Science+Business Media (www.springer.com)

In memory of our daughter Orly who combined science with love for every living thing.

Preface

In my previous book entitled *Plant Breeding for Stress Environments* (Blum 1988) plant breeding for water-limited environments was only one chapter together with other chapters on heat, cold, mineral deficiency, mineral toxicity and salinity stress. Since that publication several major developments took place regarding plant breeding for stress environments.

Firstly, plant molecular biology emerged as a major avenue of research in plant biology and crop science. Plant molecular biology and genomics research towards plant environmental stress has grown exponentially since 1988. The number of published scientific papers specifically on the molecular biology of drought stress and drought resistance increased five-fold from about 10 per year in 1990 to at least 53 per year in 2009. The major dilemma now is how to apply molecular biology and genomics to breeding for stress conditions and specifically for water limited environments. While molecular biologists underline the potential of their discipline and whereas plant breeders underline their needs, there is too much fog lying between the two – which must be cleared up. This book offers to help.

Secondly, the plant breeding community realizes now that plant stress must be addressed in plant breeding programs in a dedicated and specific manner. Selecting for yield in diverse environments is not sufficient anymore for coping with stress problems. This approach also became too costly. There is now a growing understanding that breeding for stress conditions and specifically for water limited environments requires specific components within the general breeding program to the same extent that biotic stress resistance is approached. Qualified information on how to design and perform a breeding program component for water limited environments is in great demand.

Thirdly, concerns about the agricultural implications of climate change grew exponentially since the early 1990s to the extent that the phenomenon is now being aggressively addressed in agricultural research. Even the optimists agree that climate change cannot be completely reversed and at best it could only be slowed. The impact on crop production can already be seen by increased aridity and warmer temperatures in some regions and vicious storms and floods elsewhere. Global warming is an additional serious engine of plant stress which requires specific expertise towards breeding solutions.

Fourthly, university professors who teach plant breeding may find themselves now under increasing pressure to address abiotic stress and drought stress in their classes. It should not come as a surprise that expertise in plant stress is in demand. In a recent GCP survey among plant breeders who work towards water-limited environments (Mahalakshmi and Blum 2006), most were found to be seriously at loss on how to integrate drought resistance breeding into their program. Most were not certain about protocols for drought phenotyping and the selection methods to use. In short, many felt unqualified to deal with breeding for drought resistance and water-limited environments. Most of these breeders are the product of education in the last decade or two. Education in this plant breeding area is therefore crucial. Occasional international courses on this subject are being offered, but the basic university plant breeding curriculum still requires an upgrade in this respect.

Considering all of the above, a dedicated book on applied plant breeding for water-limited environments is deemed very timely, in demand and worth the effort. Other stresses are also important but urgency with respect to drought stress seems to be an overriding consideration. Since 2000 I have been trying to address developments in other environmental stresses (including drought) in my web site at www.plantstress.com. This is a dynamic web site with constant update of information and it should be of value to practitioners and students. In some respects it can be taken as a dynamic revision of my 1988 book, with the input of many dedicated experts.

The huge volume of information developed since 1988 on the subject of plant water-relations, drought stress and drought resistance posed a serious problem towards writing this book. This information had to be carefully scanned and sifted throughout in order to select only the relevant pieces useful for application towards plant breeding. This process was quite objective and depended on my own judgment. Furthermore, the discussion of background and basic information in plant physiology was also limited to what I considered as absolutely essential for the reader understanding of the breeding issues. This text is not the place for a comprehensive review and discussion of all information available in plant physiology as related to water in plants.

Plant breeding is a complex and costly program. Breeders are under constant pressure to deliver a new product quickly. This product must withstand competition from other programs and be accepted by the client. Rewards are not easy to come by. In public institutions the breeder often faces a situation where his/her work is evaluated by the number of publications while he/she was busy producing cultivars rather than papers. Not all public systems are enlightened enough to allow for that.

Very often the practicing breeder does not have the time or the patience for extensive reading and in these fast internet times a quick fix is what is being often sought. I am sorry to say to this reader that this book still requires time, patience and an open mind in order to achieve the gain it is intended to provide. At the same time effort is made to avoid an overload of unnecessary information which might be evident, readily available or useless to the practitioner and the student. This book also assumes that the reader already has basic knowledge in plant breeding, genetics and general plant physiology. It therefore tries to build upon that knowledge.

The reader will notice that the narrative of this book contains numerous references to "old" literature dating back as far as the 1960s. This is not a matter of nostalgia nor is it an indication that this author is not up to date on his reading. It is my deliberate intention, especially for the young generation, to point out that some hard facts that underlie much of the very contemporary knowledge in plant science and agronomy is based on very contemporary "old" literature. This might also help the emerging scientist to avoid a waste of time by re-inventing the whcel.

Plant breeding for water limited environments has been receiving increased attention in recent years as evidence by many conferences, workshops and literature reviews cited throughout this book. Rather than being another literature review, this book can be considered as a science-based breeder's manual. It describes detailed methods and protocols to be used in planning and executing a breeding program for water-limited environments, as supported by solid scientific evidence and discussion. However, the manual should not be used without understanding the principles, problem and pitfalls as discussed in the background chapters. Plant breeders are known to be resistant to change – for a good reason. A misguided change introduced into their program might hamper the program for years to come. Thus breeders justly tend to walk on the well-trodden path. The purpose of this book is to inspire the breeder to makes changes but at the same time assume very careful evaluation before adopting change.

I close this preface with the eloquent statement by Robert Gauss, plant breeder (1910):

> The dominion of climate is invincible. All who come within its range must obey its laws. It grants no pardon. It makes no compromise. Compliance with the conditions imposed is the license to exist and these conditions determine the limits of activity...Let no one look with indifference upon the possibility here outlined of acclimatizing valuable crop species to arid regions or underestimate the magnitude of the achievement suggested... So vast an achievement would rank with the discovery of new continent in its enlargement of the sources of human subsistence; and well might the hope of success quicken the activities of the most sluggish and awaken ambition in the least daring.

Tel Aviv, Israel Abraham Blum

Acknowledgments While this book is based on numerous literature citations it is still unavoidable that few authors may not have been appropriately or fully recognized. My apologies and acknowledgement is extended to all. I am in debt to the many excellent scientists, practitioners and farmers whom I was fortunate to associate with during my career. This work could not have been produced without these fertile interactions in the field, the laboratory and the meeting hall. Finally, I am in a loving gratitude to my wife Dvora who patiently accepted my long hours in front of the computer and long days away from home.

References

Blum A (1988) Plant breeding for stress environments. CRC Press, Boca Raton
Gauss R (1910) Acclimatization in breeding drought-resistant cereals. Am Breed Mag 1:209–217
Mahalakshmi V, Blum A (2006) Phenotyping in the field: global capacity accessible to the GCP - Inventory of phenotyping resources and capacity for the GCP. Final report, Generation Challenge Program, El Batan

Contents

Chapter 1
The Moisture Environment

Summary Drought as an environmental factor in crop production is not taken in its catastrophic connotation. Catastrophic droughts as a problem in agriculture are a dealt with at the political, engineering or the economical levels. For the plant breeding program drought is defined as insufficient moisture supply which causes a reduction in plant production. It is the gap between crop demand for water and the supply of water. The agricultural drought on the larger scale of the region is approached at the meteorological level where the Palmer Drought Severity Index is the classical estimate. It is based on precipitations and temperature and is useful only on a longer time span such as months or weeks.

At the field level estimates of drought which affect crop production must be more precise and resolute. Here the major approach to estimates has been developed by classical accounting the energy balance of the crop. The account considers the incoming energy load on the crop brought about by solar radiation and the dissipation of this energy mainly by reflection, heat and evaporation of water (see also Chap. 2). Thus evapotranspiration and crop water use can be estimated and crop water deficit assessed in the field even on a short time scale of days or hours. The crop energy balance is driven by both environmental and crop factors. Crop factors can be genetically manipulated by plant breeding and therefore they constitute a prime subject of interest in this book.

The moisture environment in terms of its insufficiency for human and ecological sustainability can be described on several levels. Certainly catastrophic droughts were described since biblical times, but they are beyond the possible solutions through plant science and agronomy. Political, economical, engineering and social solutions are needed to cope with the catastrophe and its outcome. Where plant science is concerned, three types of drought are recognized, each involving a different scientific domain.

"*Meteorological drought*" is brought about when there is a prolonged period with less than average precipitation. Meteorological drought usually precedes the other kinds of drought.

"*Hydrological drought*" is brought about when the water reserves available in sources such as aquifers, lakes, and reservoirs fall below the statistical average.

A. Blum, *Plant Breeding for Water-Limited Environments*,
DOI 10.1007/978-1-4419-7491-4_1,

This condition can arise, even in times of average (or above average) precipitation, when increased usage of water diminishes the reserves. For example, recent discussions in the scientific and popular media describe the hydrological drought in Northern China where the demand for irrigation water exceeds the supply by the available water resources. This negative balance poses a serious threat to the future of farming in the region.

"*Agricultural drought*" is brought about when there is insufficient moisture for maximum or potential growth of crops, range or plantations. This condition can arise, even in times of average precipitation, owing to specific soil conditions, topography or biotic factors. It follows that agricultural drought can be expressed on a very wide range of plant growth reductions up to complete crop failures. It does not necessarily imply that plants must wilt or die or fail in any spectacular manner. By definition agricultural drought can cause small reductions in yield when it is mild.

Earlier classifications of the moisture environment towards the definition, assessment and even the prediction of agricultural drought were largely based on received precipitations. It still remains a simple fact that dryland (rainfed) crop performance is primarily affected by the amount and distribution of precipitation. Dryland farming is largely based on maximizing the capture of this water for crop production. Probability analysis of rainfall for selected periods during the year and the crop-growing cycle was taken as a measure of the moisture environment. It was only later recognized that the soil-moisture balance is also involved in the definition of the moisture environment and soil characteristics were introduced into the statistics. The statistical approach resulted in the treatment of drought as a meteorological anomaly characterized by a deviation from "normal" weather conditions. The main use of the statistical approach, apart from dealing with the historical perspective, was in attempting to predict severe weather anomalies. It could not seriously deal with the definition of crop-water stress, apart from defining drought in terms of duration and magnitude of water shortage.

1.1 The Palmer Drought Index

This index is a classical climatic index for drought. This is a common measurement of the moisture environment based on recent precipitation and temperature. It was developed by meteorologist Wayne Palmer during the 1960s. The Palmer Drought Index is based on a supply-and-demand model of soil moisture. Supply is comparatively straightforward to calculate, but demand is more complicated as it depends on many factors – not just temperature and the amount of moisture in the soil but also hard-to-calibrate factors including evapotranspiration and recharge rates. Palmer tried to overcome these difficulties by developing an algorithm that approximated them based on the most readily available data – precipitation and temperature. The Palmer Index is effective in determining long term drought – a matter of several months – and it is not as good with short-term forecasts (a matter of weeks). It uses a 0 as normal, and drought is shown in terms of minus or plus relative values (Table 1.1).

Table 1.1 The Palmer index scales

Index	Indication
Palmer index	
4.0 and above	Extreme moist spell
3.0 to 3.99	Very moist spell
2.0 to 2.99	Unusual moist spell
1.0 to 1.99	Moist spell
.5 to.99	Incipient moist spell
.49 to –.49	Near normal
–.50 to –.99	Incipient drought
–1.0 to –1.99	Mild drought
–2.0 to –2.99	Moderate drought
–3.0 to –3.99	Severe drought
–4.0 and below	Extreme drought
Palmer Z index (see text)	
3.50 and above	Extreme wetness
2.50 to 3.49	Severe wetness
1.00 to 2.49	Mild to moderate wetness
–1.24 to.99	Near normal
–1.99 to –1.25	Mild to moderate drought
–2.74 to –2.00	Severe drought
–2.75 and below	Extreme drought

Critics have complained that the utility of the Palmer index was weakened by the arbitrary nature of Palmer's algorithms, including the technique used for standardization. The Palmer index's inability to account for snow and frozen ground is also cited as a weakness. Still the Palmer index is widely used operationally, with Palmer maps published weekly by the United States Government's National Oceanic and Atmospheric Administration. It also has been used by climatologists to standardize global long-term drought analysis.

1.2 The Crop Moisture Index

The crop moisture index (CMI) is also a formula that was developed by Wayne Palmer subsequent to his development of the Palmer Drought Index. CMI responds more rapidly than the Palmer Index and can change considerably from week to week, so it is more effective in calculating short-term abnormal dryness or wetness affecting regional agriculture. This index is the sum of the evapotranspiration anomaly (which is generally negative or slightly positive) and the moisture excess (either zero or positive). Both terms are a function of the previous week and a measure of the current week. The evapotranspiration anomaly is weighted to make it comparable in space and time. If the potential moisture demand exceeds available moisture supplies, the CMI is negative. However, if moisture meets or exceeds demand the index is positive. It is necessary to use two separate legends because the resulting effects are different when the moisture supply is improving than when it is deteriorating. The stage of crop development and soil type should be considered when using this index. In irrigated regions, only departures from ordinary irrigation requirements are reflected.

A parameter obtained from the calculations is the monthly moisture anomaly which is also referred to as the Palmer Z Index (Table 1.1). This is the product of the moisture departure of the most recent 4 weeks and a climate weighting.

An analysis performed in Canada for more than 30 years evaluated various drought indices for their predictive capacity of spring wheat yields under variable moisture regimes (Quiring and Papakryiakou 2003). The analysis indicated that Palmer's Z-index was the most appropriate index for measuring agricultural drought in the Canadian prairies. The model evaluation indicated that the Z-index was best suited for predicting yield when there was significant moisture stress. There was a statistically significant relationship between the Z-index and spring wheat yield in all crop districts, but the strength of the relationship varied significantly by crop district due to the influence of factors other than moisture availability (e.g., disease, pests, storm damage, etc.).

1.3 The Conventions of Crop Water Use

A plant will enter a state of water deficit when its demand for water is not matched by supply. Therefore there is a need to understand plant and crop water use. Very early, Thornthwaite and Mather (1955) recognized that evapotranspiration (ET) was affected by vegetation properties as well as by solar radiation, the atmospheric capacity for removing water vapor from vegetation (temperature, wind speed, turbulence and vapor-pressure deficit) and soil properties. This understanding has driven the development of numerous models for the estimation of crop evapotranspiration. The various models differed in their nature (empirical or physically based), precision, type, and amount of data required for input and subsequently the degree of instrumentation.

CROPWAT (http://www.fao.org/nr/water/infores_databases_cropwat.html) is a decision support system developed by the Land and Water Development Division of FAO. It might be considered as a consensus. Its main functions are to calculate crop water requirements and it can also be used to evaluate rainfed production and drought effects as well as the efficiency of irrigation practices. Its documentation is useful for understanding the principles of crop water-use. The basis for this model is the modified Pennman-Monteith equation according to expert consultation convened in by FAO in 1990 (Allen et al. 1998).

Underlying all models are several basic physical and biological principles. Briefly, the energy load on the plant, expressed as the balance between the incoming solar radiation and the radiation outgoing from the crop, is defined as net radiation

(Rn). Net radiation is affected by the crop's albedo, which expresses the reflective properties of the crop. The energy received by the crop is dissipated according to the following relationship:

$$Rn = -(LE+H+S+M) \quad (1.1)$$

Where LE is latent energy used in evapotranspiration, H is the sensible heat exchange with the air, S is the soil-heat flux, and M is the total energy exchanged in photosynthesis and other metabolic processes. The magnitude of M is considered negligible. The basic energy balance reflected in the above expression considers mainly the vertical transfer of energy.

However, in the field there is also a horizontal transfer of water and heat which is defined as advection and is driven mainly by wind. Thus, the movement of wind from the edge of the field into the center of the field creates a "fetch" which involves a gradient of humidity and temperature above the edge of the crop until equilibration occurs further down wind. Therefore, the fetch also involves a gradient of plant gas exchange and plant water status from the edge of the field downwind. The length of the Fetch depends on the environment and the crop but in most normal situation it will extend for about a dozen meters or so. Fetch is an important consideration in planning field experiments, especially in a desert environment.

The environmental load that determines the crop water requirement can also be roughly represented by the potential evapotranspiration (PE), which can be measured (e.g., by class-A pan evaporation) or calculated (Allen et al. 1998). PE is sometimes referred to as "evapotranspirational demand" and it reflects the integrated effect of the various weather variables (the "demand") that drive water from the evaporating surface to the atmosphere. ET approximates PE if the crop (with a full ground cover) is able to evapotranspire (ET) freely as the surface of free water or as a wet wick.

The initial environmental indication of a possible crop water deficit situation is the increase in PE over precipitation at any given period during crop growth (barring high soil moisture storage). Long-term data on precipitation and PE may provide a rough estimate of the expected level and frequency of crop water deficits.

A closer approximation to recognizing possible crop moisture deficit is by considering actual evapotranspiration (ET) as related to demand (PE). Plant water deficit is related (not simply) to the disparity between demand for water and actual use of water, when normalized for leaf area or ground cover. This is commonly expressed in the ratio of ET to PE. Despite its shortcomings, the ET/PE ratio represents well the difference between crop moisture environments. This difference may also be represented by integrating the ET/PE curve into one seasonal value. In dryland farming this value can range from 0.1 to 1.00 depending on locality and climate. The ET/PE ratio has been worked out experimentally to the extent that it is often used for irrigation scheduling.

Further refinements in defining the moisture environment involve only ET. Given an optimal water supply, a crop will evapotranspire to a given maximum rate. Whenever moisture supply is short of the full demand, due to either soil or atmospheric factors, ET will be reduced from the maximum. The crop moisture environment can therefore be defined by the ratio of actual ET (ET_{actual}) to maximum ET ($ET_{maximum}$) rather than to PE. Plant water deficit is proportional to the rate of reduction in this ratio below a value of unity. It should be remembered that this is an indirect estimate of plant water deficit. The real estimate of plant water deficit is obtained by measuring plant water status.

Whereas water stress is defined by the inability of the crop to meet its evapotranspirational demand, a further analysis of the reasons for a reduction in the ratio of ET_{actual} to $ET_{maximum}$ is important for the definition of the moisture environment. The ability of the given crop to maximize the use of available rainfall is largely dependent on the soil properties. The physical and chemical properties of soil are responsible for the various soil-water constants, such as rate of infiltration, water-holding capacity (water release curve), surface evaporation, etc. The amount of rainfall entering the soil depends on the infiltration properties of the soil, as well as topography. A low soil-surface infiltration rate, typical of soils inflicted with surface crusting, is a serious problem. The problem is amplified in many of the semiarid regions where rainfall is often received in high-intensity storms greater than around 20 mm h^{-1}. A poor soil water-holding capacity, typical of light, sandy soils or soils of poor structure, has an apparent major effect on the plant's water relations. The classical "soil available water content" may range from about 70 to 240 mm m^{-1} depth in different soils (Rutter 1975). Poor soil water-holding capacity is, for example, a primary reason for the commonality of brief periods of water stress that may occur even in the humid tropics in crops growing on lateritic soils. Fukui (1982) argued, partly because of this reason, that the critical amount of total annual rainfall for upland rice was about 2,000 mm, and upland rice crops were often subjected to brief water stress even at a total annual rainfall of 1,200–1,800 mm. The soil's water-holding capacity also determines to a large extent the carryover efficiency of pre-seasonal moisture for crop use. This is reflected in the large disparity among sources in estimates of the fallow-system efficiency in conserving moisture, within a range of 5–40% of the total rainfall received.

Even where the soil water-holding capacity is reasonable, the effective soil volume accessible to roots may be restricted. Physical impediments to root growth and extension are common to soils with hardpans of various sorts that result from poor tillage practices or from soil horizons that contain a very high amount of clay. A clear crop response to deep tillage may be taken as a simple indicator of such problems. Poor root development in the soil may also result from soil-chemical impediments, such as excessively high or low soil pH, toxic concentration of metals or soil salinity. Soil biotic factors such as soil inhabiting pathogens or nematodes can also inhibit root development. The plant-root response to soil impediments would therefore determine the effective amount of soil moisture available to the crop, whatever is the total seasonal rainfall. The importance of deep soil moisture towards crop yield remains a contemporary issue (e.g., Kirkegaard et al. 2007).

While the total seasonal crop-water balance is important for defining the moisture environment, the seasonal components of the water balance are also crucial. Precipitation, as an intermittent phenomenon, becomes infrequent in the semiarid zone. At the same time, year-to-year and seasonal variations in amounts of rainfall become greater with the increased aridity. Thus, the "reliability" of rainfall, in terms of the annual total or the seasonal distribution, has been treated extensively by various statistical methods which were already mentioned briefly above. Such analysis is beyond the scope of this chapter, but it should be consulted for any given breeding-target environment. For a given soil and crop, the seasonal and annual variations in rainfall is critical for the formulation of a predictable profile of water stress which can be addressed by the breeder. While often one is perplexed by the variation, with the inadvertent conclusion that a stress profile cannot be predicted, climatologists are able to show that long-term data can result in a given pattern in almost any locality. The design of a "predictable" stress profile is therefore possible not in conceptual terms but rather in terms of a probability analysis of long-term rainfall data against PE and ET data, if possible. After such analysis, a conclusion may indeed be drawn that the variation in rainfall is too large for the formulation of any systematic approach to the breeding program.

"LocClim" is a free database and software for obtaining estimated climatic data for any point on the globe and it is available at the FAO web site through http://www.fao.org/sd/locclim/srv/locclim.home. A copy of the full application is available to run on your computer. Data include mean temperatures, precipitation, potential ET, sunshine, wind and humidity.

"AquaCrop" decision support system (Steduto et al. 2009), has a significantly small number of parameters and a better balance between simplicity, accuracy and robustness as compared with other previous models. Root zone water content is simulated by keeping track of incoming and outgoing water fluxes at its boundaries, considering the soil as a water storage reservoir with different layers. Instead of leaf area index, AquaCrop uses canopy ground cover. Canopy development, stomatal conductance, canopy senescence and harvest index are the key physiological crop responses to water stress. ET is simulated as crop transpiration plus soil evaporation and the daily transpiration is used to drive the daily biomass gain via the normalized biomass water productivity of the crop. The normalization is for reference evapotranspiration and CO_2 concentration to make the model applicable to diverse locations and seasons, including future climate scenarios.

The breeder must be aware first and foremost that knowledge of the moisture environment at the breeding target environment is crucial for planning any breeding project. Secondly it is underlined that "crop drought stress" as a working definition, is not to be taken in its catastrophic connotation, which implies a total yield loss. Drought stress may occur at any sub-potential yield level that is induced by the inability of the crop to meet its evapotranspirational demand. The improvement of yield at any of these levels is the purpose of breeding for water-limited environments. Since the reduction of the ratio of ET_{actual} to $ET_{maximum}$ involves both environmental and plant factors, the recognition of the reasons for the reduced ratio in the breeder's target environment is essential for developing a non-empirical and comprehensive breeding program.

However, the moisture regime at the target environment can involve also various ecological and crop management considerations which can have a decisive implication on the breeding programs and the plant traits concerned. An example can be seen in the following conclusion on reasons for the genetic diversity for osmotic adjustment (OA) in a rice population (Babu et al. 1999): "rice cultivars that were normally grown in lowland (wetland) conditions tended to express moderate to high OA capacity (with one exception). On the other hand, traditional upland cultivars, such as Azucena or Moroberekan, appeared to lack in OA." Moroberekan and Azucena are well known for their deep root development (Champoux et al. 1995; Steele et al. 2007). It is possible that dryland cultivars were inadvertently selected mainly for deeper roots (e.g., Price et al. 1997) and the ability to explore deep soil moisture, a trait that would promote avoidance of plant dehydration in a drying soil. Hence, no selection pressure was imposed toward the evolution of OA. Wetland rainfed cultivars growing in paddy soils, which have shallow roots and are subject to severe dehydration when exposed to short periods of drought stress, may have been subjected to an effective selection pressure for high OA capacity as a major possible mechanism of drought resistance. In these materials dehydration avoidance by deep roots was no option. This conclusion is further supported by a study (Babu et al. 2001) of *japonica* and *indica* type rice accessions indicating an association between good capacity for OA and poor root system.

Defining the target moisture environment and regime in the context of the breeding program will be further discussed in Sect. 3.7, mainly by considering the plant.

References

Allen RG, Pereira LS, Raes D et al (1998) Crop evapotranspiration – guidelines for computing crop water. Food and Agriculture Organization of the United Nations, Rome

Babu RC, Pathan MS, Blum A et al (1999) Comparison of measurement methods of osmotic adjustment in rice cultivars. Crop Sci 39:150–158

Babu RC, Shashidhar HE, Lilley JM et al (2001) Variation in root penetration ability, osmotic adjustment and dehydration tolerance among accessions of rice adapted to rainfed lowland and upland ecosystems. Plant Breed 120:233–238

Champoux MC, Wang G, Sarkarung S et al (1995) Locating genes associated with root morphology and drought avoidance in rice via linkage to molecular markers. Theor Appl Genet 90:969–981

Fukui H (1982) Variability of rice production in tropical Asia. In: Drought resistance in crops with emphasis on rice. International Rice Research Institute, Los Banos

Kirkegaard JA, Lilley JM, Howe GN et al (2007) Impact of subsoil water use on wheat yield. Aust J Agric Res 58:303–315

Price AH, Tomos AD, Virk DS (1997) Genetic dissection of root growth in rice (*oryza sativa* L) 1. A hydrophonic screen. Theor Appl Genet 95:132–142

Quiring SM, Papakryiakou TN (2003) An evaluation of agricultural drought indices for the Canadian prairies. Agric Forest Meteorol 118:49–62

Rutter AJ (1975) The hydrological cycle in vegetation. In: Monteith JH (ed) Vegetation and the atmosphere, vol 1, Principles. Academic, New York

Steduto P, Hsiao TC, Raes D et al (2009) AquaCrop – the FAO crop model to simulate yield response to water: I. Concepts and underlying principles. Agron J 101:426–437

Steele KA, Virk DS, Kumar R et al (2007) Field evaluation of upland rice lines selected for QTLs controlling root traits. Field Crops Res 101:180–186

Thornthwaite CW, Mather JR (1955) The water balance. Publications in Climatology. Drexel Institute of Technology, Centerton

Chapter 2
Plant Water Relations, Plant Stress and Plant Production

Summary Plant water deficit is initiated as the crop demand for water exceeds the supply. The capacity of plants to meet the demand and thus avoid water deficit depends on their "hydraulic machinery." This machinery involves firstly the reduction of net radiation by canopy albedo, thus reflecting part of the energy load on the plant. Secondly, it determines the ability to transport sufficient amount of water from the soil to the atmosphere via the stomata (which take in CO_2) in order to provide for transpiration, transpirational cooling and carbon assimilation. Water is transported by way the SPAC (soil-plant-atmosphere continuum). SPAC is largely controlled by the resistances in the continuum as determined by root, stem, leaf, stomata and cuticular hydraulic resistances. Resistances are generally a function of the plant basic anatomy, development and metabolism. Some resistance such as those of stomata is also variable depending on plant responses and environment effects.

The primary force driving water against plant resistances is the soil-to leaf gradient of water potential which is expressed in reduced leaf water potential. Reduced leaf water potential may induce osmotic adjustment which helps maintain leaf hydration at low leaf water potential. As plants enter a state of water deficit, hormones, mainly ABA are produced in the root and the shoot, causing an array of responses, most of which cannot be defined as productive in the agronomic sense. Thus, the combination of hydraulic stress and hormonal metabolism carry various impacts on plant adaptation to stress on one hand and reductions in growth and productivity on the other. The most susceptible growth stage to water deficit is flowering and reproduction, which in many crop species cannot be recovered upon rehydration. Some (not all) of the heritable plant traits and adaptive responses to water deficit can be counterproductive in term of allowing high yield potential.

2.1 The Initiation of Plant Water Deficit

Crop evapotranspiration (ET) is affected by both the environment and the crop. Crop factors that affect ET are mainly associated with the dynamics of leaf-area development and senescence and the resistances to water flux developed in the

A. Blum, *Plant Breeding for Water-Limited Environments*,
DOI 10.1007/978-1-4419-7491-4_2, © Springer Science+Business Media, LLC 2011

soil-plant-atmosphere continuum (SPAC). When actual ET is close or equal to maximum ET, the environment exercises most of the control over ET. A reduction in actual ET below maximum ET is associated with the development of a gradient of potentials between the soil and the transpiring organs, leading towards a situation defined as plant water deficit. At the same time, the relative role of the plant in affecting ET becomes greater. Therefore, the role of breeding for water limited environments is anywhere in the domain where $ET_{actual}/ET_{maximum} < 1$. It must be made very clear already at this point that breeding programs for water limited environments can be quite different if this ratio is closer to 1 or closer to 0.1, namely if plant water deficit is small or large.

In the dryland agricultural domain where plant production is a major consideration the ultimate purpose is for the plant to deliver water from the soil to the leaves thus allowing sustained leaf gas exchange and the delay of leaf death. The best plant to achieve this purpose is one that is equipped with the appropriate "hydraulic machinery" (Sperry et al. 2002) as well as additional traits to relieve the energy load on the plant as well as manage an effective use of water.

2.2 The Soil-Plant-Atmosphere Continuum (SPAC)

During the day the plant is under heavy energy load (net radiation, Rn) (Sect. 1.3). While a small fraction of this energy is used in photosynthesis, most of it must be dissipated. If this energy is still absorbed by the canopy to its fullest extent then leaves can reach a killing temperature of 40° to 50°C or more. This energy load is dissipated via three physical channels: (1) the "albedo" which is determined by the total reflectivity of the leaf as affected by its optical characters and its architecture; (2) "sensible heat" which is the radiation emitted from the canopy as heat; and (3) the "latent energy" which is dissipated by plant transpiration. In the narrow sense SPAC relates only to water movement through the system (channel 3 above) but it is most relevant towards the subject of this book to discuss all channels of energy dissipation by the plant under this heading.

2.2.1 The Albedo

The Albedo is the ratio of reflected to incident radiation. It is a unit-less measure indicating the diffuse reflectivity of any surface or body. The word is derived from *albus*, a Latin word for "white." The crop albedo is different from the leaf albedo in that the former is determined by the spectral properties of the exposed soil and the crop leaf canopy. The soil reflective properties are determined largely by its color and wetness, where greater "whiteness" increases the albedo.

The optical characteristics of the single leaf (Fig. 2.1) are determined by leaf pigments, leaf anatomy, leaf age (which is partly expressed in its pigmentation), leaf water status and leaf surface properties. A study of 45 plant species revealed

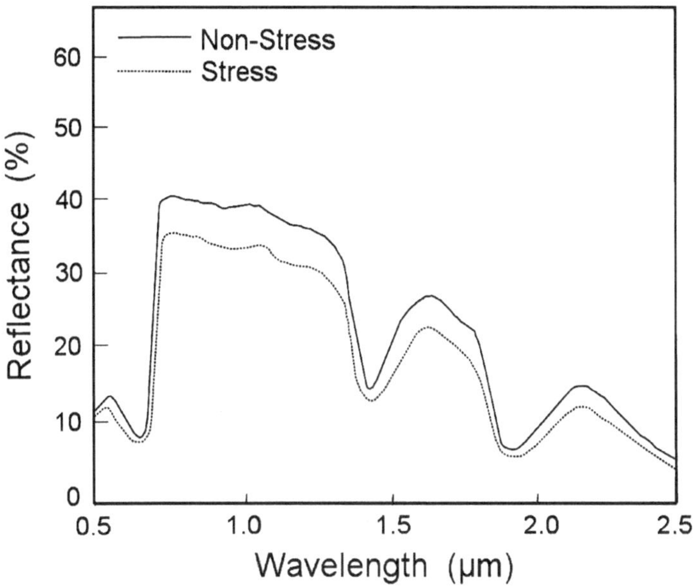

Fig. 2.1 A typical spectral reflectance curve of a typical non-stressed and a drought stressed leaf ranging from short visible to long infrared wavelength drawn as an average according to several sources of data

that that both pubescence (presence of hairs) and glaucousness (presence of a thick epicuticular wax) had marked effects on total leaf reflectance (Holmes and Keiller 2002). Pubescent leaves tended to be more effective in reflecting longer wavelengths than the ultraviolet. Surface waxes are very effective reflectors of both UV and longer wavelength radiation. As can be seen in Fig. 2.1, drought stress tends to reduce leaf reflectivity throughout the spectrum.

Research on the spectral properties of leaves was also performed in relations to remote sensing development for vegetation and crops, either from satellite platforms or from the ground. The understanding of the optical properties of leaves and how they change with leaf characteristics and the effect of the environment led to the development of remote sensing techniques which allow to sense plant drought, mineral deficiency and various biotic stresses. Multispectral signatures of crops are now being used to estimate crop growth and even yield, in conjunction with or without crop simulation models. Some of these methods as applied to breeding are discussed in Chap. 4, section "Indirect Methods (Remote Sensing)."

2.2.2 *The Water Flux*

The hydraulic system within intact plants acts as a true continuum. Water will move from the soil into the plant, through the plant and into the atmosphere in response to a water potential gradient. Water flows along a gradient of decreasing water potential.

Water potential is measured in units of negative pressure such as bars or Mega Pascals (MPa). Free water is defined to have a potential of zero. Water that contains solutes will have a negative potential and it will attract free water across a semi permeable membrane. When water is held by force as the case may be in the pores of soil, the water potential is determined by the force which is required to move this water to a state of free water. This is also the case for water held in the plant.

The physical model of water flux through the SPAC has been developed under the influence of soil physics and with the involvement of soil physicists and crop climatologists. It is still the basis of our understanding of plant water relations. However, as will be seen below there are also "metabolic" or "active" components added to this model more recently.

The movement of water through plants obeys an Ohm's law analogy, i.e., current equals driving force (the electrical potential gradient) divided by electrical resistance. Thus, water flux is more clearly understood if it is regarded as being driven by a difference in water potential, against a resistance.

Under steady-state conditions, flow through each segment of the SPAC is described as follows:

$$Waterflux = \frac{\Psi_s - \Psi_r}{r_m} = \frac{\Psi_r - \Psi_l}{r_r + r_x} = \frac{\Psi_l - \Psi_a}{r_s + r_a} \tag{2.1}$$

Where, r_m is the resistance due to the soil matrix, r_r is the root resistance, r_x is the resistance through the xylem in plant stems, r_s is the stomatal resistance, and r_a is the aerial resistance. Ψ_s, Ψ_r, Ψ_l and Ψ_a are the water potential of the soil, root, leaf and air, respectively. Resistances are additive in a series. Figure 2.2 provides a graphical schematic representation of SPAC.

The energy driving water flux through the SPAC is by and large that part of the solar irradiance which is not reflected by the canopy or dissipated as sensible heat. Therefore it must be remembered that water flux through SPAC under the Ohm's law analogy responds primarily to the seasonal, daily and hourly march of solar radiation and R_n. Other environmental factors are also in effect, such as air humidity (vapor pressure deficit), air temperature and wind. Consequently, plant water status and most prominently leaf water status vary extensively during the day in correspondence to the march of the atmospheric environment. Even passing clouds will affect transpiration on a time scale of few minutes. Normally, leaf water potential will decrease (become more negative) from sunrise towards solar noon with lowest values at or just after solar noon. As the sun begin to set, leaf water potential will increase towards full or almost full recovery at night. It is generally accepted that relatively little transpiration occurs at night but exceptions were noted (Caird et al. 2007). Towards dawn leaf water potential almost completely equates with soil water status, unless the plant is at or close to permanent wilting.

The highly dynamic state of water flux, transpiration and the associated leaf water potentials pose a problem for the comparative measurement of water fluxes or plant water status during the day. There is no problem in hooking up a single plant to various sensors and measuring its daily response from dawn to dusk.

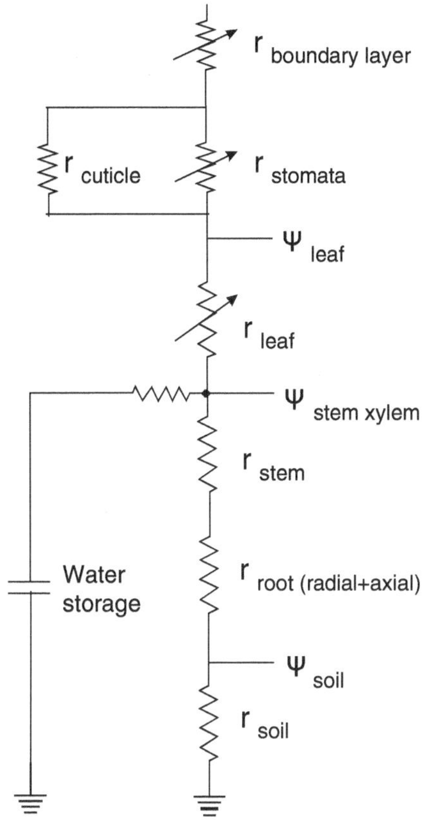

Fig. 2.2 A schematic representation of the soil-plant-atmosphere continuum as an Ohm's law analogy (see text). Arrows on resistance icons represent variable resistances. This is not to say that other resistances are absolutely static under all conditions. (Ψ= water potential; r = hydraulic resistance)

However when different genotypes must be phenotyped and compared for plant water status, it should be done on a reasonable short time span and when the plant is in a relatively stable hydraulic state under a relatively stable environment. Extensive experience shows that plants are relatively stable in terms of water flux and plant water status at dawn and for about 1 to 2 h after solar noon when there is a small plateau in the daily march of transpiration and plant water status. At dawn the plant is under a minimal water deficit while at midday it is at peak stress.

In a well hydrated plant the greatest hydraulic resistance is in the leaf and the smallest resistance is in the stem. These values may vary to some extent in absolute and relative term in different plant species and under different conditions. It is therefore important to understand the dynamics of the various components of plant resistance affecting the SPAC since they are the initial and main controls of plant water status and plant water stress, which we strive to manipulate genetically.

2.2.3 Root Resistance

The root is the most crucial organ for meeting transpirational demand at a reasonable high leaf water status, on the condition that water is available anywhere in the root horizon. Total root conductivity which is the inverse of root resistance ($K_r = 1/R_r$) is positively related to *root length density* in the soil and the *hydraulic conductivity* of the single root axe. High root length density increases the number of contact points between root and soil. This is crucial for water uptake in a drying soil. In order for the root to absorb water, water must be available at the root-to-soil interface. For this situation to occur the root must grow towards water or water must flow towards the root. Water flow in a drying soil towards the root is subjected to very high resistance. High root length density reduces the impact of soil resistance to water flow towards the root. Experience gained with drip irrigation and associated research demonstrated that well watered plants in a reasonably good soil can meet transpirational demand in full even when the root system is small.

The root axe hydraulic resistance is partitioned into radial and axial (longitudinal) resistances. Axial resistance refers mainly to water flow through root xylem vessels. It has been shown that as soon as early metaxylem vessels mature, the axial hydraulic resistance within the xylem is usually not rate-limiting (Steudle and Peterson 1998). However under certain conditions (see below) xylem resistance can increase. Radial resistance is the important component of root resistance, in addition to the root-soil interface resistance.

To pass from the soil solution into root vascular tissues, water must flow radially across a series of concentric cell layers. These layers include the epidermis, the exodermis in roots where it is differentiated, several layers of cortex cells, the endodermis, pericycle, xylem parenchyma cells and finally into the xylem vessel. Three pathways co-exist for radial water transport across living root tissues: through the cell walls (apoplastic path), from cell to cell, along the symplasm through plasmodesmata (symplastic path) or across membranes (trans-cellular path). The cell walls of exo- and endodermal cells possess a particular structure, the Casparian band (or "strip"), which consists of a deposit of suberin and/or lignin. It has been shown that in the exodermis, this structure represents an effective impediment to water flow. It is generally accepted that the Casparian band creates a tight apoplastic barrier to solutes and prevents their backflow from the stele.

It has been argued (Stirzaker and Passioura 1996) that sometimes the sum of the resistances in the plant and the soil was too small to account for the fall in water potential between the leaf and the soil, especially when plants grow in sandy soils, which are prone to dry rapidly. The root-soil interface resistance was suggested to be responsible due possibly to poor root contact with the soil or due to accumulation of solutes at the root interface. It was later shown by White and Kirkegaard (2010) that root contact as driven by extensive root branching and long root hairs is a prime determinant of moisture extraction from dry soil. Accordingly, radial root resistance and root-soil interface resistance (also involving root hairs) in series can be considered as the major resistance of the single root to water uptake.

Whereas radial water flux through the trans-cellular path is important as compared with the apoplastic path, water channels that control water movement through cellular membranes become important controls of radial root resistance. In the presence of heavily suberized roots, the apoplastic component of water flow may be small. Under these conditions, the regulation of radial water flow by water channels becomes dominant. Since water channels are under "metabolic" control, this component represents an "active" element of water transport regulation (Steudle 2000).

Zhu et al. (2010) proposed an interesting hypothesis for enhanced root growth and root-length density. They hypothesized that root cortical aerenchyma (RCA) reduces root respiration in maize by converting living cortical tissue to air volume. This should reduced root metabolic cost and release more energy for root growth. Their data for maize lines of low and high RCA show that high RCA was associated with appreciable increase in root-length density and depth.

Aquaporins are water channel proteins expressed in various membrane compartments of plant cells, including the plasma and vacuolar membranes (Javot and Maurel 2002). While their role in root water uptake and plant water status is well recognized, there are wider implications of aquaporins in plant physiology and plant response to stress (see further below). These membrane proteins belong to the major intrinsic protein (MIP) family, with members found in nearly all living organisms. Plants appear to have a particularly large number of MIP homologues. The complete genome of *Arabidopsis thaliana* has 35 full-length MIP genes. Based on sequence homology, plant MIPs cluster into four subgroups which to some extent reflect different subcellular localizations. Members of the two major subgroups, the plasma membrane intrinsic proteins (PIPs) and the tonoplast intrinsic proteins (TIPs) have been initially localized in the plasma membrane and in the tonoplast, respectively.

Mercury ($HgCl_2$) acts as an efficient blocker of most aquaporins and has been used to experimentally demonstrate the significant contribution of water channels to overall root water transport. Aquaporin-rich membranes may be needed to facilitate high-rate water flow across the trans-cellular path. Aquaporins are considered to be crucial for radial water transport in roots (Bramley et al. 2007). Roots show a remarkable capacity to alter their water permeability over the short term (i.e., in a few hours to less than 2–3 days) in response to many stimuli, such as day/night cycles or nutrient deficiency. These rapid changes can be mostly accounted for by changes in root cell membrane. The processes that allow perception of environmental changes by root cells and subsequent aquaporin regulation are basically unknown. It seems however that both MIPs and PIPs can be down-regulated or up-regulated by drought stress in *Arabidopsis*, depending also on plant part (Alexandersson et al. 2010). Drought resistance was not promoted by overexpression of PIP1 and PIP2 in Eucalyptus (Tsuchihira et al. 2010).

Abscisic acid (ABA) is a well-recognized plant hormone which accumulates in plants under drought and other stresses. It will be extensively discussed further in this and other chapters. ABA mediates many known plant responses to drought stress. It has been shown (Hose et al. 2000; Quintero et al. 1999) that exogenous

ABA enhanced root conductivity. In one detailed study (Hose et al. 2000) ABA applied at concentrations of 100–1,000 nM increased the hydraulic conductivity of excised maize roots both at the organ level (by a factor of 3 to 4) and the root cell level (by a factor of 7 to 27). It was concluded that ABA acts at the plasmalemma, presumably by interacting with aquaporins (Kaldenhoff et al. 2008). Some of the above experiments used exogenous application of ABA. Studies with transgenic plants expressing high endogenous ABA also indicate that ABA promotes hydraulic root conductivity in whole plants, such as the case for tomato (Thompson et al. 2007). ABA therefore facilitates the cell-to-cell radial water flux (Parent et al. 2009) and the uptake of water into the root as soil start drying and transpiration is reduced and when the apoplastic path of water transport is largely excluded.

The involvement of aquaporins and ABA in controlling root conductivity introduce a "metabolic" or an "active" component into the seemingly pure physical model of water flux through the SPAC. It was therefore suggested, for example, that PIP may regulate water transport across roots such that transpirational demand is matched by root water transport capacity (Sade et al. 2009; Vandeleur et al. 2009). As our understanding of aquaporins and their interaction with ABA will develop it might become possible to genetically design root conductivity to improve plant performance under drought stress. There is an apparent need for this option as can be deduced from the example of rice which has an inherently poor root conductivity causing sometimes a plant water deficit even when roots are in water (Miyamoto et al. 2001).

The majority of vascular plants form root associations with fungi to increase their absorption of mineral nutrients. Fungi, which live by absorbing nutrients from their surroundings, are ideal organisms for such associations. There are both *endomycorrhizae and ectomycorrhizae* associations. Endomycorrhizae penetrate cells of the root cortex with their hyphae. Mycorrhizae function as sophisticated root hairs; plants that associate with ectomycorrhizae often do not produce root hairs.

It has long been observed that *Arbuscular mycorrhizal* (AM) symbiosis with plant roots enhance plant water status and growth under drought stress as compared with non AM plants (Auge et al. 2001; Davies et al. 2002a; Ortega et al. 2004; Porcel and Ruiz-Lozano 2004; Ruiz-Lozano et al. 2001). The effect of AM in this respect has been traced at least partially to increased root conductivity in drought stressed or non-stressed plants (Aroca et al. 2007). The effect of AM in this respect seems to be genetically independent of the effect of root aquaporins. Studies with lettuce indicated that AM symbiosis enhanced plant tolerance to the depressing effect of exogenous ABA treatment on biomass production (Aroca et al. 2008), again suggesting a positive role of AM on root conductance and its interaction with ABA especially under drought stress. Further discussion of the role of AM symbiosis under drought stress with special reference to maize is available in Boomsma and Vyn (2008). Beyond AM, it is now recognized that various rhizosphere and root inhabiting rhizo-bacteria can impact root and plant hormone signaling pathways by producing ABA, auxins and cytokinins or by mediating plant ethylene levels (Dodd 2009). These can have important but yet unresolved effects on root hydraulic resistance and plant water relations.

Hydraulic lift is the passive movement of water from roots into the dry top soil layer, while other parts of the root system in deeper moist soil layers are absorbing water. Soil water absorbed by deep roots can be released in the upper dry soil profile at night or during periods or low irradiance. Hydraulic lift was first observed in native vegetation and later also in crop plants. In sorghum (Xu and Bland 1993) efflux of water into the dry top soil could first be detected at a dry soil water potential of about 0.55 MPa, Outflow was 5–6% of daily transpiration during periods of highest water use. More water was found to be exuded from roots in the top soil layer in a drought resistant maize hybrid than in a susceptible maize hybrid (Wan et al. 2000). The sizable amount of water from hydraulic lift allowed the resistant hybrid to reach a peak transpiration rate 27–42% higher than the drought-susceptible hybrid on days when the evaporative demand was high. There were two to threefold more primary roots in the deep moist soil in the resistant than the susceptible hybrid. Genetic variation in water transported by hydraulic lift were also found in cotton and ascribed to possible differences in root conductance (McMichael and Lascano 2010).

Large quantities of water, amounting to an appreciable fraction of daily transpiration, can be lifted at night. This temporary partial rehydration of upper soil layers provides a source of water, along with soil moisture deeper in the profile. Nutrients are usually most abundant in the upper soil layers which under dryland conditions become dry. Lifted water may provide moisture to facilitate nutrient availability, microbial processes, and the acquisition of nutrients by topsoil roots. Hydraulic lift was especially noted for P- efficient canola genotypes and it was found to enhance P and K uptake from the top dry soil (Rose et al. 2008). Lifted water into the upper soil zone might also extend root survival in the dry top soil (Bauerle et al. 2008).

2.2.4 Stem Resistance

The classical SPAC model accepts that axial hydraulic resistance of the stem is the smallest relative to that of stomata, leaf and root especially when crop plants and common fruit trees are considered. Understandably an efficient system of conduits must have been developed through evolution to allow plants and trees to meet large transpirational demand of leaf canopies against gravitational force and soil water deficit. The discussion of xylem conductivity is especially unique for the stem with its long conduits. Understandably, this topic has been discussed more extensively for trees than for herbage plants. It is however interesting still to note that improved stem hydraulic conductance was regarded as a reason for the drought resistance of a specific maize hybrid (Li et al. 2009).

The *Cohesion/tension theory* for long distance ascent of water in the xylem (mainly in trees) is based on the fact that water is a polar molecule. When two water molecules approach one other they form a hydrogen bond. The negatively charged oxygen atom of one water molecule forms a hydrogen bond with a positively charged hydrogen atom in another water molecule. This attractive force has several manifestations. Firstly, it causes water to be liquid at room temperature, while other lightweight molecules would be in a gaseous phase. Secondly, it is (along with other intermolecular forces) one of the principal factors responsible for the occurrence of surface tension in liquid water. This attractive force between molecules allows plants to draw water from the root and then pull it through the xylem (via capillary action) to the leaf.

Recent pressure probe and NMR results often challenge the frequent belief that tension is the only driving force. This seems to be particularly the case for plants faced with problems of height, drought, freezing and salinity as well as with cavitation of the tensile water. Other forces come into operation when exclusively tension fails to lift water against gravity due to environmental conditions. Possible candidates are longitudinal cellular and xylem osmotic pressure gradients, axial potential gradients in the vessels as well as gel- and gas bubble-supported interfacial gradients. Zimmermann et al. (2004) criticized the arguments developed in support of the cohesion/tension theory as an explanation of water ascent in tall trees. This was then followed by a letter of response to the journal signed by no less than 46 scientists, defending the theory against this criticism. Hence, the cohesion/tension theory became a hot issue towards which this review is not making judgement. The controversies were eminent also before the publication of Zimmermann et al. (2004) (e.g., Sperry et al. 2003). An important reason for the controversy is that that the xylem is "vulnerable" being sensitive to cavitation and embolism. If air enters the continuous column of water in the xylem, resistance to flow is created.

Rather than embolism being essentially irreversible, it also appears (Sperry et al. 2003) that there is a dynamic balance between embolism formation and repair throughout the day and that daily release of water from the xylem via cavitation may serve to stabilize leaf water balance by minimizing the temporal imbalance between water supply and demand. Sperry et al. (2003) concluded that although the cohesion–tension theory for xylem transport withstood recent challenges, a number of gaps remain in our understanding of xylem hydraulics. These include the extent and mechanism of cavitation reversal and thus hysteresis in the vulnerability curve and the structural basis for differences in air entry pressure (cavitation pressure) for different xylem types.

When various poplar (*Populus* spp.) and willow (*Salix* spp.) clones were tested for cavitation vulnerability (Cochard et al. 2007) it was found that variation in vulnerability to cavitation across clones was poorly correlated with anatomical traits such as vessel diameter, vessel wall strength, wood density and fibre wall thickness; however, a striking negative correlation was established between cavitation resistance and aboveground biomass production, indicating a possible trade-off between xylem safety and growth potential. However, the association between anatomical and structural features of the stem and cavitation vulnerability is apparently still an open issue (Cochard et al. 2009). Further discussion of cavitation vulnerability in relations to drought resistance is presented in Chap. 3, section "Stem Xylem Cavitation."

Water storage in plants (predominantly in stems) can serve as a buffer against transitional insufficient supply of water from soil. It is more common in cacti and trees, For example, in tropical forest trees (Stratton et al. 2000) it was found that seasonal and diurnal variation in leaf water potential were associated with differences among species in wood-saturated water content (a measure of water storage in trees). The species with higher wood-saturated water content were more efficient in terms of long-distance water transport, exhibited smaller diurnal variation in leaf water potential and higher maximum photosynthetic rates. The role of water storage in crop plants has not been well investigated and it is assumed to be generally small.

2.2.5 *Leaf Resistance (Excluding Stomata and Cuticle)*

The partitioning of resistances within the leaf among petiole, major veins, minor veins, and pathways outside the xylem is variable across species. Hydraulic resistances occur both in the leaf xylem as well as in the flow paths across the mesophyll to evaporation sites. Resistance therefore largely depends on the architecture of the specific leaf. Aquaporins may also be involved. A detailed discussion of leaf hydraulics has been published by Brodrib et al. (2010).

The decline in leaf conductivity in response to lower LWP arises from increase in xylem resistance due to cavitation or collapse, and/or from changes in the conductivity of the pathways outside the xylem such as the mesophyll. As leaf conductivity decreases due to dehydration stomata will close when, or before a low LWP becomes damaging. In droughted plants such a mechanism operates in tandem with chemical signals from the roots to close the stomata (discussed below).

Generally, leaf resistance is relatively lowest in crop plants and highest in conifers (Sack and Holbrook 2006).

2.2.6 *Stomatal Resistance*

Stomata affect leaf resistance by way of stomatal density and stomatal activity. High stomatal density has a role in enhancing leaf conductivity mainly under well watered conditions. As stress develops, stomatal closure becomes the main controls of resistance.

Stomata can be regarded as hydraulically and chemically driven valves in the leaf surface, which open to allow CO_2 uptake and close to prevent excessive loss of water. Movement of these valves is regulated by environmental cues, mainly light, CO_2 and atmospheric humidity. Stomatal response to humidity is of special interest with respect to plant water use in harsh environments (Fletcher et al. 2007). Stomata guard cells can sense environmental signals and they function as motor cells within the stomatal complex. Stomatal movements are controlled by the stomatal guard cells. Turgor changes in the guard cells regulate their movement. Water movement into the guard cell is driven by osmosis. Accumulation of solutes in the guard cell cytoplasm lowers guard cell water potential. Given a high hydraulic conductivity of the plasma membrane, water will flow into the guard cell and the water potential of the guard cell will equilibrate with that of the apoplast. The inflow of water will cause the turgor pressure to rise and the guard cells to swell. The increase in volume of both guard cells causes opening of the stomatal pore. Stomatal opening depends on the import of K^+ and sometimes also sugar into guard cells. Plasma membrane and vacuolar membrane ion channels and transporter proteins are involved in regulating ion status of guard cells and subsequently the dynamics of their turgor. Ca^{2+} and its interaction with aquaporin are also involved in stomatal regulation (Li et al. 2004).

Stomata open more fully at low CO_2 concentrations. When CO_2 concentration in the sub-stomatal cavity is reduced by mesophyll photosynthesis, stomatal conductivity increases. Thus CO_2 signalling of stomatal activity links the demand for CO_2 to its supply via stomata. However, stomata are similarly sensitive to CO_2 concentration outside the leaf. As a consequence of climate change, more studies are being performed recently on the effect of atmospheric CO_2 concentration on crop plant response. It was found, for example (Wall et al. 2006) that an experimental increase in atmospheric CO_2 improved wheat water status under drought stress due to the increase in daily stomatal resistance.

Light stimulates stomatal opening. Initially it was thought that the effect was transduced via the enhancement of photosynthesis by light. It was later found that the effect was achieved via blue light-specific and photosynthetic-active radiation dependent pathways. This response to blue light has been assigned to the activity of the PHOT1 and PHOT2 blue light receptors located in the plasma membrane. Light sensitivity is high and stomata will respond to shading almost instantaneously. One must remember this when leaning over the plant with a porometer in order to measure stomatal conductance.

Stomata respond to abscisic acid (ABA) by closure. ABA concentration in the leaf tissues increases as the plant sense water deficit. The guard cell receptor for ABA is unknown to the same extent that it is still an enigma for any other plant response to this hormone (e.g., Christmann and Grill 2009). It may involve Ca^{2+} signalling and regulation of plasma-membrane ion transport. Calcium, protein kinases and phosphatases, and membrane trafficking components have been shown to play a role in ABA signalling of guard cell movement, as well as ABA-independent regulation of ion channels by osmotic stress (Luan 2002). Stomata also sense the water status of distant tissues such as roots via the long-distance transport of ABA in the xylem. It is therefore believed now that stomatal activity is regulated by both hydraulic and chemical ABA signals (e.g., Christmann et al. 2007; Schachtman and Goodger 2008).

Aquaporins are also implicated in the control of stomatal conductance not only to water but also to CO_2 (Miyazawa et al. 2008). Deactivation of aquaporins was suggested to be responsible for the significant reduction in the diffusion conductance of CO_2 from the intercellular air space to the chloroplasts (internal conductance) in plants growing under long-term drought.

Stomata are therefore very effective but complex variable resistors in the SPAC which respond to the atmospheric environment on one hand and to plant water status and stress responsive plant chemical signals on the other. The consequences of stomatal resistance towards photosynthesis and the relationship to water-use and plant productivity are discussed in Chap.3, section "Stomatal Activity and Dehydration Avoidance."

2.2.7 Cuticular Resistance

In parallel to the stomata the cuticle offers a second plant surface hydraulic conductance pathway. Relative to stomata, cuticular resistance is basically non-variable on a short time-scale. When stomata are tightly closed, the cuticle remains the major resistance to transpiration at the leaf surface. If the cuticle is conductive then the effectiveness of the stomata in controlling transpiration is impaired.

The cuticle is a thin (0.1–10 μm thick) continuous membrane consisting of a polymer matrix (cutin), polysaccharides and associated solvent-soluble lipids (cuticular waxes) (Riederer and Schreiber 2001). Cuticular waxes are embedded in the cuticle and are deposited over the cuticle as "epicuticur wax" (EC). Upon the formation of the cuticle and EC, the passage of wax components through the cell wall and cuticle probably occurs via diffusion, possibly in a solvated form enabled by cell wall associated transport proteins. Lipid transfer proteins (LTPs) are thought to be involved in the transfer of lipids through the extracellular matrix. A six-fold increase of free tobacco *LTP* gene transcripts was observed after three drought events (Cameron et al. 2006).

In the following discussions "Cuticular resistance" or "non-stomatal resistance" refer to the resistance of the layer comprising of the epidermis, the cuticle and the EC. EC is a general term for complex mixtures of homologue series of long chain aliphatics like alkanes, alcohols, aldehydes, fatty acids, and esters with the addition of varying proportions of cyclic compounds like pentacyclic triterpenoids and hydroxycinnamic acid derivatives. EC can take various shapes according to plant species and plant organ, ranging from amorphous layer to ribbons, filaments, tubes and plates which can produce impressive photographs by scanning electron microscopy. EC morphology is influenced more by the physicochemical properties of the constituents rather than by the underlying cuticular membrane or the means of delivery to the surface. The shape of the wax deposit can also affect hydraulic resistance. Temperature, light intensity and humidity influence wax morphology via their effect on wax composition and probably the rate of deposition.

The hydraulic resistance of the cuticle varies. Generally it is low in tropical plants and high in xerophytic plants, indicating evolutionary adaptation to water limited conditions. Studies of EC mutants (e.g., Zhang et al. 2005; Burow et al. 2008) and experimental removal of EC by mechanical or chemical means (e.g., Araus et al. 1991) indicate that the presence of EC is very important in increasing cuticular resistance.

Stress affects EC load and cuticular resistance on a time scale of few days (Shepherd and Wynne 2006). High irradiance increase EC load. The response is very likely derived from the role of wax in reflecting excess radiation, including UV. The spectral properties of leaves are affected by EC. This has been well documented by numerous publications since that of Blum (1975). Low air humidity increases EC load and sometimes it affects the shape of the deposits. It is a well known phenomenon that plants grown from tissue culture at high humidity have little wax and tend to wilt due to excessive cuticular transpiration. Plant water deficit increases EC load (e.g., Cameron et al. 2006; Shepherd and Wynne 2006). Cuticle-associated gene transcripts in leaves were altered in *Arabidopsis* leaves subjected to drought stress and were associated with increased cuticle thickness and abundance of cuticular lipids (Kosma et al. 2009).

It is therefore evident that the full phenotypic expression of EC deposition potential of any given genotype is realized after plants are exposed to an inductive environment, such as drought, low humidity and high irradiance.

ABA promotes EC deposition. ABA treatment of Jojoba shoots resulted in increased EC load on leaves (Mills et al. 2001). CER6 condensing enzyme is involved with epicuticular wax production and it was found that ABA enhanced CER6 transcript accumulation (Hooker et al. 2002).

2.3 Plant Size and the Development of Water Deficit

Besides the factors controlling transpiration at the single leaf level, a most dominant factor in controlling whole plant and crop transpiration is total leaf area. Any amateur gardener knows that a large plant grown in a pot will require irrigation more frequently than a smaller one for the same pot volume. The disregard for the role of plant size in plant water relations has become a prevalent pitfall in pot experiments (Sect. 4.1.5.1). A major avenue by which plant evolution impacted plant adaptation to dry environments was by reduced plant size and growth rate, typical of many xerophytic and native arid land plants. It is also a common observation that when sever water deficit develops lower (older) leaves are desiccated and die first so as to reduce leaf area and plant water use.

At the crop level the demand for water as affected by plant size is controlled by leaf area index (LAI), which is the total area of live leaves per unit ground surface. Crop evapotranspiration (ET) increases with LAI until LAI reaches a maximum threshold beyond which ET does not increase. As the crop matures and leaves senesce, LAI is reduced and so does ET. Plant size and leaf area are important variables in breeding for crop adaptation to water-limited environments (Sect. 3.6).

2.4 Plant Water Status and Plant Stress

Cellular water potential is determined by several components important for cells and their surroundings. These components are derived from the effects of solute, pressure, solids (matrix), and gravity. The effect of gravity is negligible. Accordingly, cell water potential and its components are expressed as follows:

$$\psi_w = \psi_s + \psi_p + \psi_m \qquad (2.2)$$

where the subscripts s, p, and m represent the effects of solute, pressure and matrix. Each component is additive algebraically according to whether it increases (positive) or decreases (negative) the ψ_w as compared to the reference potential which is pure, free water. Whereas for free water ψ_w is null, plant cell ψ_w is always negative.

Solute lowers the chemical potential of water by diluting the water and decreasing the number of water molecules able to move compared to the reference, pure water. In the simplest terms, solutes hold the water in the cell against external pull, such as a water potential gradient developed by transpirational demand. In a similar way, wettable matrices have surface attraction that lowers the chemical potential of water. Since solutes and matric force reduce the chemical potential of water below that of free water their sign is negative. The balance in a plant cell is ψ_p (turgor potential or turgor pressure) which is positive as long as all other components allow it. In most whole-plant and crop physiology studies matric potential is neglected and the major dynamics of tissue water status is considered as the interplay and balance between ψ_w, ψ_s and ψ_p. It can be immediately seen that for a given ψ_w if ψ_s will decrease (become more negative) due to solute accumulation, ψ_p (turgor) will increase. Although some experimental results attempted in the past to show that turgor potential can sometimes be negative, this is a very debatable point. It is unresolved if negative turgor is physiologically possible or it is an apparent result of small errors in the measurement of the other components of water status.

Figure 2.3 demonstrates the most crucial facets of these relationships with special reference to the subject of this publication. The reader should concentrate very carefully on this figure and its discussion because here most mistakes are being made (e.g., Blum et al. 1996) in the interpretation of plant water status, turgor and osmotic adjustment.

As soil moisture is being used and water is transpired SWP and LWP (ψ_w) are reduced (becoming more negative). When soil moisture is abundant (high SWP) water will flow through the root and into the leaf with only a small reduction in LWP. When soil becomes drier, its water potential (SWP) is reduced further and LWP must be further reduced in order to create the necessary gradient differential, which would drive (pull) the water up from the drying soil to the leaf through all the soil and plant resistances in between.

The leaf cells contain various organic and inorganic solutes, which determine leaf OP (ψ_s). Therefore OP is lower (more negative) than LWP and the difference between the two is turgor potential (ψ_p). Turgor is lost (null value) when LWP=OP. Two theoretical cultivars are presented in this figure. Both cultivars have the same OP when the leaf is fully hydrated on the day of irrigation. In both cultivars OP is reduced as LWP is reduced. OP reduction is due to the loss of water from the leaf (*concentration effect*) and due to active solute accumulation in cells (*osmotic adjustment*) (OA). For the same LWP OP of cultivar S is reduced less than cultivar R. Therefore in cultivar S turgor is lost (reaching null) at about LWP of –3 MPa 8

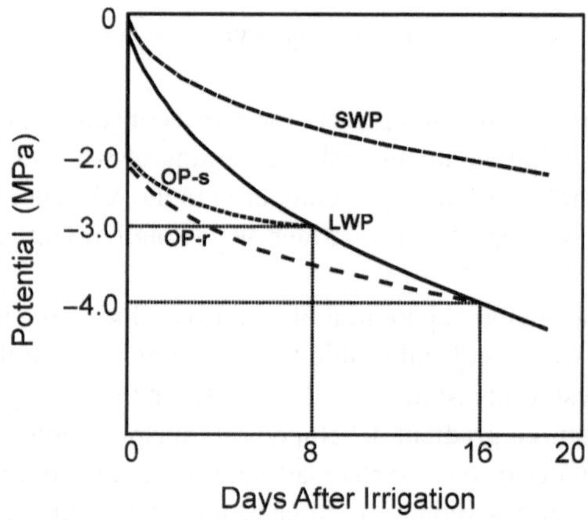

Fig. 2.3 Hypothetical schematic representation of the components of leaf water status during a soil drying cycle. *SWP* – soil water potential; *LWP* – leaf water potential (ψ_w); OP-s and OP-r represent two different cases of change in osmotic potential (ψ_s) with the reduction in LWP. See text

days after irrigation while in cultivar R turgor is lost at –4 MPa at 16 days after irrigation. Cultivar R is able to maintain turgor (and delay its wilting) for a longer period of time due to solute accumulation by OA. It can also be seen that due to its better capacity for OA, cultivar R can continue to draw water from the soil to a lower SWP, as compared with cultivar S. Cultivar R can therefore be defined as relatively drought resistant as compared with cultivar S due to a respective difference in OA. A relatively lower OP in itself is not an indication of better drought resistance or OA because it can result only from a concentration effect without any net solute accumulation.

One of the very first cellular response to water deficit is *cell wall hardening*, physically expressed in the decreased plastic extensibility and increased elastic modulus of the cell wall. This wall hardening appears to be biochemically related to decreased wall acidification and increased cross linking by phenolic substances such as lignins and diferulate bridges (Fan and Neumann 2004; Fan et al. 2006). Cell wall hardening and tightening around the cytoplasm help maintain turgor as water is lost from cells. However, hardening of the cell wall and its reduced extensibility will diminish and even stop cell growth. There is therefore a certain trade-off between cellular growth and turgor maintenance via cell wall hardening (Neumann 1995). In comparison, osmotic adjustment OA cast little if any direct cost in terms of cellular growth in turn for its effect on turgor maintenance. There is even a notion that plant species which are more capable of maintaining turgor via cell wall hardening tend to lack in OA (Barker et al. 1993).

Cell wall hardening under drought stress is reversible, depending on the rate of cellular dehydration. Chazen and Neumann (1994) claimed that the signal for cell wall hardening under water deficit was totally hydraulic in their study. However, other signals cannot be overruled, such as may be the case for ABA (e.g., Wu et al.

1994). Because of the trade-off between cell wall hardening and cell growth it is not quite clear whether greater cell wall hardening and its greater sensitivity to water deficit and turgor loss would constitute an advantage or a disadvantage in terms of affecting whole plant drought resistance (Marshall and Dumbroff 1999). It seems that crop species might differ in cell wall extensibility response to water deficit (Barker et al. 1993; Lu and Neumann 1998) so that genetic manipulation towards optimized responses under stress is theoretically imaginable. This issue still remains open in terms of application to breeding.

Leaf wilting is a symptom of turgor loss. Hence wilting is an important simple phenotypic expression of a critical stage in plant water status under drought stress and it is used extensively by breeders for phenotyping during selection under drought stress (Sect. 4.2.2.1). Wilting is displayed by various leaf presentations. In the cereals wilting is expressed by leaf rolling (Fig. 2.4). Gradual leaf movement into the rolled configuration is activated by the loss of turgor in special bulliform cells situated between veins along the axis of the leaf. When these cells loose turgor they initiate leaf curvature until tight rolling is reached at zero turgor. Leaf rolling is very sensitive to leaf turgor changes. Plants may present a daily march in leaf rolling according the daily march of plant water status and turgor. Maximum rolling is seen at about or just after solar noon. Cereals leaves roll as a defence mechanism to reduce net radiation load on the leaf. Rolling reduces transpiration and leaf water use and was found to protect PSII functionality from damage (Nar et al. 2009). As such it is an important adaptive trait for a leaf approaching zero turgor, but it is still a symptom of plant stress. When different genotypes are compared on a given day under drought stress, those with advanced leaf rolling are at a relatively lower water status than those that do not express leaf rolling on that day. *Genotypes expressing relatively delayed leaf rolling might have relatively better access to soil water or better osmotic adjustment.* Therefore, in terms of comparative performance under drought stress, delayed leaf rolling is the preferred phenotype.

2.4.1 Osmotic Adjustment (OA)

OA maintains cell water contents by increasing the osmotic force that can be exerted by cells on their surroundings and thus increasing water uptake. For the same leaf water potential, more water is held in leaf cells with greater OA resulting in higher turgor as compared with leaves with less OA (Fig. 2.3).

The adjustment results from compatible organic solutes accumulating in the cytoplasm which decreases the osmotic potential of the cytosol. Typical compatible solutes are sugars, amino acids such as proline or glycinebetaine, sugar alcohols like mannitol, and other low molecular weight metabolites. Inorganic ions may also drive OA as the case is for potassium in wheat (Morgan 1992). When plants are challenged by salinity the cellular accumulation of sodium can also be used for OA, especially if it is balanced by the accumulation of potassium. However there is a critical high sodium concentration that will toxify the cell.

Fig. 2.4 Symptoms of wilting in four plant species. From left to right: tobacco, cotton, rice and sorghum

Small cells require less solute for the same rate of osmotic adjustment (Cutler et al. 1977). The smaller size and smaller leaves typical of xerophytic plants can be partly ascribed to smaller cells and a better capacity for OA.

Some of the solutes used for OA, especially those produced by photosynthesis and used for growth are subjected to a dynamic balance between the demands by the two sinks: growth and OA. Since cell growth (expansion) is reduced by water deficit before photosynthesis (see below), there is an initial availability of carbon for OA when water deficit develops. The increase in OA allows sustaining cellular hydration and thus support continued photosynthesis and growth at slow rate under stress. When solutes used for OA are not those under heavy demand for growth (e.g., potassium, glycinebetaine), OA is relatively non-competitive to growth.

Cellular dehydration is the signal for active solute accumulation and OA generally increases with the reduction in leaf water potential. This is all too often not understood and can cause serious misinterpretation of experimental results concerning OA and drought resistance. This and other issues pertaining to OA and its role in drought resistance are discussed in Chap. 3, section "Osmotic Adjustment."

2.4.2 Abscisic Acid (ABA)

ABA was first discovered as an endogenous compound causing fruit abscission and it was named "Abscisin-ii" (Ohkuma et al. 1963). Later during the 1960s additional research by others found that this endogenous hormone also caused dormancy and was found in large amounts in wilting leaves. Subsequently it was found that ABA also induced stomatal closure. It was later described as a "stress hormone" because it was produced in plants subjected to various abiotic stresses including salinity, cold and heat all of which can involve cellular dehydration. ABA synthesis and accumulation is highly responsive to tissue water status and it increases with reduction in leaf water potential, turgor or relative water content (RWC). However there is no consensus water status threshold for ABA accumulation in plant tissues.

It is not clear how cellular water deficit induces ABA biosynthesis. The signal may constitute od cellular pressure, membrane modification, solute concentration or cell wall tension. ABA is also produced by roots in response to a drying soil (see below). Plants under drought stress contain significant amounts of ABA in their xylem. Therefore, ABA can potentially reach any plant part which is connected via the xylem. Furthermore, ABA is produced without any stress signaling in certain ripening fruit and developing seed. Detailed analyses of drought affected transcript profiles and comparisons with other studies (Huang et al. 2008) revealed that the ABA-dependent pathways are predominant in the drought stress responses. These comparisons also showed that other plant hormones including jasmonic acid, auxin, cytokinin, ethylene, brassinosteroids, and gibberellins also affected drought-related gene expression, of which the most significant was jasmonic acid. There is also extensive cross-talk between responses to drought and other environmental factors including light and biotic stresses. These analyses suggest that ABA-related stress responses are modulated by various environmental and developmental cues.

The involvement of ABA in stress perception, signaling and gene response has been reviewed and discussed by Zhang et al. (2006), Shinozaki and Yamaguchi-Shinozaki (2007), and Nakashima et al. (2009). The later reviews provide more detail but the general scheme and primary outline remain, as presented in Fig. 2.5.

2.4.2.1 ABA as a Non-Hydraulic Long-Distance Root Signal

ABA is produced in roots when they are exposed to a dry and hard soil. ABA is then found at high concentration in the xylem sap ascending from the root. ABA transported in the xylem signals the various known ABA responses in the shoot (Davies et al. 2005). Xylem sap pH is involved with xylem sap ABA activity where high pH generally enhances ABA effectiveness in the shoot. Soil drying has been shown to increase xylem sap pH. ABA solubility, transport, concentration and activity in different plant organs and cellular compartments are affected by pH. At the same time it was found that xylem sap alkalization under the effect of soil drying is not universal across all species tested (Sharp and Davies 2009). Hence, the role of pH in controlling ABA signaling is not clear and subjected to various theories (e.g., Zhang et al. 2006; Davies et al. 2005; Sharp and Davies 2009).

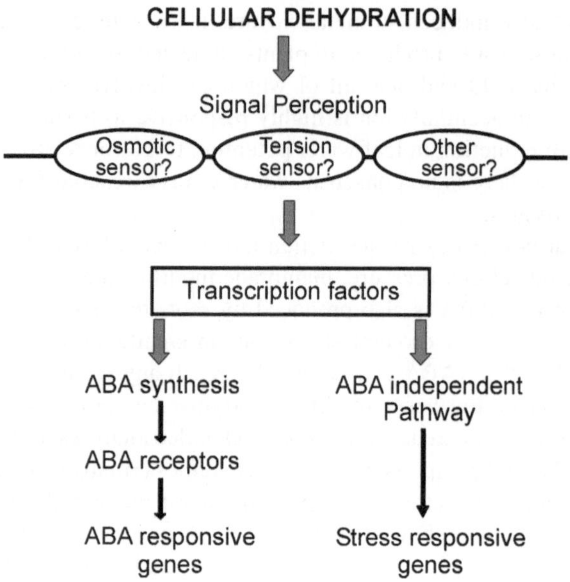

Fig. 2.5 A schematic representation of the molecular basis of drought stress perception and signaling pathways

ABA produced in the root is defined as a "long distance non-hydraulic root signal." At the very early stages of soil drying when some of the roots are exposed to a drying soil, other (deeper) roots are well supplied with water. Under such conditions the hydraulic status of the shoot is favorable. However, at the same time ABA originating from the drying roots reaches the shoot and cause stomatal closure, arrested growth and other consequences of ABA signaling. This scenario seems to contradict the Ohm's law analogy of the SPAC in the control of plant water status and its consequences in the plant, as described above. However after some debate in the literature a wide consensus has apparently been reached. The role of the hormonal signal is important at the earlier stages of drought stress while the hydraulic signal comes into full control when stress increases (e.g., Christmann et al. 2007). Galmés et al. (2007) offered an interesting concept where as soil moisture is reduced aquaporins help maintain hydraulic homeostasis while the reduction in soil water status already induces a hormonal signal to close stomata.

This is a very reasonable *modus operandi* for the native plant. The hormonal root signal serves as an "early warning system." This alarm, coming up from the root via the xylem, causes the most important early effects of ABA in the shoot, namely stomatal closure and retarded leaf growth. Both of these consequences are the effective controls of plant water use. They serve to delay the expected hydraulic signal and the consequent reduction in plant water status and turgor. It should be pointed out however that a hormonal root signal was not seen in all experiments designed to identify a hormonal root signal impacted by a dry soil (e.g., Christmann et al. 2007; Whalley et al. 2006). It could be possible that this hormonal root signal might have been perturbed by plant domestication and selection in certain cases.

The main question towards application of this knowledge in plant breeding for water-limited conditions is what should be the preferred response model – a sensitive or an insensitive "early warning system?" The "early warning system" is not so fine-tuned towards the control of water deficit response in the plant since other factors besides drought induce ABA production in roots and an increase in its mobility to the shoot, such as certain nutrient relations (Jeschke and Hartung 2000), soil strength salinity (Shaterian et al. 2005) and even certain soil biota (Dodd 2009). Furthermore, other hormones were also found to interact with ABA or with the effect of ABA on the shoot, such as the case for ethylene which can also be produced in the root (Sharp 2002). Beyond and above all these issues it is still not perfectly clear whether ABA in the shoot is a positive or a negative presence when plant production and yield under drought stress is concerned. Thus the role of the hormonal root signal in rainfed dryland crop production is not clear and the prediction of its effect on yield is not forthcoming. Genetic differences for hormonal root signal may exist, such as the case for grapevine (Beis and Patakas 2010). Blum and Sinmena (1995) tried to obtain some answers by isolating ABA insensitive variants of wheat and by studying their function under root signal promoting conditions (through partial root drying). However that study did not produce conclusive results (unpublished). A study with wheat grown in soil in pots (Xiong et al. 2007) concluded that a hormonal root signal produced at high soil moisture content was desirable for drought resistant wheat, thus favoring early stomatal closure at the onset of drought stress. This is in contrast to the current consensus that sustained stomatal conductance and transpiration under stress will support yield (Sect. 3.5.1).

Crop simulation models can be a potentially effective tool to assess if, when and where a hormonal root signal is desirable towards plant production under drought stress. Modeling efforts in this direction were initiated (e.g., Gutschick and Simonneau 2002) but they are still not perfected to the stage where they can be used as a decision support system, especially not with regard to breeding.

It should however be noted that the current emerging consensus among practicing dryland crop plant breeders is that sustained stomatal conductance and transpiration under stress will support yield (Araus et al. 2002; Blum 2009; Munns and Richards 2007). This view is compatible with plants which are less sensitive to ABA in general and to a hormonal root signal specifically.

The only available application of knowledge on hormonal root signal is being made in certain irrigation schemes. Irrigation methods that involve partial root drying (PRD) induce a hormonal signal while the plant remains well hydrated (Davies et al. 2002a, b). Irrigation of part of the roots can be technically achieved in crops normally irrigated by drip or subsurface irrigation. Crops (mainly fruit trees, olives and grapevine) irrigated by PRD show an improvement in irrigation water use efficiency for yield. Yield is often reduced somewhat in comparison with full irrigation but water use decreases proportionally more. Fruit quality has been found to improve with PRD and this might be a decisive factor in adopting PRD in choice fruit orchards. Still, fruit quality in this context is not a simple characteristic. For example, drought stressed peach produced less but larger fruit. Larger fruit fetches better price but at the same time the fruit was of poor taste (Lopez et al. 2010).

2.4.2.2 ABA Effects in Plants

Several reviews deal with the wide array of effects that ABA causes in plants (Sharp 2002; Wilkinson and Davies 2005; Liu et al. 2005; Zhang et al. 2006). Whereas ABA is often defined as a "stress hormone" which ascribes "drought tolerance" to plants, it is very important to recognize the pros and cons of high ABA concentration in various plant organs. This should allow weighing the different effects and their sum totals under given stress scenarios and given agricultural ecosystem. In view of the huge and growing literature on ABA the reader should be well informed if, where and when ABA is a blessing or nuisance towards plant production in water-limited environments. The fact that ABA is part of the stress response transcription network (Fig. 2.5) does not necessarily imply that it is a positive component of drought resistance in the agronomic perspective.

It has long been established that the most prominent effect of ABA besides stomatal closure is general *shoot growth retardation*. Figure 2.6 is a simple visual representation of the growth inhibition of wheat plants caused by an increasing physiological concentration of ABA in the root medium. Inhibition mainly involved reduced leaf size and tillering, accompanied by some loss of chlorophyll at the higher concentration. When heat stress was applied (right panel) the effect of ABA on growth reduction was amplified. It appears that heat stress and ABA effects were additive. The apparent effect of ABA under heat stress was partly caused by stomatal closure. Plants with closed stomata were less capable of transpirational cooling, causing leaf temperatures to rise to lethal levels at the highest ABA concentration. The highest concentration in itself was not lethal (left panel).

In another experiment (Fig. 2.7) wheat was grown in aerated nutrient solution. PEG was added to impose drought stress. Roots were separated from the nutrient solution by a semi-permeable membrane so as to avoid direct contact of PEG with the root (Sect. 4.2.3.2). ABA reduced growth by about 65% and PEG reduced growth by about 40%, as compared with the control. It can be seen that ABA in the nutrient solution did not provide any protection to wheat growth under drought stress. Rather, it affected growth in an additive fashion to drought stress.

Growth retardation by ABA can be caused by stomatal closure and reduced photosynthesis. In the short term growth retardation by ABA results from the inhibition of both cell expansion and cell division. The retardation of cell division seems to be caused by reduced DNA synthesis through inactivating some DNA-replication origins resulting in a lengthening of the replicon size (Jacqmard et al. 1995). Reduced tillering under the effect of ABA has long been observed in the cereals (e.g., Harrison and Kaufman 1980) and its effect may be assigned to the

Fig. 2.6 Effect of ABA on wheat growth under two temperature regimes. Plants were grown in the growth chamber in aerated nutrient solution at 15°/25°C (night/day) (control – *left panel*). Two weeks before this photo was taken the plants in the *right panel* were transferred to 25°/37°C (chronic heat stress). At the same time ABA was added to the nutrient solution in the different pans, at concentrations of 0–10 μm. Author's unpublished experiment

Fig. 2.7 The effect of 50 μm ABA on wheat grown in aerated nutrient solution with or without polyethylene glycol (PEG8000). Wheat was grown in pure nutrient solution until 2 weeks before this photo was taken when PEG was added in three daily increments to reach a final solution water potential of –0.5 MPa so as to impose drought stress. ABA was then added. Wheat was grown in vermiculite in vials where a semi-permeable membrane separates the roots from the nutrient solution (Sect. 4.2.3.2). Author's unpublished experiment

inhibition by ABA of kinetin. There are indications that ABA might enhance growth when endogenous ethylene accumulation is the cause of growth retardation (Sharp and LeNoble 2001). Xylem and apoplastic pH can affect the way in which ABA regulates stomatal activity and leaf growth.

It has long been established that ABA promote *root growth* (e.g., Munns and Sharp 1993). ABA also increase root hydraulic conductivity (see above), presumably by enhancing root aquaporins activity. Since ABA also reduces leaf area, the result is the often observed increase in the root/shoot dry matter ratio under the effect of drought stress. Such higher ratio has a major impact on the crop SPAC and the maintenance crop water status when drought stress develops. Thus, ABA involvement in root growth and function can enhance plant performance under drought stress in the field. However, this effect should be considered within the complete array of ABA effects on the plant.

ABA cause *flower abscission* (Aneja et al. 2004) in tune with the initial discovery of its effect in plants (Ohkuma et al. 1963). Elevated concentration of ABA in the reproductive structures may inhibit embryonic cell division and subsequently impair *fruit and seed set and development*. In soybean, drought-induced increase in xylem ABA concentration, and not pod water potential, was found to control pod growth (Liu et al. 2003). In wheat, grain set was negatively correlated with the endogenous ABA concentration under drought (Westgate et al. 1996). Application of ABA to wheat leaf sheath of well watered plants inhibits floret development, and decreases the number of fertile florets and grain set (Morgan 1980; Wang et al. 2001). The effect of ABA towards flower sterility is largely mediated by pollen dysfunction (Oliver et al. 2007) following the inhibition of pollen germination and pollen tube growth (Frascaroli and Tuberosa 1993). There is therefore compelling evidence indicating that ABA seriously hampers plant reproduction, which in grain and fruit crops translates into yield reduction.

In the cereals, ABA inhibits *endosperm cell division* on one hand (Mambelli and Setter 1998) and on the other hand it promotes starch accumulation and *grain filling* in wheat and rice (Yang et al. 2004, 2006). This was attributed mainly to the enhanced sink activity by regulation of key enzymes involved in starch synthesis. In contrast to the above results in wheat and rice it has been shown in maize (Cheikh and Jones 1994) that the reduction of grain growth under heat stress involved the inhibiting effect of ABA.

ABA seems to have a role in enhancing *stem reserve* mobilization into the growing grain in rice and wheat (Yang et al. 2001, 2003), which is linked to accelerated leaf senescence. Kinetin delayed senescence and reduced stem reserve mobilization to the grain.

ABA enhances plant *senescence* in contrast to kinetin. This has been seen in the known expressions of senescence such as chlorophyll breakdown (Figs. 2.6 and 2.7) and specific changes in cell ultra structure (e.g., chromatin condensation, thylakoid swelling, plastoglobuli accumulation) and metabolism (e.g., protein degradation, lipid peroxidation) (Munne-Bosch and Alegre 2004). There is therefore compelling evidence that ABA is involved in the breakdown and transport of storage materials from senescing leaves into the developing grain while kinetin acts to conserve leaf viability.

ABA treatment of *Poa bulbosa* L a summer perennial grass geophyte (Ofir and Kigel 1998), resulted in cessation of leaf and tiller production and in the development of typical features of *dormancy*, namely bulbing at the base of the tillers and leaf senescence. Photoperiodic induction and heat stress, both of which are known to induce dormancy in this plant were accompanied by an increase in endogenous ABA concentration at the tiller base. ABA induced grape bud dormancy and the rate of dormancy was proportional to ABA concentration (Or et al. 2000). Dormancy of rose buds cultured in vitro could be broken by fluoridone an inhibitor of ABA synthesis. Dormancy was regained by constant ABA application (Le Bris et al. 1999). High endogenous ABA or high seed embryo sensitivity to ABA

Table 2.1 A summary of ABA effects and consequences in the plant

Trait	Effect
General growth	Decrease
Cell division	Decrease
Cell expansion	Decrease
Germination	Decrease
Tillering	Decrease
Root growth	Increase
Root hydraulic conductance	Increase
Flower abscission	Increase
Pollen viability	Decrease
Seed and fruit set	Decrease
Grain and fruit growth	Decrease
Starch synthesis in cereal grains	Increase
Plant reserve mobilization to the grain	Increase
Leaf senescence	Increase
Dormancy	Increase

retained embryo dormancy in maturing seed of sorghum (Steinbach et al. 1997) and wheat (Rasmussen et al. 1997).

When all of the above results pertaining to the cereal grain are taken together it appears that ABA reduces grain size but enhances stored assimilates transport into the grain and starch synthesis in the grain. If ABA supply to the grain is sustained, it will induce dormancy upon maturation.

Table 2.1 offers a concise summary of the positive and negative consequences of ABA, in terms of the final effect on plant production under drought stress. It helps explain the reduction seen in yield of wheat (Quarrie 1991) and maize (Sanguineti et al. 1996) lines selected for a constitutive capacity for high leaf ABA content. Near isogenic maize lines constitutively producing high or low leaf ABA content were developed by backcrosses. The difference in ABA accumulation was mainly due to one major QTL. The effect of this QTL was evaluated in testcrosses subjected to drought stress and non-stress conditions in the field (Landi et al. 2007). The effect of the high leaf ABA QTL was seen in lower yield under both water regimes indicating a basic negative effect of ABA accumulation on maize yield. Selection for low leaf ABA resulted in higher yielding maize under non-stress and moderate stress conditions (Landi et al. 2001). On the other hand Kholova et al. (2010) found that pearl millet lines resistant to terminal drought stress had constitutively higher leaf ABA content. They argued that water-saving due to apparent moderate water-use under the effect of high ABA was beneficial for sustaining the final stages of growth and grain filling under drought stress. It might be added here also that constitutively high ABA content could perhaps has enhanced stem reserve utilization for grain filling under terminal stress (see above).

It can therefore be speculated very reasonably that ABA evolved as a life conserving mechanism when the plant enters a stress situation. Where drought stress is concerned the first consequences of ABA activity are to reduce water use and conserve plant hydration via reduced shoot growth, reduced stomatal conductance and promoted root growth and its hydraulic conductance. As stress increases ABA serve to reduce the sink load (see below) on the stressed plants by reducing the number of the developing fruit and/or seed. However, few remaining seed are still retained and filled well. When total plant assimilate production is limited by stress it would be a reasonable strategy to limit the number of sinks in order to produce at least a few viable seeds. Filling of the remaining seed in the cereals is supported by ABA induced stem reserve mobilization. Dormancy is then affected in order to conserve the seed until the next season.

This *survival* strategy is extremely important to the plant in terms of its ontogenicity and evolution. However, when this plant is used for the farmer's livelihood, other considerations can be more important and they may not fit the above built-in strategy of ABA regulation (Table 2.1). If we understand this, the way for manipulating ABA signaling towards sustained plant production under drought stress will open. However, one thing must be absolutely clear: ABA cannot be arbitrarily defined (as sometimes seen in the literature) as a "drought resistance hormone." It is a stress hormone.

2.5 Growth and Water Deficit

Cell growth depends on turgor and cell wall extensibility. The relationship is described by the classical Lockhart equation (Lockhart 1965). Expansion rate of a cell equals to m (P – Y), where rn is the extensibility of the cell wall, P is the turgor pressure, and Y is a minimum value of P below which the cell will not grow. Passioura and Fry (1992) argued that Y (and sometimes m) may vary in response to changes in P on a time scale of about 10 min. The result is that, apart from the transient responses, cell expansion rate is often maintained at an approximately steady value despite changes in P. This has been later supported by data of others, such as Serpe and Matthews (2000) indicating it to be the case at least for moderate decrease in turgor. Cell wall growth therefore accounts for how m and Y may vary to maintain a constant growth rate despite moderate changes in turgor.

During growth, plant cells secrete proteins called "expansins," which unlock the network of cell wall polysaccharides, permitting turgor-driven cell enlargement. For example, expansins were implicated in the drought responses of maize seedlings, where maintenance of root growth involved increased expansin activity in the growing region (Wu et al. 1996). Drought increases the expression of expansin genes in a spatial and temporal pattern that closely matches the changes in expansin protein activity (Cosgrove 2000).

Inhibition of cell expansion under drought stress involves both the reduction in turgor and the loss of cell wall extensibility. Loss of cell wall extensibility also involves changes in polysaccharide content and structure in the cell wall. In the resurrection plant *Myrothamnus flabellifolius* (Moore et al. 2006) constitutive presence of high concentration of arabinose in cell walls provide the necessary structural properties to be able to undergo repeated periods of desiccation and rehydration. Genetic engineering of specific cell wall properties was suggested by Cosgrove (2000) as a potential option for drought resistance improvement. However there are vast complexities that still exist in attempting to understand how cells grow especially under environmental cues such as water deficit.

Cell division can occur only after cells reach a certain size. Old views considered that cell growth and enlargement was more sensitive to water deficit than cell division. More recently it has been found for sunflower leaves that cell division and enlargement were similarly affected by water deficit (Granier and Tardieu 1999). In another study by the same group (Tardieu and Granier 2000) it was shown that water deficit reduced the final cell number in leaves by way of increasing cell cycle duration. More studies are requires before a universal rule can be established regarding the relative sensitivity to water deficit of cell division and cell enlargement. One must also consider that cell division take place in certain growth regions of the young leaf while cell enlargement take place in various parts of young and old leaves. The specific cellular position and environment within the leaf can have a decisive effect on cell sensitivity to measured bulk leaf water deficit.

The integrated and final effect of both cell enlargement and division on leaf growth under stress is the important issue in terms of the whole plant in the field. It has been argued on the basis of experimental work with maize (Reymond et al. 2003) that a single leaf growth under drought stress can be predicted and its genetic background might be resolved. Basic growth process of plant tissues might be under a universal genetic control, whether under non-stress or stress conditions (Welcker et al. 2007). This is certainly an attractive proposition implying that plant growth under drought stress might perhaps be amenable to simple genetic manipulation despite the plant's apparent complexity.

However there are still major plant structural and physiological components to consider where whole plant growth under drought stress is considered. Whole plant structural and morphological features are relatively stable under drought stress as compared with features of dynamic organ expansion. Meristem and organ differentiation seem to be relatively resilient as compared with expansion growth. Any experienced agronomist will confirm that determinate plants subjected to drought stress will nearly always maintain the same number of leaves but leaves become smaller. Hence, differentiation and expansion growth must be treated differently in order to understand and manipulate whole plant response to drought stress.

The leaf canopy constitutes a major control over transpirational demand of the crop as well as the crop light interception. Canopy development and size at any given day in the field is determined by the expansion of all of the growing leaves as well as by leaf number and the senescence of older leaves. Plants subjected to soil moisture deficit develop a gradient of water potential such that leaves at a higher insertion are at lower water potential than leaves at lower insertion – with all the consequences of leaf water potential, turgor and their effect on growth. Leaf expansion is reduced by water deficit before leaf photosynthesis is inhibited. Hence, photosynthate that has been normally used for leaf expansion is now available for either osmotic adjustment or translocation. Light distribution in the canopy and the extinction coefficient also impose a variable vertical profile of photosynthesis in addition to leaf water potential and leaf age.

Taken together, all this translates into the fact that at any given time each leaf in the canopy is very unique in its own physiology and microenvironment and the response to soil water deficit. If one considers also a flux of ABA (and possibly xylem sap pH) ascending in the xylem along this gradient, then we have a very complex system for simulation. Furthermore, in terms of the purpose of simulation towards plant breeding one has to consider the interpretation of the model. For example, high rate of leaf death under stress will usually be taken as stress inflicted damage to the crop and its productivity. This however is not necessarily always the case as seen in sorghum, a relatively drought resistant plant. When drought stress develops not all green viable leaves respond similarly in stomatal closure. Rather, older leaves senesce and die while upper younger leaves retain full turgidity and open stomata. Thus, whole plant water use is reduced but leaf gas exchange is retained in the most viable and light-exposed part of the canopy (Blum and Arkin 1984). Furthermore, leaf senescence under stress can also be linked to enhanced stem reserves mobilization into the grain as discussed above.

It therefore seems that designing a plant that can sustain growth and productivity when its tissues are dehydrated is not forthcoming. What appears to be the solution at the present state of our knowledge is to design a plant that can avoid dehydration. It also appears that plant reserve mobilization into the growing grain is a powerful resource for enhancing grain yield under stress during grain filling.

2.6 Root Growth Under Drought Stress

When drought stress develops, root-to-shoot ratio increases in terms of final dry matter weight. Total root dry matter very rarely increase in absolute terms under drought as compared with non-stress conditions. However this change in ratio also indicates that root-length density per unit live leaf area generally increases. Root-length density at deep soil may increase relative to root length density at shallow soil.

Four factors are behind the relative (or in rare cases the absolute) increase in root growth under drought stress. These are not totally independent and certain interactions between factors in affecting root growth were noted.

Firstly, Root growth is less sensitive than leaf growth to the same tissue low water potential (Hsiao and Xu 2000). The reason is in the greater osmotic adjustment in the extension region of roots as compared with leaves (Ober and Sharp 2007). In the apical few millimeters of the primary root of maize seedlings, proline concentration increased dramatically under water deficit. It could contribute up to 50% of osmotic adjustment (Sharp et al. 2004). However besides proline certain photosynthetic products also serve as osmoticum in roots. Since leaf expansion is arrested before photosynthesis is affected by shoot water deficit, some of the excess carbohydrates are assumed to be diverted to the root, supporting osmotic adjustment and root growth. Even shoot osmotic adjustment can drive deeper soil moisture extraction (e.g., Chimenti et al. 2006).

Secondly, it has long been established that ABA promotes root growth while it inhibits shoot growth (see above). The role of ABA accumulation in roots in enhancing root growth in a drying soil has been clearly proven by the use of fluoridone (an ABA synthesis inhibitor) and by two ABA knockdown mutants (Sharp et al. 2004; Ober and Sharp 2007). Loss of ABA synthesis capacity hindered root growth only under drought stress. However there were difference in root growth response to drought stress and ABA between the root tip and the immediate growth zone above it, indicating a complex control of ABA function in roots subjected to water deficit. Cytokinins are involved in inhibiting root branching and enhancing primary root growth (Havlová et al. 2008) by preventing the formation of an auxin gradient that is required to pattern lateral root primordia (Laplaze et al. 2008).

Thirdly, cell wall expansion is an important factor in enhancing root growth in a drying soil. The importance of expansin proteins and the expression of expansin genes in this respect have already been discussed above.

Fourthly, plant morphological and developmental interactions can greatly modify root growth in a drying soil and determine root distribution in the soil, especially in the cereals. In sorghum, crown (adventitious) roots are formed in a distinct temporal cycle from buds in the basal stem internodes. When the top soil is wet the initiated crown roots penetrate into the soil, grow and constitute the major part of the root system that occupies the top wet soil (Fig. 2.8). If the

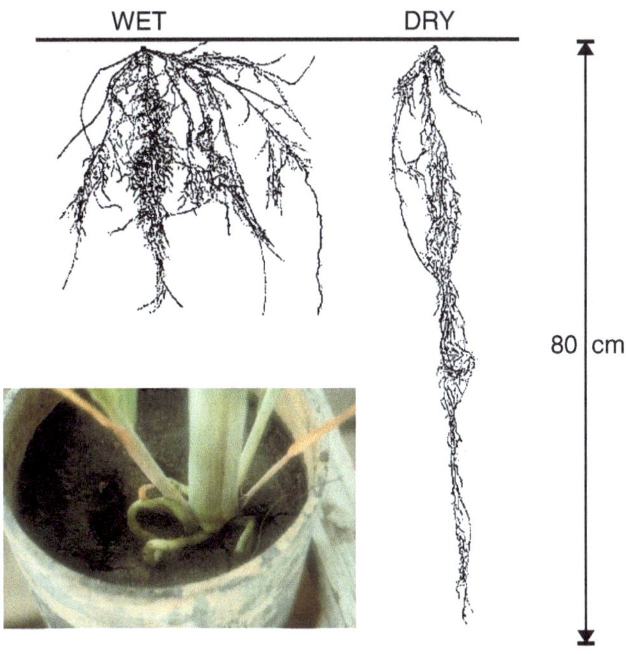

Fig. 2.8 Computer enhanced display of sorghum adventitious (*crown*) roots grown in fully wet soil (*left*) and in soil in which only the top was dry but was wet deeper (*right*) (Blum and Ritchie 1984; with permission). Inset: initiated crown roots that could not penetrate the hard top-soil (see text)

top-soil is dry and hard (while deep soil is still wet), these crown roots do not penetrate the soil (Blum and Ritchie 1984). Photosynthate that would have been used for the growth of these new roots is diverted towards the growth of existing roots, which grow deeper into the soil. Therefore, by limiting crown root number per plant, a dry soil surface causes compensatory growth in existing roots, which subsequently reach deeper soil layers. Different root distribution profile results from the balance between crown root number and crown root growth. Similar results were found by Troughton (1980) in perennial ryegrass where crown roots were associated with tillering. Consequently in a tillering crop plant such as wheat, a drying soil was found to limit root growth at the top 30 cm while promoting root extension and growth into depth (Asseng et al. 1998). Once re-watered, plants reverted to fast root growth at the top soil at the expense of deep soil.

2.7 The Formation of Yield and Drought Stress

Crop biomass production is linearly related to crop transpiration, or water use. The equation first proposed by de Wit (1958) still stands: $B = mT/E_0$, Where B = total crop biomass, m = crop constant, T = crop transpiration and E_0 = free water (potential) evaporation. This equation is the foundation for our understanding that biomass production is linked to transpiration. The primary consideration for enabling total plant production under drought stress is sustained transpiration All other considerations as much as they may be important are secondary when production is concerned.

The development of a relationship between economic yield (e.g., grain, fruit, fiber, or tuber) and water use is far more complex whereas economic yield is not equated with total biomass. A first approximation is achieved through the introduction of the "harvest index" (HI) to the calculation, where a given crop-specific fraction of the total dry matter is partitioned into economic yield. This approximation is imperfect, as the harvest index changes with the water regime especially when drought stress occurs towards the end of the crop season. As such HI is a complex result and balance of genetic and environmental effects when different genotypes are compared. *HI is not an explanation. It is a result.* HI is useful tool for the analysis of results rather than a tool for obtaining results in breeding. The most well known analysis involving HI is that which is done repeatedly for various crops since the first study in wheat by Austin et al. (1980). They showed that most of the historical genetic progress in grain yield was obtained by a *de facto* increase in HI rather than by an increase in biomass production at a given HI (with few exceptions). However, HI does not help explain the basis of this change in ratio in the course of historical modern plant breeding. The speculation offered here is that selection for yield alone (as done historically) put a selective pressure on morphology and assimilate partitioning process but not on basic plant production (biomass). We have no idea what would have resulted in the historical perspective if both biomass and yield were persistently selected during the breeding process. Certainly, the occasional voiced or written recommendations to select for HI as a conclusion from the historical analysis in this or other cases is an example of a misguided conclusion based on this ratio. The lesson learned from the historical perspective is that biomass and yield should be selected for while retaining HI.

Most research on yield formation has been done in the cereals. A useful approach to understand yield has already been developed years ago by defining *yield components*. Hence, yield of wheat, barley, sorghum, millet or rice is the multiplication of the number of inflorescences per unit area of land, by the number of grains per inflorescence, by single grain weight. Even the analysis of panicle weight components allowed better understanding of yield formation and heterosis in sorghum (Blum 1970, 1977). The reader can derive the definition of the yield components of maize, sunflower, pulses or cotton etc.

All yield components taken together constitute the "sink" while all assimilate contributing parts of the plant are considered the "source." Whereas certain yield components can be developmentally interactive, such as grain weight and the number of grains per inflorescence in sorghum, component compensation is an important developmental mechanism for reconstituting yield under or upon recovery from stress - to a limit. For example, if tiller number is reduced by stress, grain mass per inflorescence can increase upon recovery via grain number or grain weight – depending on source activity and sink structure. It is not uncommon to observe an increase in sorghum grain weight under drought stress, due to a decrease in grain number per panicle, or an increase in grain number per panicle in compensation for a decrease in panicle number (Blum 2004).

The plant has a large potential for the creation of yield sinks, beyond what is realized even under a non-stress conditions. Cotton produces more flower buds and wheat produces more tillers or more florets that will ever bear fruit to maturity. Despite constant breeding for a more efficient cereal plant, excessive tillering and the natural degeneration of a proportion of the tillers have been apparently retained in present cultivars. This may have been the result of the selection pressure for stability of yield across different environments. Plants without a capacity for plastic development may lack in adaptation to variable growing condition.

Drought stress can reduce yield by affecting the sink or the source. Source capacity is reduced under drought stress as a result of stress effects on leaf area, gas exchange and carbon storage available for grain filling as well as from an increase in leaf senescence and the increase in rate of certain developmental processes.

The reduction in sink capacity under drought stress is caused by arrested organ differentiation as well as by the dysfunction of the differentiated reproductive organs. Thus, for example, drought stress reduces the number of tillers either by stopping their sequence of differentiation or by death of growing or grown tillers. The number of flowers (or florets) in the inflorescence will be reduced by arrested differentiation or by abortion and degeneration of developed flowers under stress. The reduction in the number of grains developed from a given number of flowers in the inflorescence can be affected by induced sterility of female or male organs as well as by stress induced abortion of embryos.

There is a very large volume of evidence that the most drought stress sensitive plant growth stage is flowering. This can be seen in the classical presentation by O'Toole (1982) with contribution from TC Hsiao (Fig. 2.9) where yield of rice is reduced most when stress occurs during plant reproduction. Peak stress sensitivity is at anthesis and fertilization. This presentation for rice represents well most if not all other cases of grain and fruit bearing crops.

A well demonstrated case for a non-cereal crop has been described for chickpeas where drought stress at the reproductive growth stage caused flower abortion, as well as pistil (or stamen) and pollen failure causing a reduction in total seed number per plant (Fang et al. 2010). The specific sensitivity of reproduction to drought stress is compounded by the fact that plants at flowering are large and pose a heavy demand for water. Reproductive failure is basically irreversible unless non-determinate crop plants are considered. There the failed reproductive organs cannot re-grow but they can be replaced upon recovery by new growth and the differentiation of new reproductive organs. Depending on their inherent drought resistance, non-determinate crop species offer better probability for recovering some yield under later season drought stress.

During its differentiation and early growth the flower or the inflorescence is usually protected by other tissues against excessive water loss, at least in comparison

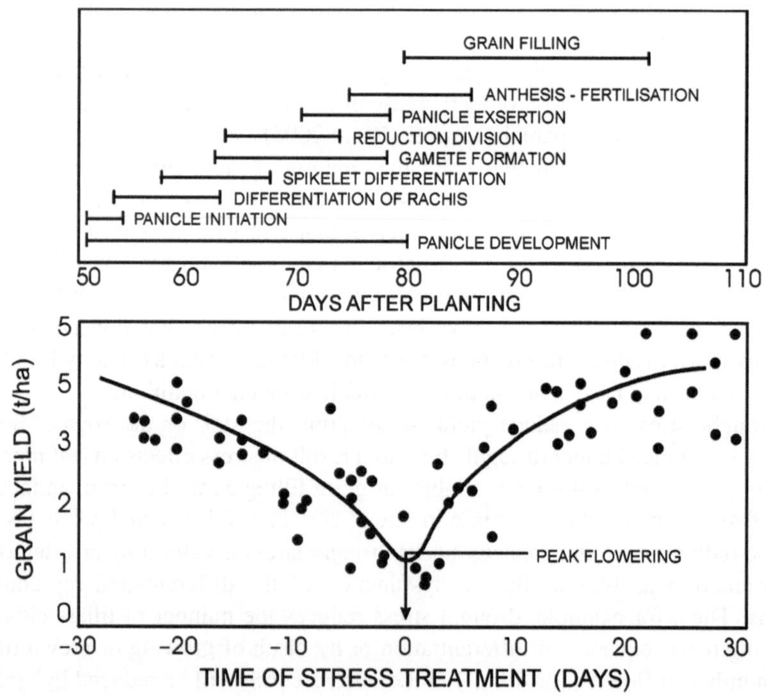

Fig. 2.9 The effect of drought stress applied at different rice growth stages on grain yield (Adapted by O'Toole from personal data by T.C. Hsiao and as presented in O'Toole (1982). With permission) *Bottom panel*: yield when stress occurred at different times during growth. *Top panel*: the respective stages of reproductive development when stress occurred

with exposed leaves. In the cereals and some other crop plant species the inflorescence is relatively protected against evaporation by protective surfaces such as thick cuticle and heavy epicuticular wax load. For these reasons, at least, the water status of the inflorescence may be expected to be better than that of the leaf. Still, a reduction in inflorescence or flower water status under severe stress can occur, thereby causing reproductive failure. However, even in a hydrophyte species such as rice, manipulations to reduce panicle transpiration and improve its water status did not increase its fertility under conditions of soil moisture stress (Garrity et al. 1986). The increase in shoot ABA concentration under drought stress, whether produced in situ or imported from the root (see Sect. 2.4.2.2) is a most likely reason for reproductive failure – irrespective of inflorescence water status.

In Maize, grain water status was reported to be stable under varying conditions (Borras et al. 2003) and ovary turgor under drought stress was the same as under non-stress conditions (Schussler and Westgate 1995). However, in wheat grain filling processes under stress conditions were partly limited by low grain water status as well as by the reduced assimilate supply to the grain (Ahmadi and Baker 2001). It seems that grain water status or sensitivity of grain growth to grain water status is not the only factor controlling grain growth. Each of the above sink-limiting dysfunctions which are caused by drought stress can be mediated by the direct effect of the water deficit of the organ in question, by plant hormones or by reduction in the supply of carbohydrates to the organ. The first two factors were already discussed above. The third one warrants further consideration.

In maize, female florets were more sensitive to reduction in water potential than pollen (Boyer and Westgate 2004). It was further found that invertase activity was inhibited and starch content was diminished in the ovaries. Sucrose infused to the stems of droughted maize rescued many of the ovaries otherwise destined to abort. Sucrose feeding restored some of the ovary starch and invertase activity. These studies indicate that sugar deficiency in maize ovaries was an important cause of abortion under drought stress. Invertase is implicated as a limiting enzyme step for grain yields during drought stress. On the other hand the ovary was more resilient than pollen in drought stressed wheat (Ji et al. 2010) while carbohydrate availability supported anther resilience.

The role of carbohydrate deficiency in pod abortion under drought stress was also evidenced in soybeans (Fulai et al. 2004). Sherson et al. (2003) concluded that the hydrolysis of sucrose by cell-wall invertase and the subsequent import of hexose into target cells appear to be crucial for appropriate metabolism, growth and differentiation in plants.

Sugar concentration in plant tissues constitutes an important signal, and sugar responsive genes have a role in the response of plants to drought stress (Koch 1996; Smeekens 1998). Sugar responsive genes participate in the control of resource distribution among tissues and organs. Carbohydrate depletion up regulates genes for photosynthesis, remobilization, and export, while decreasing mRNAs for storage and utilization. A role for plant hormones (particularly ABA) in sugar-response pathways was found by using various ABA mutants of *Arabidopsis thaliana*. ABA might be important also in regulating tissue response to sugar.

An additional important source for grain filling is carbohydrate reserves stored in the stem in the form of starch or fructan. Whenever the demand by the sink for assimilates grows beyond the supply by the current source, available stem reserves may be used for grain filling. Stem reserves can also be used in tandem with the current assimilate supply by photosynthesis. A large amount of information is available on the importance of stem reserve utilization (SRU) for grain filling especially under drought and heat stress. SRU was found to be important for grain filling in wheat, barley, triticale, rice, maize, sorghum, pearl millet, safflower, sunflower, chickpeas. SRU was ineffective in lupine (Palta et al. 2007). It was concluded that an important physiological component of the increase in wheat yield in the UK from 1972 to 1995 was a larger source for grain filling through increases in stem carbohydrate reserves (Shearman et al. 2005). SRU contribution to grain mass depends on the amount stored and the capacity for remobilization of storage to the grain. Further discussion SRU role in drought resistance is presented in Chap. 3, section "Stem Reserve Utilization for Grain Filling."

Finally, in the biological sense the sink constitute a load on the source; whereas a large sink dictates a high rate of assimilate demand from the source. Experiments with partially de-grained wheat plants indicated that a large sink signals higher stomatal conductance and gas exchange in the flag leaf (Blum et al. 1988). Under stress conditions this effect brought about a significant reduction in flag leaf water status and a reduction in its capacity for osmotic adjustment. Lower flag leaf stomatal conductance was observed in millet plants after removal of their panicles (Henson and Mahalakshmi 1985). Higher stomatal conductance involved greater transpiration and water-use. This is compatible with the fact that high yielding wheat cultivars (having a large sink) could be identified by their higher rate of transpiration (Reynolds et al. 1994). A large sink load on the source would lead to earlier leaf senescence under stress, as compared with a plant of smaller sink (Khanna-Chopra and Sinha 1988).

Thus, sink load and its effect on plants under stress may be taken as one example of the fact that a high yield potential is basically not compatible with sustainable yield under severe drought stress. Further discussion of this important point with regard to breeding is presented in Sect. 3.3.

The sensitivity of plant reproduction to drought stress must have evolutionary roots. The plant apparently constantly monitors its status with respect to sugar pool and ABA signaling. Besides ABA, other hormonal signals are possibly involved in the monitoring of sink-source relationships. By means which were discussed above the plant under stress can adjust its reproduction in response to water status, hormones and sugars all of which signal the amount of available assimilates (current source). What an agronomist may define as a reproductive failure is in the evolutionary sense a method of survival under drought stress. The reduction in sink size allows the survival of few seeds in tune with the small source. Maintenance of a large sink demand under stress would have resulted in the total failure of reproduction in the face of assimilate shortage.

While this is a most appropriate strategy of survival for natural vegetation, it may not be suitable for crop plants. Agriculture is based on the idea that the farmer and its supporting research make the decisions on the basis of knowledge and experience pertaining to plants, inputs, soils, water, climate and economics. This is one of the domains where plant genetics should modify and adapt plant responses and signaling systems to the specific agroecosystem in order to economically optimize and stabilize plant production. Philosophically, the farmer is prepared to and capable of taking a greater risk than evolution when it comes to plant reproduction under stress.

References

Ahmadi A, Baker DA (2001) The effect of water stress on grain filling processes in wheat. J Agric Sci 136:257–269

Alexandersson E, Danielson JÅH, Råde J et al (2010) Transcriptional regulation of aquaporins in accessions of *Arabidopsis* in response to drought stress. Plant J 61:650–660

Aneja M, Gianfagna T, Ng E (2004) The roles of abscisic acid and ethylene in the abscission and senescence of cocoa flowers. J Plant Growth Regul 27:149–155

Araus JL, Febrero A, Vendrell P (1991) Epidermal conductance in different parts of durum wheat grown under Mediterranean conditions – the role of epicuticular waxes and stomata. Plant Cell Environ 14:545–558

Araus JL, Slafer GA, Reynolds MP et al (2002) Plant breeding and drought in C3 cereals: what should we breed for? Ann Bot 89:925–940

Aroca R, Porcel R, Ruiz-Lozano JM (2007) How does arbuscular mycorrhizal symbiosis regulate root hydraulic properties and plasma membrane aquaporins in *Phaseolus vulgaris* under drought, cold or salinity stresses? New Phytol 173:808–816

Aroca R, Vernieri P, Ruiz-Lozano JM (2008) Mycorrhizal and non-mycorrhizal *Lactuca sativa* plants exhibit contrasting responses to exogenous ABA during drought stress and recovery. J Exp Bot 59:2029–2041

Asseng S, Ritchie JT, Smucker AJM et al (1998) Root growth and water uptake during water deficit and recovering in wheat. Plant Soil 201:265–273

Auge RM, Kubikova E, Moore JL (2001) Foliar dehydration tolerance of mycorrhizal cowpea, soybean and bush bean. New Phytol 151:535–541

Austin RB, Bingham J, Blackwell RD et al (1980) Genetic improvement in winter wheat yields since 1900 and associated physiological changes. J Agric Sci 94:675–689

Barker DJ, Sullivan CY, Moser LE (1993) Water deficit effects on osmotic potential, cell wall elasticity, and proline in five forage grasses. Agron J 85:270–275

Bauerle TL, Richards JH, Smart DR et al (2008) Importance of internal hydraulic redistribution for prolonging the lifespan of roots in dry soil. Plant Cell Environ 31:177–186

Beis A, Patakas A (2010) Differences in stomatal responses and root to shoot signalling between two grapevine varieties subjected to drought. Funct Plant Biol 37:139–146

Blum A (1970) Nature of heterosis in grain production by the sorghum panicle. Crop Sci 10:28–31

Blum A (1975) Effect of the BM gene on epicuticular wax deposition and the spectral characteristics of sorghum leaves. SABRAO J 7:45–52

Blum A (1977) The basis of heterosis in the differentiating sorghum panicle. Crop Sci 17:880–882

Blum A (2004) Sorghum physiology. In: Nguyen HT, Blum A (eds) Physiology and biotechnology integration for plant breeding. CRC Press, Boca Raton

Blum A (2009) Effective use of water (EUW) and not water-use efficiency (WUE) is the target of crop yield improvement under drought stress. Field Crops Res 112:119–123

Blum A, Arkin GF (1984) Sorghum root growth and water-use as affected by water supply and growth duration. Field Crops Res 9:131–142
Blum A, Ritchie JT (1984) Effect of soil surface water content on sorghum root distribution in the soil. Field Crops Res 8:169–176
Blum A, Sinmena B (1995) Isolation and characterization of variant wheat cultivars for ABA sensitivity. Plant Cell Environ 18:77–83
Blum A, Mayer J, Golan G (1988) The effect of grain number (sink size) on source activity and its water-relations in wheat. J Exp Bot 39:106–114
Blum A, Munns R, Passioura JB et al (1996) Genetically engineered plants resistant to soil drying and salt stress: how to interpret osmotic relations? Plant Physiol 110:1051
Boomsma CR, Vyn TJ (2008) Maize drought tolerance: potential improvements through arbuscular mycorrhizal symbiosis? Field Crops Res 108:14–31
Borras L, Westgate M, Otegui ME (2003) Control of grain weight and grain water relations by post-flowering source-sink ratio in maize. Ann Bot 91:857–867
Boyer JS, Westgate ME (2004) Grain yields with limited water. J Exp Bot 55:2385–2394
Bramley H, Turner DW, Tyerman SD et al (2007) Water flow in the roots of crop species: the influence of root structure, aquaporin activity, and waterlogging. Adv Agron 96:33–196
Brodrib TJ, Feild TS, Sack L (2010) Viewing leaf structure and evolution from a hydraulic perspective. Funct Plant Biol 37:488–498
Burow GB, Franks CD, Xin Z (2008) Genetic and physiological analysis of an irradiated bloomless mutant (epicuticular wax mutant) of sorghum. Crop Sci 48:41–48
Caird MA, Richards JH, Hsiao TC (2007) Significant transpirational water loss occurs throughout the night in field-grown tomato. Funct Plant Biol 34:172–177
Cameron KD, Teece MA, Smart LB (2006) Increased accumulation of cuticular wax and expression of lipid transfer protein in response to periodic drying events in leaves of tree tobacco. Plant Physiol 140:176–183
Chazen O, Neumann PM (1994) Hydraulic signals from the roots and rapid cell-wall hardening in growing maize (*Zea mays* l) leaves are primary responses to polyethylene glycol-induced water deficits. Plant Physiol 104:1385–1392
Cheikh N, Jones RJ (1994) Disruption of maize kernel growth and development by heat stress – role of cytokinin abscisic acid balance. Plant Physiol 106:45–51
Chimenti CA, Marcantonio M, Hall AJ (2006) Divergent selection for osmotic adjustment results in improved drought tolerance in maize (*Zea mays* L) in both early growth and flowering phases. Field Crops Res 95:305–315
Christmann A, Grill E (2009) Are GTGs ABA's biggest fans? Cell 136:21–23
Christmann A, Weiler EW, Steudle E et al (2007) A hydraulic signal in root-to-shoot signalling of water shortage. Plant J 52:167–174
Cochard H, Casella E, Mencuccini M (2007) Xylem vulnerability to cavitation varies among poplar and willow clones and correlates with yield. Tree Physiol 27:1761–1767
Cochard H, Holtta T, Herbette S et al (2009) New insights into the mechanisms of water-stress-induced cavitation in conifers. Plant Physiol 151:949–954
Cosgrove DJ (2000) Loosening of plant cell walls by expansions. Nature 407:321–326
Cutler JM, Rains DW, Loomis RS (1977) The importance of cell size in the water relations of plants. Physiol Plant 40:255–260
Davies FT, Olalde-Portugal V, Aguilera-Gomez L et al (2002a) Alleviation of drought stress of Chile ancho pepper (*Capsicum annuum* L cv San Luis) with arbuscular mycorrhiza indigenous to Mexico. Sci Hort 92:347–359
Davies WJ, Wilkinson S, Loveys B (2002b) Stomatal control by chemical signalling and the exploitation of this mechanism to increase water use efficiency in agriculture. New Phytol 153:449–460
Davies WJ, Kudoyarova G, Hartung W (2005) Long-distance ABA signaling and its relation to other signaling pathways in the detection of soil drying and the mediation of the plant's response to drought. J Plant Growth Regul 24:285–295
de Wit CT (1958) Transpiration and crop yields. Versl Landabouwk Onderz 64:1–88

Dodd IC (2009) Rhizosphere manipulations to maximize 'crop per drop' during deficit irrigation. J Exp Bot 60:2454–2459

Fan L, Neumann PM (2004) The spatially variable inhibition by water deficit of maize root growth correlates with altered profiles of proton flux and cell wall pH. Plant Physiol 135: 2291–2300

Fan L, Linker R, Gepstein S et al (2006) Progressive inhibition by water deficit of cell wall extensibility and growth along the elongation zone of maize roots is related to increased lignin metabolism and progressive stelar accumulation of wall phenolics. Plant Physiol 140:603–612

Fang X, Turner NC, Yan G et al (2010) Flower numbers, pod production, pollen viability, and pistil function are reduced and flower and pod abortion increased in chickpea (*Cicer arietinum* L) under terminal drought. J Exp Bot 61:335–345

Fletcher AL, Sinclair TR, Allen LH Jr (2007) Transpiration responses to vapor pressure deficit in well watered 'slow-wilting' and commercial soybean. Environ Exp Bot 61:145–151

Frascaroli E, Tuberosa R (1993) Effect of abscisic acid on pollen germination and tube growth of maize genotypes. Plant Breed 110:250–254

Fulai L, Christian RJ, Mathias NA (2004) Drought stress effect on carbohydrate concentration in soybean leaves and pods during early reproductive development: its implication in altering pod set. Field Crops Res 86:1–13

Galmés J, Pou A, Alsina MM et al (2007) Aquaporin expression in response to different water stress intensities and recovery in Richter-110 (*Vitis* sp): relationship with ecophysiological status. Planta 226:671–681

Garrity DP, Vidal ET, O'Toole JC (1986) Manipulating panicle transpiration resistance to increase spikelet fertility during flowering stage water stress. Crop Sci 26:789–795

Granier C, Tardieu F (1999) Water deficit and spatial pattern of leaf development Variability in responses can be simulated using a simple model of leaf development. Plant Physiol 119:609–620

Gutschick VP, Simonneau T (2002) Modelling stomatal conductance of field-grown sunflower under varying soil water content and leaf environment: comparison of three models of stomatal response to leaf environment and coupling with an abscisic acid-based model of stomatal response to soil drying. Plant Cell Environ 25:1423–1434

Harrison MA, Kaufman PB (1980) Hormonal regulation of lateral bud (tiller) release in oats (*Avena sativa* L). Plant Physiol 66:1123–1127

Havlová M, Dobrev PI, Motyka V et al (2008) The role of cytokinins in responses to water deficit in tobacco plants over-expressing trans-zeatin O-glucosyltransferase gene under 35S or SAG12 promoters. Plant Cell Environ 31:341–353

Henson IE, Mahalakshmi V (1985) Evidence for panicle control of stomatal behaviour in water-stressed plants of pearl millet. Field Crops Res 11:281–290

Holmes MG, Keiller DR (2002) Effects of pubescence and waxes on the reflectance of leaves in the ultraviolet and photosynthetic wavebands: a comparison of a range of species. Plant Cell Environ 25:85–93

Hooker TS, Millar AA, Kunst L (2002) Significance of the expression of the CER6 condensing enzyme for cuticular wax production in *Arabidopsis*. Plant Physiol 129:1568–1580

Hose E, Steudle E, Hartung W (2000) Abscisic acid and hydraulic conductivity of maize roots: a study using cell- and root-pressure probes. Planta 211:874–882

Hsiao TC, Xu LK (2000) Sensitivity of growth of roots versus leaves to water stress: biophysical analysis and relation to water transport. J Exp Bot 51:1595–1616

Huang D, Wu W, Abrams SR et al (2008) The relationship of drought-related gene expression in *Arabidopsis thaliana* to hormonal and environmental factors. J Exp Bot 59:2991–3007

Jacqmard A, Houssa C, Bernier G (1995) Abscisic acid antagonizes the effect of cytokinin on DNA-replication origins. J Exp Bot 46:663–666

Javot H, Maurel C (2002) The role of aquaporins in root water uptake. Ann Bot 90:301–313

Jeschke WD, Hartung W (2000) Root-shoot interactions in mineral nutrition. Plant Soil 226:57–69

Ji X, Shiran B, Wan J et al (2010) Importance of pre-anthesis anther sink strength for maintenance of grain number during reproductive stage water stress in wheat. Plant Cell Environ 33:926–942

Kaldenhoff R, Ribas-Carbo M, Flexas J et al (2008) Aquaporins and plant water balance. Plant Cell Environ 31:658–666

Khanna-Chopra R, Sinha SK (1988) Enhancement of drought-induced senescence by the reproductive sink in fertile lines of wheat and sorghum. Ann Bot 61:649–653

Kholova J, Hash CT, Lava Kumar P et al (2010) Terminal drought-tolerant pearl millet [*Pennisetum glaucum* (L.) R. Br.] have high leaf ABA and limit transpiration at high vapour pressure deficit. J Exp Bot 61:1431–1440

Koch K (1996) Carbohydrate-modulated gene expression in plants. Annu Rev Plant Physiol Plant Mol Biol 47:509–540

Kosma DK, Bourdenx B, Bernard A et al (2009) The impact of water deficiency on leaf cuticle lipids of *Arabidopsis*. Plant Physiol 151:1918–1929

Landi P, Sanguineti MC, Conti S et al (2001) Direct and correlated responses to divergent selection for leaf abscisic acid concentration in two maize populations. Crop Sci 41:335–344

Landi P, Sanguineti MC, Liu C et al (2007) Root-ABA1 QTL affects root lodging, grain yield, and other agronomic traits in maize grown under well-watered and water-stressed conditions. J Exp Bot 58:319–326

Laplaze L, Benkova E, Casimiro I et al (2008) Cytokinins act directly on lateral root founder cells to inhibit root initiation. Plant Cell 19:3889–3900

Le Bris M, Michaux-Ferrière N, Jacob Y et al (1999) Regulation of bud dormancy by manipulation of ABA in isolated buds of Rosa hybrida cultured in vitro. Aust J Plant Physiol 26:273–281

Li Y, Wang G-X, Xin M et al (2004) The parameters of guard cell calcium oscillation encodes stomatal oscillation and closure in Vicia faba. Plant Sci 166:415–421

Li Y, Sperry JS, Shao M (2009) Hydraulic conductance and vulnerability to cavitation in corn (*Zea mays* L) hybrids of differing drought resistance. Environ Exp Bot 66:341–346

Liu F, Andersen MN, Jensen CR (2003) Loss of pod set caused by drought stress is associated with water status and ABA content of reproductive structures in soybean. Funct Plant Biol 30:271–280

Liu F, Jensen CR, Andersen MN (2005) A review of drought adaptation in crop plants: changes in vegetative and reproductive physiology induced by ABA-based chemical signals. Aust J Agr Res 56:1245–1252

Lockhart JA (1965) An analysis of irreversible plant cell elongation. J Theor Biol 8:264–276

Lopez G, Behboudian MH, Vallverdu X et al (2010) Mitigation of severe water stress by fruit thinning in 'O'Henry' peach: implications for fruit quality. Sci Hort 125:294–300

Lu ZJ, Neumann PM (1998) Water-stressed maize, barley and rice seedlings show species diversity in mechanisms of leaf growth inhibition. J Exp Bot 49:1945–1952

Luan S (2002) Signalling drought in guard cells. Plant Cell Environ 25:229–237

Mambelli S, Setter TL (1998) Inhibition of maize endosperm cell division and endoreduplication by exogenously applied abscisic acid. Physiol Plant 104:266–272

Marshall JG, Dumbroff EB (1999) Turgor regulation via cell wall adjustment in white spruce. Plant Physiol 119:313–320

McMichael BL, Lascano RJ (2010) Evaluation of hydraulic lift in cotton (*Gossypium hirsutum* L) germplasm. Environ Exp Bot 68:26–30

Mills D, Genfa Z, Benzioni A (2001) Effect of different salts and of ABA on growth and mineral uptake in jojoba shoots grown in vitro. J Plant Physiol 158:1031–1039

Miyamoto N, Steudle E, Hirasawa T et al (2001) Hydraulic conductivity of rice roots. J Exp Bot 52:1835–1846

Miyazawa S-I, Yoshimura S, Shinzaki Y et al (2008) Deactivation of aquaporins decreases internal conductance to CO_2 diffusion in tobacco leaves grown under long-term drought. Funct Plant Biol 35:553–564

Moore JP, Nguema-Ona E, Chevalier L et al (2006) Response of the leaf cell wall to desiccation in the resurrection plant *Myrothamnus flabellifolius*. Plant Physiol 141:651–662

Morgan JM (1980) Possible role of abscisic acid in reducing seed set in water-stressed wheat plants. Nature 285:655–657
Morgan JM (1992) Osmotic components and properties associated with genotypic differences in osmoregulation in wheat. Aust J Plant Physiol 19:67–76
Munne-Bosch S, Alegre L (2004) Die and let live: leaf senescence contributes to plant survival under drought stress. Funct Plant Biol 31:203–216
Munns R, Richards RA (2007) Recent advances in breeding wheat for drought and salt stresses. In: Jenks MA, Hasegawa PM, Mohan Jain S (eds) Advances in molecular breeding toward drought and salt tolerant crops. Springer, Dordrecht
Munns R, Sharp RE (1993) Involvement of abscisic acid in controlling plant growth in soils of low water potential. Aust J Plant Physiol 20:425–437
Nakashima K, Ito Y, Yamaguchi-Shinozaki K (2009) Transcriptional regulatory networks in response to abiotic stresses in *Arabidopsis* and grasses. Plant Physiol 149:88–95
Nar H, Saglam A, Terzi R et al (2009) Leaf rolling and photosystem II. Efficiency in *Ctenanthe setosa* exposed to drought stress. Photosynthetica 47:429–436
Neumann PM (1995) The role of cell wall adjustment in plant resistance to water deficits. Crop Sci 35:1258–1266
Ober ES, Sharp RE (2007) Regulation of root growth responses to water deficit. In: Jenks MA, Hasegawa PM, Jain S (eds) Advances in molecular breeding towards drought and salt tolerant crops. Springer, Dordrecht
Ofir M, Kigel J (1998) Abscisic acid involvement in the induction of summer-dormancy in *Poa bulbosa*, a grass geophytes. Physiol Plant 102:163–170
Ohkuma K, Lyon JL, Addicott FT et al (1963) Abscisin II, an abscission-accelerating substance from young cotton fruit. Science 142:1592–1593
Oliver SN, Dennis ES, Dolferus R (2007) ABA regulates apoplastic sugar transport and is a potential signal for cold-induced pollen sterility in rice. Plant Cell Physiol 48:1319–1330
Or E, Belausov E, Popilevsky I et al (2000) Changes in endogenous ABA level in relation to the dormancy cycle in grapevines grown in a hot climate. J Hort Sci Biotechnol 75:190–194
Ortega U, Duñabeitia M, Menendez S et al (2004) Effectiveness of mycorrhizal inoculation in the nursery on growth and water relations of *Pinus radiata* in different water regimes. Tree Physiol 24:65–73
O'Toole JC (1982) Adaptation of rice to drought prone environments. In: Drought resistance in crops with emphasis on rice. International Rice Research Institute, Los Banos
Palta JA, Turner NC, French RJ et al (2007) Physiological responses of lupin genotypes to terminal drought in a Mediterranean-type environment. Ann Appl Biol 150:269–279
Parent B, Hachez C, Redondo et al (2009) Drought and abscisic acid effects on aquaporin content translate into changes in hydraulic conductivity and leaf growth rate: a trans-scale approach. Plant Physiol 149:2000–2012
Passioura JB, Fry SC (1992) Turgor and cell expansion: beyond the Lockhart equation. Aust J Plant Phys 19:565–576
Porcel R, Ruiz-Lozano JM (2004) Arbuscular mycorrhizal influence on leaf water potential, solute accumulation, and oxidative stress in soybean plants subjected to drought stress. J Exp Bot 55:1743–1750
Quarrie SA (1991) Implications of genetic differences in ABA accumulation for crop production. In: Davies WJ, Jones HG (eds) Abscisic acid: physiology and biochemistry. Bios Scientific Publishers, London
Quintero JM, Fournier JM, Benlloch M (1999) Water transport in sunflower root systems: effects of ABA, Ca^{2+} status and $HgCl_2$. J Exp Bot 50:1607–1612
Rasmussen RD, Hole D, Hess JR et al (1997) Wheat kernel dormancy and plus abscisic acid level following exposure to fluridone. J Plant Physiol 150:440–445
Reymond M, Muller B, Leonardi A et al (2003) Combining quantitative trait loci analysis and an ecophysiological model to analyze the genetic variability of the responses of maize leaf growth to temperature and water deficit. Plant Physiol 131:664–675

Reynolds MP, Balota M, Delgado MIB et al (1994) Physiological and morphological traits associated with spring wheat yield under hot, irrigated conditions. Aust J Plant Physiol 21:717–730
Riederer M, Schreiber L (2001) Protecting against water loss: analysis of the barrier properties of plant cuticles. J Exp Bot 52:2023–2032
Rose TJ, Rengel Z, Ma Q et al (2008) Hydraulic lift by canola plants aids P and K uptake from dry topsoil. Aust J Agric Res 59:38–45
Ruiz-Lozano JM, Collados C, Barea JM et al (2001) Arbuscular mycorrhizal symbiosis can alleviate drought-induced nodule senescence in soybean plants. New Phytol 151:493–502
Sack L, Holbrook NM (2006) Leaf hydraulics. Annu Rev Plant Biol 57:361–381
Sade N, Vinocur BJ, Diber A et al (2009) Improving plant stress tolerance and yield production: is the tonoplast aquaporin SlTIP2;2 a key to isohydric to anisohydric conversion? New Phytol 181:651–661
Sanguineti MC, Conti S, Landi P et al (1996) Abscisic acid concentration in maize leaves – genetic control and response to divergent selection in two populations. Maydica 41:193–203
Schachtman DP, Goodger JQD (2008) Chemical root to shoot signaling under drought. Trends Plant Sci 13:281–287
Schussler JR, Westgate ME (1995) Assimilate flux determines kernel set at low water potential in maize. Crop Sci 35:1074–1080
Serpe MD, Matthews MA (2000) Turgor and cell wall yielding in dicot leaf growth in response to changes in relative humidity. Aust J Plant Physiol 27:1131–1140
Sharp RE (2002) Interaction with ethylene: changing views on the role of abscisic acid in root and shoot growth responses to water stress. Plant Cell Environ 25:211–222
Sharp RG, Davies WJ (2009) Variability among species in the apoplastic pH signalling response to drying soils. J Exp Bot 60:4363–4370
Sharp RE, LeNoble ME (2001) ABA, ethylene and the control of shoot and root growth under water stress. J Exp Bot 53:33–37
Sharp RE, Poroyko V, Hejlek LG et al (2004) Root growth maintenance during water deficits: physiology to functional genomics. J Exp Bot 55:2343–2351
Shaterian J, Georges F, Hussain A et al (2005) Root to shoot communication and abscisic acid in calreticulin (CR) gene expression and salt-stress tolerance in grafted diploid potato clones. Environ Exp Bot 53:323–332
Shearman VJ, Sylvester-Bradley R, Scott RK et al (2005) Physiological processes associated with wheat yield progress in the UK. Crop Sci 45:175–185
Shepherd T, Wynne GD (2006) The effects of stress on plant cuticular waxes. New Phytol 171:469–499
Sherson SM, Alford HL, Forbes SM et al (2003) Roles of cell-wall invertases and monosaccharide transporters in the growth and development of *Arabidopsis*. J Exp Bot 54:525–531
Shinozaki K, Yamaguchi-Shinozaki K (2007) Gene networks involved in drought stress response and tolerance. J Exp Bot 58:221–227
Smeekens S (1998) Sugar regulation of gene expression in plants. Curr Opin Plant Biol 1:230–234
Sperry JS, Hacke UG, Oren R et al (2002) Water deficits and hydraulic limits to leaf water supply. Plant Cell Environ 25:251–263
Sperry JS, Stiller V, Hacke UG (2003) Xylem hydraulics and the soil plant-atmosphere continuum: opportunities and unresolved issues. Agron J 95:1362–1370
Steinbach HS, Benech-Arnold RL, Sanchez RA (1997) Hormonal regulation of dormancy in developing sorghum seeds. Plant Physiol 113:149–154
Steudle E (2000) Water uptake by plant roots: an integration of views. Plant Soil 226:45–56
Steudle E, Peterson CA (1998) How does water get through roots? J Exp Bot 49:775–788
Stirzaker RJ, Passioura JB (1996) The water relations of the root-soil interface. Plant Cell Environ 19:201–208
Stratton L, Goldstein G, Meinzer FC (2000) Stem water storage capacity and efficiency of water transport: their functional significance in a Hawaiian dry forest. Plant Cell Environ 23:99–106

Tardieu F, Granier C (2000) Quantitative analysis of cell division in leaves: methods, developmental patterns and effects of environmental conditions. Plant Mol Biol 43:555–567
Thompson AJ, Andrews J, Mulholland BJ et al (2007) Overproduction of abscisic acid in tomato increases transpiration efficiency and root hydraulic conductivity and influences leaf expansion. Plant Physiol 143:1905–1917
Troughton A (1980) Production of root axes and leaf elongation in perennial ryegrass in relation to dryness of the upper soil layer. J Agric Sci Camb 95:533–538
Tsuchihira A, Hanba YT, Kato N (2010) Effect of overexpression of radish plasma membrane aquaporins on water-use efficiency, photosynthesis and growth of Eucalyptus trees. Tree Physiol 30:417–430
Vandeleur RK, Mayo G, Shelden MC et al (2009) The role of plasma membrane intrinsic protein aquaporins in water transport through roots: diurnal and drought stress responses reveal different strategies between isohydric and anisohydric cultivars of grapevine. Plant Physiol 149:445–460
Wall GW, Garcia RL, Kimball BA et al (2006) interactive effects of elevated carbon dioxide and drought on wheat. Agron J 98:354–381
Wan CG, Xu WW, Sosebee RE et al (2000) Hydraulic lift in drought-tolerant and -susceptible maize hybrids. Plant Soil 219:117–126
Wang Z, Cao W, Dai T et al (2001) Effects of exogenous hormones on floret development and grain set in wheat. Plant Growth Regul 35:225–231
Welcker C, Boussuge B, Bencivenni C et al (2007) Are source and sink strengths genetically linked in maize plants subjected to water deficit? A QTL study of the responses of leaf growth and of anthesis-silking interval to water deficit. J Exp Bot 58:339–349
Westgate ME, Passioura JB, Munns R (1996) Water status and aba content of floral organs in drought-stressed wheat. Aust J Plant Physiol 23:763–772
Whalley WR, Clark LJ, Gowing DJG et al (2006) Does soil strength play a role in wheat yield losses caused by soil drying? Plant Soil 280:279–290
White RG, Kirkegaard JA (2010) The distribution and abundance of wheat roots in a dense, structured subsoil – implications for water uptake. Plant Cell Environ 33:133–148
Wilkinson S, Davies WJ (2002) ABA-based chemical signalling: the co-ordination of responses to stress in plants. Plant Cell Environ 25:195–210
Wu YJ, Spollen WG, Sharp RE et al (1994) Root growth maintenance at low water potentials – increased activity of xyloglucan endotransglycosylase and its possible regulation by abscisic acid. Plant Physiol 106:607–615
Wu YJ, Sharp RE, Durachko DM et al (1996) Growth maintenance of the maize primary root at low water potentials involves increases in cell-wall extension properties, expansin activity, and wall susceptibility to expansins. Plant Physiol 111, 765–772
Xiong Y-C, Li F-M, Zhang T et al (2007) Evolution mechanism of non-hydraulic root-to-shoot signal during the anti-drought genetic breeding of spring wheat. Environ Exp Bot 59:193–205
Xu XD, Bland WL (1993) Reverse water flow in sorghum roots. Agron J 85:384–388
Yang JC, Zhang JH, Wang ZQ et al (2001) Activities of starch hydrolytic enzymes and sucrose-phosphate synthase in the stems of rice subjected to water stress during grain filling. J Exp Bot 52:2169–2179
Yang JC, Zhang JH, Wang ZQ et al (2003) Involvement of abscisic acid and cytokinins in the senescence and remobilization of carbon reserves in wheat subjected to water stress during grain filling. Plant Cell Environ 26:1621–1631
Yang JC, Zhang JH, Ye YX et al (2004) Involvement of abscisic acid and ethylene in the responses of rice grains to water stress during filling. Plant Cell Environ 27:1055–1064
Yang JC, Zhang J, Liu K et al (2006) Abscisic acid and ethylene interact in wheat grains in response to soil drying during grain filling. New Phytol 171:293–303
Zhang J-Y, Broeckling CD, Blancaflor EB et al (2005) Overexpression of WXP1, a putative *Medicago truncatula* AP2 domain-containing transcription factor gene, increases cuticular wax accumulation and enhances drought tolerance in transgenic alfalfa (*Medicago sativa*). Plant J 42:689–707

Zhang J, Jia W, Yang J et al (2006) Role of ABA in integrating plant responses to drought and salt stresses. Field Crops Res 96:111–119
Zhu J, Brownjonathan KM, Lynch P (2010) Root cortical aerenchyma improves the drought tolerance of maize (*Zea mays* L.). Plant Cell Environ 33:740–749
Zimmermann U, Schneider H, Wegner LH et al (2004) Water ascent in tall trees: does evolution of land plants rely on a highly metastable state? New Phytol 162:575–615

Chapter 3
Drought Resistance and Its Improvement

Summary Plant breeding has been successful in developing drought resistant crop cultivars. However the traditional breeding method by using yield as a selection index and performing multi-environmental yield trials has been costly and slow. Plant physiology is now incorporated into the breeding program by using physiological selection criteria relevant to the designated plant ideotype and subsequent plant performance in the target stress environment. Genomics offer a great potential for the improvement of breeding efficiency towards water limited environments. There are still inherent problems in deploying marker assisted selection and transgenic technology into breeding program for drought resistance. The potential of genomics can be realized only when it will be well synchronized with plant breeding concept, theory and methods.

It has often been voiced and published that "drought resistance" is complex and therefore its improvement is difficult. This chapter aims to diffuse some of these beliefs and demonstrate that the issue is not as complex as seen by the novice or as seen from the "gene discovery" platform.

Breeding for drought resistance can basically follow an analogy of breeding for disease resistance in terms of concept and design (with few exceptions). Drought resistance is approached in terms of its components, namely dehydration avoidance, dehydration tolerance and drought escape. The most widespread and effective mechanism of drought resistance in crop plants is dehydration avoidance, which is the ability of the plant to maintain its hydration. It is controlled by plant constitutive traits and plant adaptive traits. Dehydration tolerance which is the ability to function in a dehydrated state is rare but can sometimes be important. It is shown that when stress physiology, plant genetics and knowledge of the target environment are combined it is possible to design an appropriate plant ideotype to be used as guide in breeding for the specific water limited environment.

3.1 Genetic Gains Achieved in Plant Breeding for Drought Resistance

A good number of years ago at the time when research in genomics of plant abiotic stress resistance was just emerging this author received for review a research proposal dealing with functional genomics of plant stress tolerance. In the description

A. Blum, *Plant Breeding for Water-Limited Environments*,
DOI 10.1007/978-1-4419-7491-4_3,

of the goal and aims of the research it was stated that "plant breeding has not produced varieties suitable for use in stressful environments." The proposal went on to explain how this void will be filled by the genomics research done under that proposal. At that time this statement was taken as an amusing glitch from an inexperience young scientist. However, as time went by and as more such opinions were voiced and published by reputable scientists (e.g., the Preface in Jenks et al. 2007) it became clear that education in this respect was missing and it is time to correct that deficiency.

When all the information available is considered, plant breeding was generally more effective in achieving gains in yield under non-stress than under stress conditions. However, the fact remains that plant breeding achieved gains in yield under conditions of drought stress.

Perhaps the most remarkable conclusion was made by Lynch and Frey (1993). Frey was at the time one of the deans of plant breeding in both theory and practice, specializing in oat breeding. From their research on the historical yield gains made in oat breeding they concluded that "oat breeding primarily has improved the ability of oat cultivars to perform in stressful environments."

Experiments with historical cultivars of various crops are the common method for assessing progress in yield improvement by plant breeding. Following are some of these results when breeding for stress conditions was evaluated.

Selection of maize for high yield potential in itself has led to consistent increases in yield in both stress and non-stress conditions (Castleberry et al. 1984). This phenomenon will be further discussed below. On the other hand, a study of maize hybrids developed in Ontario Canada from the1950s to the late 1980s (Tollenaar and Wu 1999) showed that genetic yield improvement was 2.5% per year and that most of the genetic yield improvement could be attributed to increased stress resistance. Improvement in stress resistance in recent over old maize hybrids have been shown for high plant population density, weed interference, low night temperatures during the grain-filling period, low soil nitrogen and low soil moisture. Comparisons of 36 widely grown successful hybrids released at intervals from 1934 to 1991 in the USA showed continuing improvements in tolerance to abiotic stresses such as heat and drought, excessively cool and wet weather, low soil fertility, high density planting (Duvick 1997). Eighteen maize hybrids that were released between 1953 and 2001 in the USA were tested under drought stress at different growth stages (Barker et al. 2005). It was found that the genetic gain in yield when stress occurred during flowering and mid grain filling was 124 and 91 kg ha^{-1} yr^{-1}, respectively. Genetic gain without stress was higher (211 kg ha^{-1} yr^{-1}), but the point was well made that drought resistance in maize reproduction was improved.

Historical winter wheat cultivars dating from 1870s to 1987 were evaluated in experiments in Kansas (Cox et al. 1988). Genetic gain in yield was 0.6 and 0.4% per year under non-stress and stress conditions, respectively. Historical Australian wheat cultivars representing a series from the 1860s to 1982 were grown in dryland field trials over 4 years in the wheat belt of Western Australia (Perry and D'Antuono 1989). Yields have increased from 1,022 kg ha^{-1} in 1884 to 1,588 kg ha^{-1} in 1982, representing a gain of 0.57% per year. The yield potential (non-stress) of wheat in

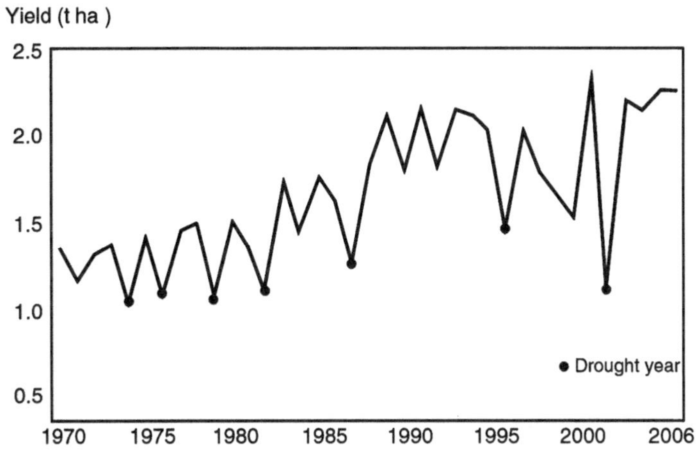

Fig. 3.1 Trends in rice yield and major drought years during 1970–2006 in the Eastern India state of Orissa (From Pandy and Bhandari (2008). With permission)

1982 was estimated at about 7,000 kg ha^{-1}. Therefore this study was performed under the harsh Australian dryland conditions.

Improved crop cultivars developed under dryland conditions were seen to possess distinct physiological adaptive traits which supported yield under stress. Modern soybean cultivars maintained higher leaf water potential as compared with older cultivars when subjected to drought stress (Boyer et al. 1980). Recent maize hybrids had relatively better capacity than older ones to access soil water under drought stress (Hammer et al. 2009). A study of historical wheat cultivars in the UK indicated that modern cultivars had a relatively high capacity for stem reserve storage available at flowering (Shearman et al. 2005) which support grain filling under stress. This is an important dehydration tolerance mechanism (section "Stem Reserve Utilization for Grain Filling"). Finally, ample genetic variation in dehydration avoidance was found in advanced high yielding wheat breeding germplasm (Blum et al. 1981).

In conclusion, plant breeding has been successful in continuously improving drought resistance of dryland and rainfed crops for decades. It is impossible to imagine how economically viable crop production could be sustained and continuously improved under the dryland farming systems in semi-arid lands such as Australia, the American mid-West, the Canadian prairie or the Mediterranean region without the development of drought and heat resistant cultivars. Even ancient farmers selecting seed for next season sowing in their harsh environment were apparently developing drought resistant landraces of sorghum, pearl millet (Blum and Sullivan 1986), wheat (Blum et al. 1989) and barley (Ceccarelli et al. 1998). Those were the first pre-scientific era plant breeders which achieved progress in breeding for drought resistance.

A most compelling evidence for the agronomist and the farmer is in Fig. 3.1, which represents progress in rice yield in the drought-stricken state of Orissa in

India. Mean yield level is low because of insufficient inputs and other limiting factors. Yield in drought years was severely reduced but it still increased with time in tune with the increase in yield in good years. A large part of this progress can be assigned to the genetic improvement and adoption of drought resistant rice cultivars in addition to minor improvements in crop management.

3.2 Genomics and Breeding for Drought Resistance

Since the advent of molecular biology and plant biotechnology, plant breeding as we know it has been redefined by molecular biologists as "conventional breeding" or "classical breeding." It is not clear what then by definition is "unconventional breeding" or "modern breeding." These could not be biotechnology or molecular biology or genomics because these disciplines cannot produce (breed) new cultivars on their own accord unless done in the context and framework of plant breeding. Hence, biotechnology and molecular biology are important disciplines which constitute tools to support and enhance plant breeding. A leading group in molecular biology of plant abiotic stress resistance recently concluded that "the collective and cooperative efforts of plant molecular biologists, physiologists, and breeders are required to generate stress-tolerant grasses through genetic engineering" (Nakashima et al. 2009). Therefore, there is no "classical" or "conventional" breeding. There is just "plant breeding," very hopefully with the support of genomics and the continuing support as ever by plant physiology, plant pathology, entomology, food technology, agricultural engineering, statistics and more.

This book does not offer a comprehensive discussion of genomics in relations to drought resistance improvement. The subject has been profusely discussed in many conferences, review papers and edited books. Here we are interested in results and proven ways to achieve them. Hence, genomics and molecular biology will be discussed here only to the extent that a contribution to plant breeding for water limited environment is prominent and established. Where basic genomics research and potential towards the future are concerned the reader is referred to the literature on genomics.

Three breeding-related major activities in genomics are recognized here: Gene expression work, molecular mapping and transgenics. These will be further discussed in specific cases below but here a few introductory comments are warranted.

3.2.1 Gene Expression and Gene Discovery

This is a popular activity in genomics, where plants are experimentally subjected to drought stress and genes that are up- or down- regulated by drought stress are recorded and compared with known genes listed in databases. Understanding the function of these hundreds of genes and transcription factors is the first step in making any practical sense of gene expression results. In unique cases some of these results were carried forward into transgenic model plants or even crop plants which were then tested for function at the whole plant level. A minimal number of cases were carried forward into field studies and productivity assessments under conditions of limited water supply. These can be regarded as pre-breeding materials and potential genetic resources for breeding and are therefore discussed in Sect. 5.4.

Regretfully, some of the initial steps in gene expression research still suffer from lacking methodology with regard to stress testing. A popular approach is to grow plants in the growth chamber or the greenhouse in pots and subject them to rather short stress cycle typical of root-restricted pot-grown plants. Barker et al. (2005) showed in maize that 27% of the expressed genes were regulated by a short 5 days drought stress applied in pots while only less than 1% of the genes were regulated by a slow (realistic) 5 weeks stress in the field.

The persisting question is which of the stress responsive or regulated genes have a role in plant resistance to drought and a value for breeding towards improved production under water limited conditions. For example, a gene expression study in rice cultivars which differed in their established drought resistance (Degenkolbe et al. 2008) found that more genes were drought regulated in the sensitive than in the tolerant cultivars. Furthermore seen that certain genes regulated in the susceptible or resistant cultivars caused a shift towards dilapidated responses such as senescence and reduced photosynthesis.

Hence gene expression studies are important but in the breeding perspective they are also very primordial steps which must be extended into functional genetics studies and value assessment.

3.2.2 Marker-Assisted Selection (MAS) for Drought Resistance

Molecular mapping for drought resistance has been here since the early 1990s. It has been addressing drought related plant traits ranging from yield under stress to root architecture and osmotic adjustment. At this time at least 70 mapping exercises concerning drought stress were performed just in rice, using at least 12 different populations. In contrast to this extensive activity there is just one well documented case of a drought-adapted rice cultivar released in India from a breeding program using MAS (Steele et al. 2006, 2007). This cultivar ('Birsa Vikas Dhan 111') was selected for larger root by MAS from the cross Kalinga-III × Azucena. The target root QTLs were first identified by Adam Price and Brigitte Courtois under a collaborative partnership between Bangor University, UK; Gramin Vikas Trust, Ranchi, Jharkhand, India; and Birsa Agricultural University (BAU), Ranchi, Jharkhand, India.

There might be other undocumented cases in the private seed industry. There are at this time a few publicly documented examples where MAS is approaching application in breeding for drought resistance traits which are discussed below.

It is therefore evident that at this time MAS for drought resistance traits has not yet developed into an important component of breeding for water-limited environments.

Collins et al. (2008) concluded that "MAS has contributed very little to the release of improved cultivars with greater tolerance to abiotic stresses." They further added that "a given QTL can have positive, null, or negative additive effects depending on the drought scenario. This complication has considerably slowed the utilization of QTL data in breeding. Additionally, predicting the performance of a particular genotype when considering multiple QTLs is complex."

Several reasons are recognized for the present low utility of MAS for drought resistance breeding, in contrast to the progress made in applying MAS in other breeding disciplines.

1. The quality of drought field phenotyping work during the mapping exercise is often unsatisfactory. Reasons are poor understanding of what traits constitute effective drought resistance in the target environment; low standards of the field work, mainly involving insufficient control of drought stress in the experiment; and insufficient control of other abiotic or biotic stresses. Thus phenotyping of drought resistance in the field has become a major bottleneck to high quality molecular mapping (Xu and Crouch 2008).
2. Drought resistance as reflected in the final yield under stress is affected by constitutive as well as by adaptive plant traits (Blum 1997, Sect. 3.6). QTLs for constitutive traits are reasonably consistent and stable across environments. However QTLs for adaptive traits are expressed in response to stress, its severity and timing during a given plant growth stage. This implies relative inconsistency in mapping drought resistance QTLs across environments and populations. Often there is a call for high resolution molecular mapping. High resolution is also required in phenotyping and in pinpointing the roles of constitutive and adaptive trait in ascribing resistance. For example (Liu et al. 2006), rice sterility under drought stress at flowering can be caused by reduced anther dehiscence. Two properties may contribute to high anther dehiscence: (1) constitutively superior development of fibrous structures in the endothecium at the anther apex and base and (2) better maintenance of constitutive pollen size. At the same time dehydration avoidance and maintenance of high plant water status as adaptive or constitutive processes can also help avoid sterility (e.g., Yue et al. 2005). Thus, molecular mapping of panicle sterility which is commonly phenotyped by counting sterile spikelets under drought stress will confound all the above different factors which may cause variable sterility and QTLs in different populations or environments. The desirable drought resistance QTLs should be identified at a low level of plant organization and at the determinant physiological causes of the trait involved. In this respect, "yield" as the most complex trait is actually a "moving target" for molecular mapping, with very few exceptions.
3. Mapping populations are by definition very diverse genetically. In most cases they are also diverse for phenology. Whereas drought stress for phenotyping purposes is applied to the population at a given date, the result is that early flowering genotypes receive stress later in their development while late flowering genotypes receive it at an earlier growth stage. This is especially crucial if stress is targeted to the flowering stage. Recently, a solution to the problem is attempted by developing diverse mapping populations that are not variable in flowering time.

In conclusion, genotyping QTLs is straight forward due to constant polishing and enhancement of theories, methods and equipment. It is also becoming more affordable. However, most of the problems in applying MAS to drought resistance breeding can be traced to *poor phenotyping* of drought resistance and the failure to address the most relevant traits which ascribe drought resistance for the given crop and environment – namely being able to target the required drought resistant ideotype (Sect. 3.7).

3.2.3 Transgenic Plants

The technology for moving genes across plant species has been perfected. It does not present any serious problem with respect to implanting the desirable genetic factor into most target genotypes. The number of reported transgenic plants claimed to express genes for drought resistance is increasing steadily (Fig. 3.2). By the end of 2009 about 110 genes were reported to putatively ascribe drought resistance in transgenic plants. Of these only very few were tested for yield under drought stress in the field.

Gene discovery research and the subsequent identification of the physiological function of the gene in the whole plant and under field conditions have been subjected to criticism by plant scientists and breeders. Criticism has been directed at the test protocols used and their value in predicting plant drought resistance. The first open criticism (Blum et al. 1996) was followed by additional critique to the

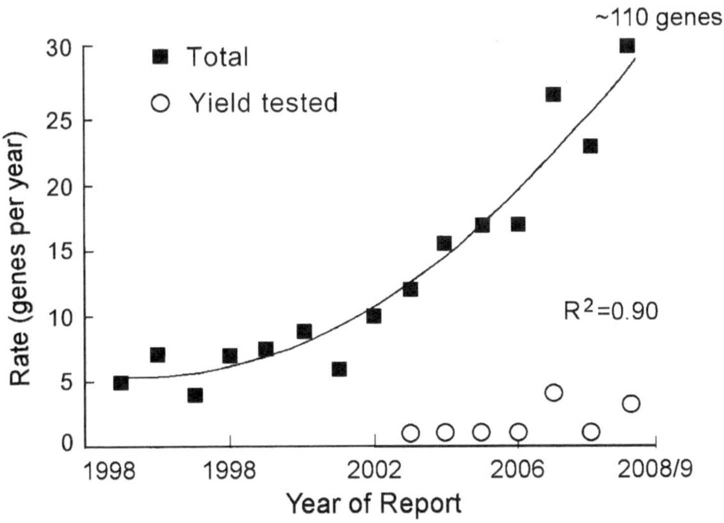

Fig. 3.2 Rate of production of transgenic plants for putative drought resistance genes as reported in the literature and the number of these transgenic plants tested for yield under drought stress in the field

extent that plant breeders tended to doubt the real contribution of transgenic technology towards breeding for water limited environments.

Beyond objective problems such as gene expression in the mutant or regulatory issues, the primary shortcoming is the quality of testing for function when drought resistance is claimed. In public research institutes transgenic plants are often not subjected to serious or even qualified evaluations and the scientific reports are sometimes satisfied with minimal laboratory testing. On the other hand there are very interesting cases which passed field testing for yield under drought stress and appear to carry a value for breeding (e.g., Wang et al. 2005a, b; Huang et al. 2007; Xiao et al. 2007; Castiglioni et al. 2008).

This area of research has a great potential for improving crop drought resistance. The subject is further discussed in Sect. 5.4 in terms of application to breeding.

3.3 Drought Resistance in Terms of Yield

3.3.1 Drought Resistance and Yield Potential: The Crossover Interaction

L.P. Reitz who was a leading plant breeder in the mid twentieth century wrote that "breeders worship the yield column in their field-books" (Reitz 1974). Breeders are right in doing so, and it has been producing results. He also warned against following a mirage in a selection program. When yield is tested at the end of several years of selection, only then mistakes are revealed – and there is no return. Where yield is the traditional target of the breeding program it is also the final test and almost always the main selection index. Reitz stated the main corollaries of crop plant breeding at the time, namely that crop cultivars fall into three categories: (1) those with uniform superiority over all environments; (2) those relatively better in poor environments; and (3) those relatively better in favorable environments. As yield potential increased since that time, category (1) should be changed to "those with uniform superiority over all moderate environments." More on this modification and the reasons behind it (Blum 2005) will be further discussed in the subsequent sections of this chapter.

Figure 3.3 represents a realistic (e.g., Blum and Pnuel 1990; Ceccarelli and Grando 1991) schematic linear regressions of yield of two different cultivars in response to the water regime. Water regimes are represented by the "Environmental Index" which is the mean yield over many cultivars in locations differing in their water regime.

Yield of the two cultivars is reduced with increasing drought stress. Cultivar A is a high yielding cultivar of high yield potential. Cultivar B is a cultivar which does not perform as well under well watered conditions but it performs relatively well at low water regimes. By definition cultivar B is relatively drought resistant, having a lower yield potential. There is a crossover interaction between the two cultivars at an environmental index of around 3–4 ton ha^{-1}, which is typical of several cereal crops (Ceccarelli 1989; Blum and Pnuel 1990; Ceccarelli and Grando 1991; Fischer et al. 2003). Apparently, cultivar B possesses traits non-existent in A, which allows it to crossover and perform better than A when drought stress is severe. These same traits might also be responsible for its poor performance under potential conditions.

The persisting question and the continuous debate among breeders is whether the crossover is unavoidable if yield is to be improved at the low end, involving a loss in yield potential. There are reports where yield improvements have been shown to occur on both ends, where the new improved cultivar performed better than A in all environments (which would present a parallel line with respect to A in Fig. 3.3). This is the typical response involved with early stage breeding which sometimes proceeds from landraces or "local" cultivars to improved cultivars. This type of improvement can be found in technical reports and conference presentations coming from breeding programs in developing countries and it basically represents the genetic impact of a "green revolution." Namely, the improvement of inherent and innate features of the crop plant which affect its performance in any environment. These might involve optimization of phenology, reduced height, reduced lodging, increased harvest index, better threshability, better canopy structure, and sometimes even disease resistance. This phase of yield progress across all environments can also be seen in some reports on the yield improvement across historical varieties. For example, Castleberry et al. (1984) demonstrated that the genetic improvement of maize hybrids yield in the USA was similar under both dryland and irrigated conditions from the 1930s to the 1970s. Yield improvement with the 1980s hybrids already produced a crossover interaction with respect to the 1970s hybrids.

Where yield is initially reduced by drought stress from an environmental index of 8 to about 5 ton ha^{-1} (Fig. 3.3) the high potential of cultivar A appears to continue

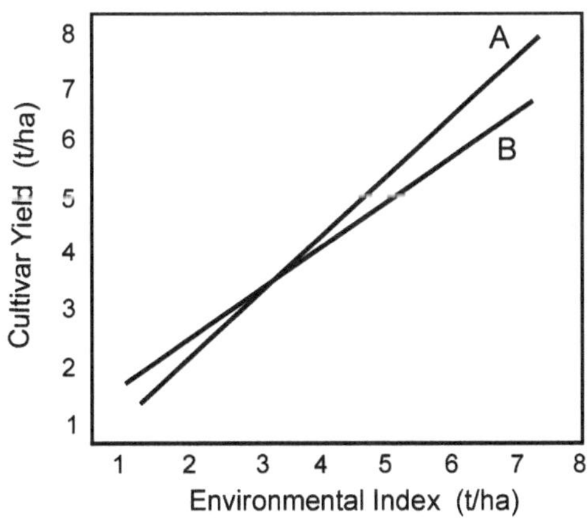

Fig. 3.3 A schematic representation of the linear regression of yield of two different cultivars on an environmental index. The environmental index is derived from the mean yield of many tested cultivars under a range of conditions which differ in their water regime. Hence low environmental index represents conditions of drought stress and vice versa for a high index (see text)

to support its relative yield advantage. It yields better than the drought resistant B despite stress, as long as stress is not severe so as to reduce yield to the point of crossover. Therefore, a high yield potential is sustaining yield under moderate drought stress. Looking at such similar crop responses (Lafitte et al. 1997; Pantuwan et al. 2002; Campos et al. 2004; Bidinger et al. 2005; Barker et al. 2005) a rule of thumb would be that *breeding for high yield potential will most likely help sustain yields under conditions where mean yield is not reduced more than about 50% of the potential.* Breeders sometimes define this as "yield potential spillover" (Barker et al. 2005). This is a crucial point to consider. It is critical for designing a selection program towards drought resistance. There are numerous publications describing selection programs for drought resistance in terms of yield where stress intensity is mild as expressed by a mean yield reduction of only 20–40% from non-stress conditions. The better performing lines selected under such mild stress conditions are described in these publications as "drought resistant" or "drought adapted," which they are not. They are most likely lines of improved yield potential (e.g., Pantuwan et al. 2002). This trade-off and cross-over between yield potential and drought resistance is well expressed also in the discussion by Ceccarelli (1989) on a scale of yields appropriate to barley in his range of target environments.

It can therefore be concluded that drought resistance (as expressed by the cross-over interaction) and high yield potential are most likely mutually exclusive or will become so at some level of yield potential improvement in the specific breeding program. This crossover has a sound biological basis as discussed in detail by Blum (2005) and as briefly explained below.

Firstly, in most past and present selection programs towards improved yield under drought stress the main selection criterion is yield under stress. When selection for yield under severe drought stress is performed there is a correlated genetic shift towards plant constitutive traits which support moderate water-use. These are mainly reduced leaf area and LAI (Blum and Sullivan 1997; Fukai et al. 1999), constitutively low leaf transpiration rate (Kholov et al. 2010), early flowering (e.g., Blum and Pnuel 1990; Yadav and Bhatnagar 2001) and reduced tillering in the cereals. Low constitutive (potential) total plant dry matter was associated with better water status and yield under drought stress in rice (Fukai et al. 1999). QTL-based pearl millet hybrids selected for drought resistance (Bidinger et al. 2005) were relatively higher yielding in a series of stress environments. However, this gain under stress was achieved at the cost of a lower yield in the non-stressed environments. Their particular stress adaptation was consistent with early flowering, limited tillering and low biomass. Reduced leaf area and LAI, reduced tillering capacity and earlier flowering are basically not compatible with a high yield potential phenotype.

Secondly, high grain yield cultivars have a large sink. Large sink constitutes a load on the plant in terms of its water-use and water status (Sect. 2.7). Under drought conditions this load can be expressed in low plant water status.

Thirdly, a deep root is a major attribute of dehydration avoidance. It is common knowledge in rice breeding for example (Fischer et al. 2003) that deep rooted genotypes such as typical upland rice tend to have limited tillering. Extensive tillering which is typical of high yielding rice cultivars involves extensive adventitious root production which can limit growth and extension of existing roots into deep soil. However this does not necessarily hold across all cases, such as the case for heterosis (see below).

In conclusion, the crossover interaction is a consequence of some basic biological attributes of crop plants especially observed in the cereals. As yield potential increases it becomes more difficult to achieve drought resistance in the highest yielding genotype. Reported parallel improvement of both yield potential and yield under stress as compared with an older standard cultivar can be taken as an indication that the a high yield potential has not been approached yet in the new cultivar for that growing environment (climate, soil and biota).

Having said all of the above, there are notable exceptions where drought resistance did not compromise high yield potential. These are cases where drought resistance is conditioned by a major trait that has no negative effect on yield potential and sometimes has even a positive effect. Examples for such exceptions are:

1. *Anthesis-to-silking interval* (*ASI*) is a crucial developmental trait that carries very large effect on maize fertility (section "The Effect of Phenology"). Drought stress during the reproductive stage delays silking and thus pollen shed does not coincide with silk receptivity. ASI-based drought resistant maize sustains short ASI despite stress (Bolanos and Edmeades 1996). A short ASI under drought stress does not bear a negative effect on yield potential.
2. Experiments performed under managed stress environment at IRRI (Kumar et al. 2007; Venuprasad et al. 2008) used selection for yield under severe drought stress applied at flowering. Selected drought resistant lines were also found to perform very well under irrigated conditions. Such stress impairs fertility and resistant lines must have the capacity to sustain fertility under stress. Mechanisms which support plant reproduction process under stress are not likely to compromise yield potential and may even have a positive effect on plant performance under non-stress conditions, where for example hot or high vapor pressure deficit conditions occur in a well irrigated crop.
3. It is theoretically very possible that selection for yield in a field stress environment can result is selection pressure for resistance to other abiotic stresses besides drought. It would be very logical to expect, for example, the improvement in both drought resistance and heat tolerance when selection is continuously performed for yield in a dry hot environment. However, evidence for such multiple resistances developed as a result of selection for yield is rare. A recent study in rice indicated that the Pup1 locus responsible for tolerance to phosphorus deficiency was found in 80% of varieties recognized as drought resistant (Hyoun Chin et al. 2010).

Hybrid varieties are a private case of high yield potential genotypes and their performance under drought stress is of interest and importance. The uniqueness of the hybrids is that their high yield potential is achieved by heterosis – a genetic-physiological phenomenon which basically still remains an enigma. The classical plant breeding texts often deal with the question of the expression of heterosis over a wide range of environments, ranging from low to high yield conditions.

Most breeders who work with hybrids are divided into those who believe that heterosis is expressed throughout the environmental spectrum and those who believe that open-pollinated varieties are superior to hybrids at the low yield end. Heterosis might present an advantage under drought stress. For example, heterosis in sorghum can be expressed also in a large root (Blum et al. 1977a, b) and better soil moisture capture. Araus et al. (2010) found that heterosis in maize was manifested in more effective water use and better plant water status which might have resulted from larger roots in the hybrids as compared with the parental lines. Root size as a driver of heterosis in plant water status or even nutrient capture is very reasonable for maize in view of the common inbreeding depression in this crop plant. This explanation is less likely in sorghum where inbreeding depression is rare or non-existent.

However my discussions with breeders experienced in hybrid breeding seem to support the general case for a crossover interaction for yield, again.

Heterosis is commonly driven by parental lines of general combining ability which support the broad expression of high yield potential. A large root in the hybrid is an example of an important component of heterosis for yield as mentioned above. As drought conditions develop and as the profile of drought becomes more specific (related to growth stage etc.) general combining ability for high yield may not be sufficient to sustain yield and the hybrid requires some additional specific drought resistance traits to cope with this environment. Such traits will be provided by the specific parental lines. Parental line selection for drought resistance becomes crucial for hybrid performance under drought stress.

This has already been observed by Jenkins (1932) who noted that during the dry weather of 1930 ten maize hybrids made with one of the lines were completely free from leaf burning, whereas those of another line ranged from some to many plants with burned leaves in the different crosses. Much later, the same conclusion is reflected in a study with maize, where the correlation in yield between hybrids and parents increased as drought stress increased (Betrán et al. 2003a). In another study with maize it was concluded that there was a need for drought resistance in both parental lines to achieve acceptable hybrid performance under severe drought (Betrán et al. 2003b). Elite × landrace pearl millet crosses which expressed good heterosis under drought stress owed their superiority to the drought resistance of the landrace parents originating from the dry Rajasthan region (Prester and Weltzien 2003). Improvement of female lines for semi arid conditions was considered essential for adapting sorghum hybrids to the dryer regions of Kenya (Haussmann et al. 1999).

It can therefore be concluded for hybrids that in most cases high yield potential is important above the crossover point while specific parental drought resistance is important for hybrid performance below that point. The differences in opinions among breeders and among publications might have resulted from the specific rate of drought stress in their test environments, whether it was above or below the point of crossover as well as the specific crop, germplasm and environment.

Finally, numerical taxonomic methods were always used by breeders to reduce their own frustration with the crossover interaction (genotype by environment interaction) in trying to achieve "a cultivar for all seasons." These methods had no biological basis or a basis for understanding the reasons for the interaction. Horner and

Frey (1957) were among the first to apply the method in their breeding program. They were able to show for oats, by a fairly simple method, that a substantial reduction of the cultivar by environment (crossover) interaction resulted from a proper division of the state of Iowa into sub-regions. Optimum subdivision into four parts resulted in a reduction of the interaction component by 30%. Similar results were later found also for winter wheat in Kansas (Liang et al. 1966). In more recent times these basic considerations were elegantly underlined and further polished (Cooper and Hammer 1996). This was followed by an attempt to develop a biological insight into resolving the problem of genotype by environment interaction by using crop simulation and genetic models (e.g., Hammer et al. 2006).

However, when water limited environments are being dealt with in the breeding program in isolation of all other environmental limitations to yield the crossover interaction can be successfully resolved. The most important lesson in this respect can be learned from the combined works of Ceccarelli and associates in rainfed barley which spanned from 1987 (Ceccarelli 1987) to 2005 (Al-Yassin et al. 2005). Taken together it can be concluded that where drought stress is severe, a cultivar adapted to the region becomes more specific and less common to other regions. This is well supported by other studies showing that wide adaptation was achieved at the cost of poorer performance in specific stress environments. Henceforth, a variety with uniform superiority over all environments (Reitz 1974) is a rare occurrence. Regretfully, for most breeders working in rainfed environments "a super cultivar for all seasons" is still the Holy Grail.

3.3.2 The Heritability of Yield and Drought Stress

Whereas yield is a common selection criterion in most field crops breeding programs, its inheritance becomes a crucial issue especially if selection for yield is performed under low yield (stress) conditions.

Yield has been dealt with quite extensively in quantitative genetics studies and later in molecular mapping experiments with the common and frequent conclusion that yield is inherited in a "complex manner" and this is why heritability for yield can be elusive, inconsistent and erratic. Sometimes "epistasis" is blamed for inconsident heritability estimates from one study to the other. It is quite amazing that we keep struggling with the inheritance of yield for nearly a century with the same recurring conclusion. Perhaps yield as such is not truly heritable?

We are still limited in our understanding of how yield is being controlled at the basic physiological level. Subsequently we are still limited in our understanding of the genetic control of yield. It is the multitude of developmental, physiological and biochemical processes and their interactions at any level of plant organization and any level of biological scale that are under genetic control. The sum total is integrated toward a final effect on yield. This sum total is then subjected to a genetic analysis by legitimate biometrical methods that attempt to compensate for the remoteness of the yield trait from its underlying, truly heritable biology. I must conclude that the study of the inheritance of yield is a mere biometrical exercise that sometimes with luck can be found to be useful for a specific breeder × crop × environment situation.

Even the simplest yield component or the simplest developmental interaction exemplifies the superficial state of the genetic analysis of yield. Already Johnson and Frey (1967) for oats and Blum (1973) for sorghum have shown that stress conditions caused an increase in kernel weight due to reduction in kernel number per panicle under the effect of stress. Similarly, heterosis in the yield of barley was not in evidence because as grain number increased under the effect of heterosis, grain weight decreased due to competition within the spike (Carleton and Foote 1968). What then would be the meaning of a genetic analysis of grain weight per panicle (a less complex trait than yield) when intra-plant competition or developmental plasticity can introduce large phenotypic variation into the analysis. This also reflects on the problem of resolving QTL by environment interaction for yield or similarly complex traits (see below). The analysis will improve biometrically and might produce predictable results when dealing with relatively stable environments. The analysis will hardly produce practical results towards non-stable environments, Environmental stability of a crop increases as all crop inputs are supplied, including abiotic and biotic stress protection.

The quantitative genetic analysis of yield for certain materials in given environments may prove useful for designing an approach to selection for that material and environment. However, in the biological sense, both the analysis and its application to selection are still empirical and loose. Extreme care is advisable towards accepting and interpreting the genetic analysis of yield at face value. This is of special concern under drought stress where field spatial variability is inherent (Sect. 4.1.1).

Heritability is defined as the ratio of additive genetic variance ("narrow sense" H) or genetic variance ("wide sense" H) to environmental variance. Decades of experience in plant breeding have shown that this is a useful parameter for evaluating the expected genetic progress in selection for complex or simple traits. Rather than deriving from H an exact number of "expected gain," the breeder is advised to regard H as an indication of probability for achieving gain and use it together with other available information. For example, H of 50% accompanied by a large budget might provide the same gain as H of 90% with a small budget.

A continuous debate takes place in the literature and in the plant breeding community on the value of H for yield when selection is performed under low yield (stress) conditions. Evidence has been accumulating to the effect that H for yield under stress conditions is reduced and selection for yield becomes relatively less effective as compared with non-stress conditions (Blum 1988). The type of stress was not important in those studies. Low yield in the selection environment was producing low H. The cause for reduced H under low yield conditions has been traced to both low genetic variance and high environmental variance for yield in stress environments. A study in beans for example (Frahm et al. 2004) concluded that drought stress and low yields contributed to high environmental coefficients of variation (CV), which reduced selection efficiency. Breeders experienced in dryland work will testify that drought stress causes increased spatial field variability in

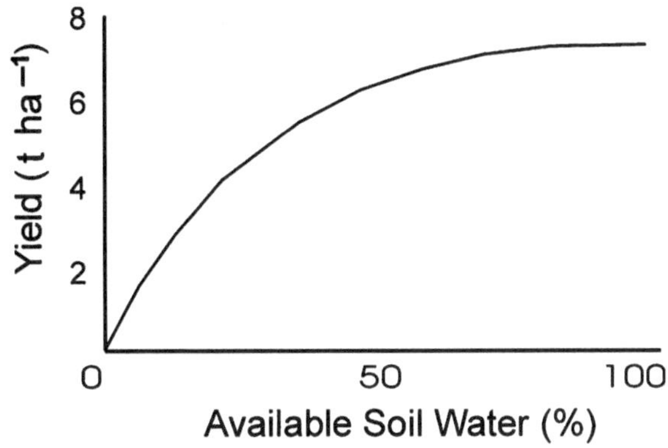

Fig. 3.4 A schematic classical yield response to water supply

crop growth and yield, as compared with a well-watered crop. The reason is very simple and it is traced to the non-linearity of the classical yield response to moisture supply (Fig. 3.4).

As water supply decrease slightly from the optimum the associated rate of yield reduction per unit water supply is small. When water supply is further reduced towards the exponential part of the curve, yield reduction per unit change in water supply becomes much greater. Therefore when a significant soil moisture deficit develops in the field small variations in soil water content or availability within the field have a relatively large effect on yield.

On the other hand Ceccarelli and associates working with barley in harsh environments (e.g., Ceccarelli and Grando 1991; Al-Yassin et al. 2005) consistently reported high heritability for yield under stress and a considerable progress in selection for yield under stress. The magnitude of heritability under stress was associated in one of their studies with the genetic distance between parents of crosses. Their use of drought adapted barley landraces in crosses (Ceccarelli et al. 1998) must have ensured that their populations contained sufficient genetic variability for drought resistance, rather than just variation for yield potential. Similarly, Venuprasad et al. (2007) found that selection for rice yield under severe drought stress was effective if the population carried high levels of drought resistance derived from a resistant donor. Hence, the genetic and physiological structure of the population in terms of drought resistance is crucial in determining H for yield under stress and the efficiency of selection for yield under stress.

Use in selection of less complex plant traits that carry a physiological relevance towards drought stress such as ASI in maize (Bolanos and Edmeades 1996) can compensate for the low heritability of yield and improve progress in selection. This is one of the corollaries of breeding and selection towards water limited environments as will be discussed in this and later chapters.

In conclusion, two important considerations can improve heritability and the selection efficiency for yield under drought stress and/or compensate for the low H for yield under stress:

1. Manage the drought field selection environment to improve homogeneity in soil moisture conditions under stress and thus reduce experimental error (Sect. 4.1.1).
2. Use populations which contain sufficient genetic variability for yield and for the specific drought resistance traits relevant to the target environment and the ideotype. You should know that some breeders describe these traits as "secondary." In this context they might be considered primary. Use these traits *together* with yield in the selection index to enhance selection efficiency (e.g., Chapman and Edmeades 1999).

3.3.3 QTLs and Yield Under Drought Stress

The difficulty of resolving the inheritance of yield and the question whether yield is a genetic entity at all is reflected also in the molecular mapping of yield QTLs and the struggling attempts to use marker assisted selection (MAS) for yield.

QTLs for yield rarely express repeatable expression for the same population in different years or locations, notwithstanding different populations. QTL × environment interaction for yield is the rule. The resolution of yield QTLs is generally low (Campos et al. 2004). Ribaut et al. (1997) concluded from their work in maize that "no major QTLs, expressing more than 13% of the phenotypic variance, were detected for any of these (yield) traits, and there were inconsistencies in their genomic positions across water regimes." This is typically reflected also in a study of rice (Zou et al. 2006) where a total of 32 QTLs for grain yield and its components were identified and the phenotypic variation explained by individual QTLs varied from 1.29 to 14.76%.

Perhaps the most typical representation of the status of the current significance and utility of mapped QTLs for yield under drought stress is represented by two different studies with the same population of barley derived from the cross Steptoe × Morex. Romagosa et al. (1999) found that four genomic regions (QTLs1–4) were associated with yield in different dryland environments. QTL1 was effective in identifying high yield genotypes in all environments. Genotypic (MAS) and tandem genotypic and phenotypic selection were at least as good as phenotypic selection. Consistent selection responses were detected for QTL1 alone and together with QTL3. Other three QTLs expressed strong interaction with the environment and were considered ineffective in selection unless the biological basis of their interaction would be resolved. Zhu et al. (1999) constructed a barley population by crossing two superior lines in terms of their yield QTLs which were selected from the same Steptoe × Morex population. They found that all target QTLs showed significant QTL by environment interaction. Digenic epistatic effects were also detected between some QTL loci.

On the background of the above discussion, the finding of a large-effect QTL on yield of upland rice is indeed an exceptional result (Bernier et al. 2007; Venuprasad et al. 2009). Similarly, a relatively stable yield QTL across diverse environments was detected in wheat (Kuchel et al. 2007). Bidinger et al. (2007) reported for pearl millet a major effect QTL which explained 13–25% of variation in grain yield over 12 environments subjected to terminal drought stress.

In rice, the effect of this QTL on grain yield increased with increasing intensity of drought stress at flowering, from having no effect under well-watered conditions to having an additive effect of more than 40% of the trial mean yield in the most severe stress treatments. However, it was later found that the effect of this allele was largely epistatic and its expression differed among populations.

It still remains that QTL by environment interaction is the rule, and a QTL main effect on yield is an exception. Given the enormous inconsistency of QTLs across environments and traits, breeders are reluctant to adopt MAS for yield under drought stress. Some help might be offered in the future by multi-environment QTL mixed models (Mathews et al. 2008), perhaps.

The conclusion made by Romagosa et al. (1999) point at the future direction in selection for yield. Empirical-numerical approach to QTL analysis and use in selection for yield and its stability is likely to produce limited results. The biological basis of yield and its response to the environment should be resolved and introduced into the analysis and the execution of the selection program, be it phenotypic or genotypic.

QTL by environment interaction can be driven by a host of environmental and biotic factors which produce different genotypic order in yield in each environment. Here we discuss the water environment exclusively and we seek genotypes that perform best in a limited water regime. If all other variables such as soil type, nutrients and biotic stresses are controlled the resolution of yield inheritance should improve and selection for drought resistance by yield as a criterion should become more effective. This is a simple version of the "eco-statistical" modeling approach proposed by van Eeuwijk et al. (2005). They recommended an approach that includes a synthesis of a multiple QTL model and an eco-physiological model to describe a collection of genotypic response curves. Their review describes an example of yield data analysis from the North American barley genome project with added environmental information. QTL by environment interaction at barley chromosome 2H was found to depend on the temperature during heading. A QTL allele substitution increased or decreased yield within 0.11 ton ha^{-1} for every degree of temperature increase. Similarly, Ouk et al. (2007) found in rice that stress at flowering and late season was driving G × E interaction.

Selection for yield under stress, whether genotypic or phenotypic, must address drought stress in isolation of any other environmental and biotic stress. This is again underlined by a multilocation study in maize (Messmer et al. 2009) where the expression of yield and yield component QTLs was quite stable under the same water regime. Their conclusion that "the best approach to breeding for drought tolerance includes selection under water stress" is not surprising at all to the same extent that selection for disease resistance under disease infection is not (Sect. 3.4.1).

3.4 Drought Resistance in Terms of Physiology

The Preface mentions the prevailing and almost persistent complaint found in the literature and expressed in conferences and research proposals, namely that "drought resistance is a very complex trait." One of the objectives of this chapter is to clarify to the reader that drought resistance is not as complex as it is being implied. Despite the danger of being regarded too deliberate I must state that the assumed complexity of drought resistance appears to result from four major possible reasons which I have personally evidenced: (1) insufficient education on the subject (2) low proficiency for translating knowledge into practical problem solving (3) serious methodological faults in phenotyping drought resistance in one's work, and (4) the genomics "gene discovery" platform where the numerous expressed genes under a single stress event (whether relevant or not) are viewed as an indication of complex "adaptation." Views taken from the whole plant platform allow to consider drought resistance as a far less complex trait.

Voiced and documented opinions (e.g., Passioura et al. 2007; Munns and Richards 2007) have been known to reject the use of "drought resistance" (or "tolerance") in an agricultural context with the argument that drought resistance implies survival and xerophytism while under limited water supply in agriculture we are interested in productivity and not just in the mere survival of the crop. According to this opinion drought resistance belongs with natural vegetation and not with crop production. This argument does not survive in the face of fact. The term "drought resistance" was rejected by these authors in favor of "water productivity" as a working definition for plant breeders in water-limited environments. In the discussion of the subject under the definition of "water productivity" Munns and Richards (2007) proceed to examine the components of "water productivity." The reader can easily recognize that these components are actually the same as for "drought resistance." For example they point out the importance of roots towards water productivity, while at the same time roots are a major component of "drought resistance" ("dehydration avoidance") (section "Roots and Dehydration Avoidance"). Similarly, they indicate that a new Australian "water productive" wheat cultivar excel in osmotic adjustment. Osmotic adjustment is a major component of drought resistance ("dehydration avoidance") (section "Osmotic Adjustment") in both natural vegetation and crop plants. It will also be shown below that osmotic adjustment is effective for both "productivity" and "survival." What these authors aim at is perhaps the "effective use of water" (EUW) (Blum 2009) as being determined by various components of drought resistance and plant developmental plasticity, thereby leading to sustained yield under drought stress. This will be further discussed in this chapter.

Even real plant survival in its xerophytic facets, which is rejected as being relevant towards plant production under stress (Passioura et al. 2007), can still be important in an agricultural context, as will be discussed in detail below. Serraj and Sinclair (2002) even went a step further and argued that osmotic adjustment (a major component of "dehydration avoidance" linked to sustained yield under drought stress; section "Osmotic Adjustment") is not important for plant production under drought stress since it just enhances survival.

In conclusion, new semantics do not really make a difference in understanding the underlying physiology of plant production under drought stress or towards helping breeders define their breeding target. Besides "water productivity" we also find terms such as "plant protection," "plant resilience," "plant defense," "plant homeostasis" etc. which are often volunteered by newcomers to this discipline. Unnecessary novel semantics can promote confusion and drive the impression that "drought resistance is complex."

The term "drought resistance" and its logical components have been used by experts in this field since its introduction by Levitt (1980). It is a most helpful approach since it allows a physiologically logical dissection of plant capacity to perform under drought stress in a manner which can be acted upon by geneticists and breeders. In order to further elaborate the relevance of "drought resistance" as a working definition towards breeding for improved yield under drought stress, an analogy to breeding for "disease resistance" (not "disease survival") is used below. Italics in this example highlight the analogies to the point that the reader will grasp the direction of the analogy.

3.4.1 The Disease Resistance Analogy

Disease resistance (*drought resistance*) as such is not defined enough to be used in actual selection and breeding. The first and primary information needed before a breeding program can be designed is which disease (*which drought*) is the target of breeding. Only when the disease is classified, characterized and defined one can plan a breeding program. The definition and characterization of the disease includes the recognition of its *expressions and symptoms* in the plant. For certain diseases there is seedling *growth stage*, adult plant stage or flowering stage infection (*stress*). Resistance at different stages can be very different physiologically and genetically.

Disease *resistance* and *tolerance* are distinctly different in terms of physiology, genetics and breeding approach. Resistance means that the plant is not infected. It is capable of disease infection *avoidance*. This is the most common mechanism dealt with in breeding for disease resistance because it is practically more effective than tolerance. Under disease epiphytotics the resistant plant is free of the disease (*free of water deficit*). Disease *tolerance* on the other hand is rare in plant pathology and plant breeding. It is the capacity of the plant to *function* when it is heavily infected by the disease (*dehydrated*). Plants may *survive* under heavy disease infection but plant production is seriously impaired. Tolerance is economically important if it sustains plant production under disease stress.

Plant pathologists are also interested in understanding how damage caused by the pathogen affect yield. Initially, phenotypic selection programs for resistance concentrate on disease symptoms and expressions in the plant. A scale and a score system for plant symptoms have been constructed as a tool for phenotyping resistance. Symptoms are not only visual but selection programs can use plant physiological symptoms and expressions in phenotyping resistance. However, many physiological symptoms are not useful in screening of large populations because physiological measurements are slow. For some diseases, remote sensing techniques which are fast have been applied to selection work.

Plant breeders rarely use yield as a selection index for disease (*drought*) resistance. It is recognized that yield is a genetically complex trait and plant resistance is in most cases genetically independent of yield. It would be expected that yield under disease stress will generally express low heritability. Therefore selection for disease resistance by addressing the disease response of the plant is far more effective than by using yield under disease infection. Certainly at some point in the breeding program selection for yield and yield potential are included in parallel or in line with selection for disease resistance and other traits such as product quality, etc. By and large the inheritance of disease resistance is basically simpler than the inheritance of yield.

Plant pathologists together with plant breeders have developed detailed *phenotyping protocols for each disease and crop* of interest. Breeders, geneticists and molecular biologists do not invent disease resistance phenotyping protocols in isolation of plant pathology knowledge. Plants are subjected to heavy and *controlled* infection by the disease under which plant reactions can be observed or measured. The main concern is to avoid any "escapes" in phenotyping work.

Phenotyping can be executed by subjecting the population to disease epiphytotics in a natural environment where the disease occurs on a permanent basis (defined as a "hot spot"). Alternatively, the disease is cultured in some form or method and spread upon the plants in the field or the greenhouse as a "*managed* disease *environment*." Care is taken to avoid any un-intentional biotic or abiotic stress in the phenotyping process so as not to bias results for disease reaction. The control and elimination of other stresses in the phenotyping site is as important as the infection by the disease.

Once genetic donors for disease resistance are identified a backcross program is commonly used to incorporate resistance into the adapted and desirable cultivar. MAS is sometimes used as an aid in breeding for disease resistance, provided that the disease resistance has been well mapped. MAS for yield under disease infection is not acceptable for reasons described above.

In conclusion, this analogy should demonstrate that breeding for drought resistance is in principle analogous to the established approach of breeding for disease resistance (with few exceptions). Genes for resistance are identified and recombined with an agronomic genotype by established methods of phenotypic or genotypic selection. The point to remember is that breeding for drought resistance is an emerging plant breeding discipline possibly standing today where plant pathology has been almost a century ago.

3.4.2 The Components of Drought Resistance

Levitt (1980) analyzed and defined the basis for plant response to environmental stress upon which the understanding of the physiology and genetics of resistance could be built upon. He presented the most serious and comprehensive terminology for environmental stress resistance. His explanation of stress and strain terms and the derived resistance or tolerance terms should serve as guideline for any individual involved in stress work. His approach is therefore adopted and used here, with one addition. The term “stress escape” was not considered by Levitt for the obvious reason that it does not involve any exposure of the plant to stress and the subsequent development of strain. However, in the agronomic domain and the related breeding work, “drought escape” is an important component of breeding solutions for limited water environments.

Previously in this chapter the term “secondary traits” has been brought up, as it is sometimes used by breeders in addressing various plant traits involved in drought resistance. Secondary traits were defined ad hoc as they emerged to be relevant in various studies. All traits defined as “secondary traits” can be easily redefined as distinct components of drought resistance or as developmental traits or as grain yield components. Therefore, wherever relevant, these “secondary traits” will be discussed under the appropriate component(s) of drought resistance. The discussion of traits as components of drought resistance rather than as “secondary traits” will clarify their physiological background, context and actual role in promoting drought resistance.

3.4.2.1 Dehydration Avoidance

Dehydration avoidance is defined as the plant’s ability to maintain a relatively high level of hydration under conditions of soil or atmospheric water stress. The basic concept of dehydration avoidance is in the fact that by retaining a high level of tissue hydration, despite exposure to the stress environment, the plant does not develop a strain as a result of stress. Namely, various physiological, biochemical, and metabolic processes that are involved in growth and yield are not exposed to plant tissues water deficit. In this sense they are “protected” from stress and avoid being strained. The plant does not “avoid drought” and therefore the term is not “drought avoidance” as erroneously mentioned in the literature sometimes.

The common measure of dehydration avoidance is the tissue’s water status under drought stress as expressed by plant (leaf) water potential or relative water content (RWC). The maintenance of turgor is also a component of dehydration avoidance. Therefore, osmotic adjustment, as a means for retaining turgor at a given tissue water potential (Fig. 2.3), is a component of dehydration avoidance.

Levitt (1980) recognized two plant types with respect to dehydration avoidance: plants that avoid dehydration by reduced transpiration and water-use (“water savers”) and plants that use means other than reduced transpiration (“water spenders”) in order to maintain dehydration avoidance. These two types of plants were also characterized by classical botanists through inappropriate anthropomorphic definitions, such as “pessimistic” and “optimistic,” respectively. The important point is the implications of each of the two phenotypes of dehydration avoidance toward plant production in an agricultural ecosystem. The available evidence in crop plants suggests that a combination of both responses may exist in one species, probably as a result of selection by man.

The corollary is that high stomatal conductance and transpiration are linked to plant production and an effective use of water by the plant is in most cases the key to sustained production in limited water environments (Blum 2009).

It is therefore important to clarify at the onset of this discussion that evolution in natural vegetation and selection by man in domesticated plants most likely ascribed a different preference to different drought resistance mechanisms. Levitt (1980) noted that a "water-saving" mechanism is common in xerophytes subjected to extreme water stress. Xerophytes, almost by definition, were evolved for survival. Their morphological features, growth behavior, and biomass production all attend to the fact that they were selected for low production in return for a high level of survival. In most (but not all) cases this is not the ideal phenotype for an agricultural ecosystem. However, as discussed above, xerophytism is not equated with drought resistance. It is only one facet of resistance.

Stomatal Activity and Dehydration Avoidance

Stomatal activity is driven by various plant adaptive responses to stress and the effect of plant constitutive traits on plant function under stress. Stomata will remain open or close depending on many factors as those discussed in Chap. 2, including hydraulic signals, hormonal signals, CO_2, light, pH, ion status and aquaporins. Thus there are potentially numerous avenues for intervention in stomatal activity by genetic and transgenic manipulations. On one hand "insensitive" mutants were obtained where stomata retained relatively open position, resulting in very high sensitivity to leaf water loss and ensuing wilting (e.g., Nagel et al. 1994). On the other hand transgenic plants were developed to express high stomatal sensitivity and quick closure in response to mild drought stress (e.g., Laporte et al. 2002; Jung et al. 2008). In most cases the modification of stomatal sensitivity to drought stress was achieve via intervention in ABA metabolism. It is interesting that in both cases of high and low sensitivity, drought resistance was claimed to be the result of the modification. Careful consideration must be given to the genetic manipulations of stomatal sensitivity. Sensitivity to which signal? Hydraulic, hormonal, light, CO_2? The delicate interplay between CO_2 substomatal concentration, stomatal conductance and transpiration already led Jones (1998) to the realistic conclusion that it is often difficult to decide whether stomata are controlling gas-exchange or vice versa. Hence, "wielding the axe" on one facet of stomatal activity can lead to unpredictable results with respect to breeding. While we seem to have the tools for genetic intervention in stomatal activity, we are not yet capable to fully grasp the nature and impact of such intervention at the whole plant and crop level in the field. As far as applied plant breeding for crop production under drought stress is concerned direct genetic intervention in stomatal activity is not on the agenda at this time.

Therefore, breeding for drought resistance might be addressing and phenotyping normal stomata functions only as a reflection of other important plant processes that sustain crop production under drought stress. Photosystem activity is relatively less sensitive to leaf water deficit as compared to stomata (Saccardy et al. 1996;

Flexas and Medrano 2002; Flexas et al. 2004). As drought stress develops most of the control over photosynthesis is by way of stomatal activity. Photosystem responds to leaf dehydration only when the latter becomes sever. Dehydration injury to the photosystem will generally occur upon or after stomatal closure. There is even some indication that certain photosystem functions can acclimate to dehydration under field conditions (Maroco et al. 2002). Such acclimation might not be of great significance to productivity when stomata already close. However, it might have a role upon recovery (Galle and Feller 2007). Therefore, sustaining stomatal conductance throughout the main part of the drying cycle is crucial for photosynthesis and plant production.

It is now very well established that high yielding genotypes of cotton, wheat and rice sustain high stomatal conductance and transpiration under drought stress (Blum et al. 1982; Izanloo et al. 2008; Sanguineti et al. 1999) or under well-watered conditions (Fischer et al. 1998; Horie et al. 2006; Lu et al. 1994; Reynolds et al. 1994), as indicated by leaf gas exchange measurements or canopy temperature measurements. The role of dehydration avoidance with respect to stomata and leaf gas exchange is to maintain water supply and sustain leaf hydration and turgidity so as to delay stomatal closure.

Non-stomatal Control of Water Use and Canopy Energy Balance

Leaf Epicuticular Wax

Epicuticular wax load (EC) has two major consequences: reduction of cuticular (non-stomatal) transpiration and the increase in leaf reflectance (albedo). Increase in reflectance results in reduced transpirational demand due to lower R_n (Sect. 2.2). When the cuticle is impermeable due to high EC, all leaf transpiration is controlled by stomata, without any "leaks."

It has been amply demonstrated that EC can be increased 30–150% by leaf water deficit. Therefore the evaluation and consideration of genetic variations in EC must be performed after plants were acclimated or hardened by drought stress.

The effect of accumulated EC on non-stomatal transpiration is finite and the question is how much EC is required to hinder the cuticle hydraulically impermeable? Exact values for each crop species are lacking. In sorghum, for example, the maximal EC load beyond which an increase in EC did not affect cuticular transpiration has been estimated between 0.7 and 1.5 mg dm^{-2}, depending on the study (Blum 2004). In wheat a range of 1.5–2.8 mg dm^{-2} was found among different cultivars (Nizam-Uddin and Marshall 1988). A general conclusion on cuticular permeability could be made for certain crop species, such as the case is for rice. Rice cuticle is considered very permeable and any increase in EC should constitute an important breeding target towards dehydration avoidance (Nguyen et al. 1997).

The effect of EC on yield has not been studies extensively. In a large field study involving common wheat, durum wheat, triticale and barley, Fischer and Wood (1979) found that drought resistance in grain yield was associated with increased

Fig. 3.5 Glaucous (*right*) and normal (*left*) common wheat genotypes

leaf glaucousness, among other traits. According to *Merriam-Webster* dictionary the definition of "glaucous" is "having a powdery or waxy coating that gives a frosted appearance and tends to rub off" (Fig. 3.5).

Johnson et al. (1983) and Richards et al. (1986) field-tested isogenic lines of durum wheat differing in glaucousness. The trait was not found to be associated with the amount of EC but with leaf reflectance which was greater in glaucous than non-glaucous lines. Evidently glaucousness influences the spectral characteristics of the leaf. The glaucous lines yielded relatively better most likely because of lower leaf temperature and longer leaf area duration. These results were later confirmed almost to the letter in barley by Febrero et al. (1998) indicating the importance of increased albedo by leaf glaucousness. Pea cultivars high in EC had relatively cooler leaves and produced higher harvest index under dry conditions (Sanchez et al. 2001). A study with 18 sesame cultivars differing in their post flowering EC content did not show any positive relationship between EC and yield under post flowering drought stress (Kim et al. 2007). Other plant factors were evidently influencing yield at this late growth stage. Selection for high EC in Altai wild rye (*Leymus angustus*) was effective but it was found that higher EC was associated with lower forage yield (Jefferson 1994). No physiological data were presented to help understand the nature of this negative association. More recently Gonzalez and Ayerbe (2010) found that barley genotypes with greater leaf EC had lower cuticular transpiration and relatively greater yield than standard cultivars under drought stress.

The genetic control of EC is often found to be relatively simple and heritability estimates are generally high (e.g., Jefferson 1994; Jordan et al. 1983; Clarke et al. 1994; Giese 1976; Haque et al. 1992). Single gene mutants and isogenic lines for EC are available in several crop plants. An extensive chemical mutagenesis study in sorghum (Peters et al. 2009) revealed 18 new epicuticular wax loci with various wax phenotypic expressions. It was concluded that epicuticular wax inheritance in sorghum was not simple.

"Glossy" is a waxy non-glaucous leaf phenotype found in sorghum and rice. Sequence analysis in rice revealed 11 homologous genes for the Glossy-1 (GL1) gene (Islam et al. 2009). Of these the OsGL1-2-over-expression in rice exhibited more wax crystallization, a thicker epicuticular layer and improved drought resistance during reproductive stage as compared with the wild type.

The WXP1 gene of *Medicago truncatula* has a major role in promoting EC deposition. It was shown that the transcript level of WXP1 is inducible by ABA treatment (Zhang et al. 2007). Several rice QTLs identified for EC co-located with QTLs linked to shoot and root-related drought resistance traits (Srinivasan et al. 2008). The CER6 enzyme of *Arabidopsis thaliana* has a role in EC production. Abscisic acid was found to enhance CER6 transcript accumulation and thus increase EC load (Hooker et al. 2002).

In conclusion, EC is a serious candidate for improving dehydration avoidance in cases where EC load is not heavy enough to hinder the cuticle impermeable or where the increase in EC or the enhancement of glaucousness is expected to increase reflectivity and reduce leaf temperature under stress.

Leaf Pubescence

Pubescence ("leaf hairiness") generally increase leaf spectral reflectance within the range of 400–700 nm, resulting in lower net radiation and lower leaf temperatures under high irradiance and drought stress. The increased reflectance in the photosynthetically active waveband is expected to reduce photosynthesis under non-stress conditions, but not under drought and heat stress conditions. Under conditions of stress, there is a tradeoff between the effect of pubescence on net radiation and its effect on photosynthesis. In such cases, relatively little effect of pubescence toward reduced photosynthesis is expected.

The role of leaf pubescence in the crop's energy balance was investigated by Baldocchi et al. (1983) using soybean isogenic lines for pubescence. Dense pubescence reduced ET without affecting CO_2 exchange, thus causing greater water use efficiency. Since net radiation was not affected by pubescence in this study, its effect on the partitioning between sensible and latent heat was explained by the greater penetration of radiation into the canopy of the densely pubescent lines. However, another study with a greater number of isogenic lines of soybeans (Nielsen et al. 1984) revealed that increased pubescence decreased net radiation by about 0.5–1.5%. The associated effect of dense pubescence on yield under stress was inconsistent.

Since the time when these studies were performed in soybean hardly any new information has been developed on the impact of leaf pubescence on crop performance under drought or heat stress. Its contribution to dehydration avoidance in terms of crop productivity is not well established. More information is needed before the trait can be seriously considered in selection for drought resistance.

The value of pubescence towards breeding for drought avoidance should perhaps be considered in tandem with its impact on frost (Maes et al. 2001) and insect resistance (Lambert et al. 1992).

The genetic control of leaf pubescence can be relatively simple such as in wheat (Taketa et al. 2002) and soybean (Specht et al. 1985). A study performed with a Chinese soybean population revealed 20 QTLs for pubescence density (Du et al. 2009).

Leaf Posture

Leaf posture has an important effect on canopy energy balance and its water regime. The radiative load on the individual leaf is maximized when solar radiation is received perpendicular to the leaf, especially around solar noon. Any deviation from the perpendicular will reduce the load. Paraheliotropic leaf movement (particularly in legumes) minimizes the amount of solar radiation received by the leaf.

Paraheliotropism may be taken also as a private case of the general effect of leaf angle as a dehydration avoidance mechanism based on reduced net radiation. "Erect leaf" is a distinct morphological feature common in the cereals and often considered in breeding as a favorable component of canopy architecture. Erect leaves allow better distribution of irradiance into the canopy (reduced "extinction coefficient") instead of just illuminating the top leaves. In this sense, erect leaves are generally at a better leaf-water status than lax leaves when subjected to drought stress (Innes and Blackwell 1983; Ludlow and Bjorkman 1984). This is most probably the reason why "erect leaf" lines yielded better than "lax leaf" lines of wheat under conditions of moisture stress. Edmeades et al. (1999) concluded that erect leaves offer some adaptive advantage to maize under drought stress, though not necessarily under well-watered conditions.

The erect leaf phenotype as a component of dehydration avoidance should be considered in cereal crop breeding in tune with other considerations such as the effect of erect leaf canopy on weed competition, etc. For example, one of the major reason for the advantage and popularity of the African 'Nerica' rice cultivars developed by introgressing wild *Oryza glaberrima* germplasm into cultivated *Oryza sativa* (Sect. 5.3) is their lax and spreading leaf canopy that help compete with weeds.

Roots and Dehydration Avoidance

A most evident control of plant water status is at the root system. Root growth and function in relation to water uptake has been reviewed extensively in the literature and discussed in Sect. 2.2.

There are known facts about root growth and function which can be applied to the planning and execution of breeding and selection work. However there are still very exciting questions that have to be resolved. This section is not a comprehensive review of roots and root function with respect to plant water relations and plant growth. Here, only the topics of interest and relevance to breeding opportunities will be discussed, including root interactions with the environment which can implicate the phenotype.

Root Architecture

The two major architectural attributes of roots concerning water uptake are root length density (total length of roots per unit soil volume) and root depth. In terms of the SPAC (Sect. 2.2), plant hydraulic conductance is positively associated with root length density. Both root depth and root length density are determined firstly by the basic (potential) root architecture and its development in time as the plant develops and as the soil dries.

Root architecture varies extensively among crop species and genotypes. It also varies extensively with soil physical, chemical and biological characteristics. There is a basic genetic control of root growth and development. Various mutants for root growth were identified and even implicated in drought resistance (e.g., Xiong et al. 2006), but the developmental interactions in the plant and plant response to the soil have a large effect on transient or final root architecture and function under drought stress. Root plasticity is still a problem for developing a realistic root module in crop simulation models even if fractal theories are considered (Lynch 1995; Manschadi et al. 2008). It is also a problem for the breeder in terms of setting a realistic target for selection. The minimum that one can project are some basic macroscopic potential root traits and functions that should be considered in the breeding program. It should be remembered that the root is a well organized system no less than the shoot.

There is a basic difference between crops with a tap root system such as cotton and the root system of the cereals. A single axial root system with branching is far more predictable than the cereal root system. Initially the cereal root system is composed of the seminal roots and the adventitious (crown) roots. Adventitious roots develop on a temporal cycle from axillary buds in the lower stem internodes, as in maize or sorghum. In highly tillering plants such as small grains or rice, adventitious rooting is associated with tiller production and establishment. Profusely tillering cultivars tend to have a profuse and relatively shallow root system while limited-tillering cultivars tend to have a less profuse and relatively deeper root system, as a general design (Sect. 2.6). Porter et al. (1986) described well these associations between seminal roots, adventitious roots and tiller development in wheat. When drought stress develops some of the smaller existing tillers tend to abort and new tiller development is arrested. This is an effective mechanism for the plant to limit water use and develop deeper roots (Fig. 3.6). Constitutive limited tillering in small grain cereals and rice can be regarded as one component of an upland rainfed plant ideotype. Limited tillering is characteristic of many spring wheat cultivars developed for dryland conditions by empirical selection for yield. Exceptions are noted. In millet, for example, limited tillering is not an advantage under pre-flowering drought stress (van Oosterom et al. 2006). Barley which is relatively more drought resistant than wheat tends to tiller quite profusely. The advantage of barley over wheat in adaptation to dry

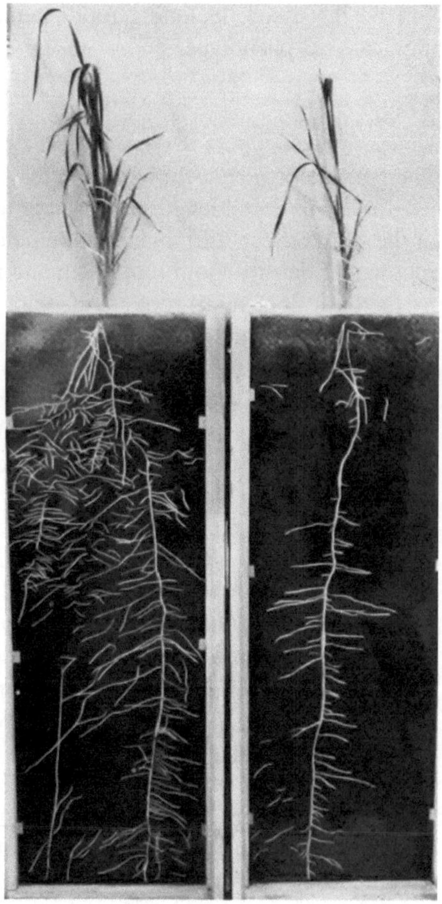

Fig. 3.6 Wheat cultivars Thatcher (*left*) and Lemhi (*right*) grown in soil in glass paneled root boxes and photographed 7 weeks after planting. Note the correspondence between tillering and adventitious root number. A difference in root depth is not seen here because of the limited depth of the boxes and because plants were well irrigated (From Hurd (1974). With permission)

conditions may lie in factors other than tillering, such as earliness or early vigor (Lopezcastaneda et al. 1995).

The effect of limited tillering on root architecture and depth should not be confused with root biomass. Restricted tillering in wheat did not result in relatively greater root biomass when plants were grown in root boxes and irrigated daily (Palta et al. 2007).

There is ample information in the literature indicating large genotypic differences in root depth or maximum root length among genotypes of cultivated crop species. The variation is found among cultivars, germplasm and exotic genetic resources such as landraces. Rice has been especially subjected to molecular mapping studies of root architecture, morphology, and growth with the recognition that deep penetrating roots are very important in this crop under drought stress.

Effect of Soil Moisture

The effect of a drying soil on root distribution and depth was already discussed above and in Sect. 2.6. Here a few points are raised with respect to breeding. There is a general agreement that soil moisture deficit rarely increase root biomass but root-to shoot dry matter ratio tends to increase. The relative biomass distribution in the root system tends to change in favor of deeper rooting at the expense of shallow rooting. As soil dries up from the top roots at the top die while existing live roots grow into depth. When rainfall or irrigation water is received growth pattern tends to revert and more root growth is observed in the wetter top soil while root growth at depth ceases. There are species differences in this respect. For example, root growth in depth upon the development of soil moisture deficit was more pronounced in Chickpeas than in soybeans (Benjamin and Nielsen 2006), both of which have a taproot system.

Past concepts of soil moisture "availability" considered a fixed low limit for soil moisture extraction set to 1.5 MPa of soil water potential (SWP). However, it has been realized that genetic variation exists in plant capacity to reduce soil moisture to lower SWP. While these additional extracted amounts may not be large, they can be very effective when the plant is under stress. For example it was calculated that each additional millimeter of water extracted during wheat grain filling generated an extra 55 kg ha^{-1} of grain yield (Manschadi et al. 2006). Plant factors controlling soil moisture extraction to low SWP are root-length density, shoot osmotic adjustment and osmotic adjustment of the root (Sect. 2.2). Survival of the shoot to lower SWP is important for enhanced soil moisture extraction (Volaire and Lelievre 2001).

There are some indications that different parts of the root respond differently to soil drying. For example, Main root extension of mustard was inhibited while lateral roots continued growing (Vartanian 1981). Lateral root growth of maize and wheat was promoted with the development of moderate water deficit but not with severe deficit (Ito et al. 2006). Lateral root growth is crucial in dryland row crops, whereas it determines the capacity for extraction of soil moisture in the spacing between rows.

Roots grow geotropically as influenced by their basic developmental architecture and by soil resistance and porosity. If wet soil is encountered extended root growth develops in that zone influenced by the low resistance of wet soil. There has always been a traditional belief that roots sense water from distance and grow towards moist soil (root hydrotropism). Solid evidence for hydrotropism has been scarce. More recently, studies with *Arabidopsis thaliana* mutants appear to support the existence of root hydrotropic response (Eapen et al. 2005) and more research in real crop environments is required before any conclusions can be reached.

The breeder must be reminded of the fact that beyond soil moisture a host of other factors can impact and hinder root growth and function. These include biotic factors such as soil inhibiting pathogens, insects and nematodes as well as abiotic factors such as soil salinity, alkalinity, acidity, waterlogging and mineral toxicity.

Effect of Soil Strength

Soil bulk density and soil strength (or hardness) affects root growth. This explains why roots can grow to an abnormally large size in aeroponics culture and why roots grow deeper in sandy soil than in dense clay soil. Thus, the advantage of a deep rooted rice genotype was well expressed when grown on sandy soil but not when grown on heavier paddy soil (Yue et al. 2005). The effect of high bulk density towards reduced root growth is achieved by way of physical resistance to root penetration and possibly certain hypoxic conditions in extreme cases. There is a wide consensus that soil strength and soil hardpans constitute a serious limitation to yield which becomes more acute when the soil is drying.

There are consistent indications in several crop plants that the effect of soil strength on shoot growth might be mediated by way of a hormonal (most likely ABA) root signal transported from root to shoot in the xylem sap. Soil compacted to high bulk density induces hormonal root signal (Sect. 2.4.2.1).

Genetic variation in plant response to and performance under high soil strength has been found. Cotton germplasm that could penetrate a hardpan were identified (May and Kasperbauer 1999). Soybean accession PI416937 had better root penetration in soil columns set to high bulk density of 1.75 g cm^{-3} than a common commercial cultivar (Busscher et al. 2000). Use of wax-petrolatum layers to experimentally simulate hardpan became a popular method for phenotyping and screening for hardpan penetration capacity (Yu et al. 1995) (Chap. 4, section "Root Penetration Ability"). Distinct differences in root penetration ability of wax layer at different soil moisture regimes were seen among wheat cultivars (e.g., Kubo et al. 2004). Variation in rice root penetration was found among rice accessions with some traditional cultivars having the highest capacity (Babu et al. 2001). Four QTLs for root-penetration ability in rice were identified (Zheng et al. 2000). A three-marker model that accounted for 34% of the variation in root penetration has been proposed (Ray et al. 1996). However, differences among rice cultivars in the ability of roots to penetrate wax layers were not related to their elongation rates through uniformly strong media (Clark et al. 2003). At the same time alleles ascribing high root penetration capacity in rice also promoted maximum root length (Price et al. 2000).

The Effect of Phenology

Not many studies were performed on the association between plant phenology and root development and function in terms of plant water relations, and especially not in recent times. The topic is important since plant phenology is a major trait in adapting genotypes to any given environment and certainly the water regime. The role of phenology on root growth and architecture is discussed below under Sect. 3.4.2.3.

The Effect of Plant Height

Ever since the "green revolution" was linked to high harvest index and reduced plant height there has been a consistent interest in the effect of plant height genes on root development and function. The concern was that while dwarf genotypes had greater yield potential they might be lacking in rooting and the capacity for soil moisture and nutrient capture. Despite this concern, there is insufficient information on the subject especially when some of it appears to be inconclusive.

Most of the available information is on the Rht height genes of wheat. Four studies with wheat isolines differing in height (Mccaig and Morgan 1993; Pepe and Welsh 1979; Holbrook and Welsh 1980; Ehdaie and Waines 1996) and one study with barley (Irvine et al. 1980) concluded that height genes did not have a clear and consistent effect on root growth, root dry matter accumulation or soil-moisture extraction. On the other hand one study with well-watered wheat (Miralles et al. 1997) found that total root length and root dry-weight per unit area at anthesis were increased with decreased plant height. Controlled studies on wheat root penetration capacity of a simulated hardpan indicated that the Rht height genes had no effect on seminal or crown root penetration capacity (Kubo et al. 2005). In sunflower hybrids and open-pollinated varieties, dwarf genotypes had reduced rooting depth and soil moisture extraction as compared with tall genotypes (Angadi and Entz 2002).

In conclusion, there is no consistent trend for an association between height genes and root growth and function.

Root Xylem Vessel Diameter

A special case for managing plant water status and dehydration avoidance has been described for wheat that grows mainly on stored soil moisture in parts of Eastern Australia. When crops are grown on limited stored soil moisture a major consideration is to limit soil moisture use in early growth stages so that some water will remain for the terminal reproductive growth stage. Under such growing conditions the top soil is dry most of the time and no adventitious roots develop and grow into the soil. The main root system of wheat then constitutes of the seminal roots which grow into deep soil. Richards and Passioura (1989) developed a program to select and breed wheat lines with reduced seminal root xylem vessel diameter in order to increase root axial hydraulic resistance and thus restrict water uptake. Wheat lines developed and tested under this program yielded up to 7% more than control cultivars under severe drought conditions (mean yield of 1–2 ton ha^{-1}) with a smaller advantage under less extreme stress (Richards et al. 2002). These lines were never released due to their susceptibility to a new strain of rust disease. It should be noted that where intermittent seasonal rainfall occurs this concept is irrelevant. It might be appropriate for cereal crops grown exclusively on stored soil moisture conditions which normally result in terminal drought stress. It should not be adopted indiscriminately for other water regimes. The author is unaware of any additional attempts to perform such a selection program.

The Effect of Root Senescence

Roots naturally grow senesce and die as a leaves do. There is a large turnover of roots whereas new ones are initiated and grow while old ones die. The casual observer of roots may not realize that not all roots present at any given time are alive and functioning. Dead roots are often brown-black in color as compared to the white off-white-pink color of live roots.

The longevity of the seminal root in cereals differs in different crops under different growing conditions. The natural longevity of the seminal root of sorghum can be relatively short (Blum et al. 1977a, b) as compared with that of the maize or the small grains where it can survive to plant maturity. In irrigated sorghum grown in soil the rate of root mortality at flowering was about two thirds of the cumulative root production by the plant to that date (Blum and Arkin 1984). The rate of root mortality was highest in the top soil and it decreased with depth.

The Effect of Mycorrhizae and Other Soil Biota

The roots of many plants are invaded by fungi that form symbiotic associations called mycorrhizae. These are of two types: endotrophic (AM, for arbuscular mycorrhizae) in which the fungus penetrates the root cells, but has little effect on external appearance, and ectotrophic in which the fungus covers the external surface and causes marked hypertrophy and extensive branching of roots. The most prevalent type is the arbuscular mycorrhizae that develop on roots of most terrestrial plants. The AM fungi are able to grow into the root cortex forming intercellular hyphae from which highly branched structures (arbuscules) originate within cortex cells. The arbuscules are responsible for nutrient exchange between the host and the symbiont, transporting carbohydrates from the plant to the fungus and mineral nutrients, especially phosphate, and water from the fungus to the plant. With respect to plant water relations, mycorrhizae has a role in capturing water and increasing root hydraulic conductance subsequently enhancing dehydration avoidance. The interested reader is referred to a concise review on the subject (Auge 2004). Additional soil microorganisms were implicated via their root associations in the augmentation of plant drought resistance (e.g., Marulanda et al. 2009). While soil biota interactions with the root hold great potential for enhancing yield under drought stress, their use in plant breeding for water limited environments is not forthcoming yet at this time.

Nitrogen Fixation in the Legumes Under Drought Stress

In their symbiotic interaction with rhizobia, legumes develop root nodules in which atmospheric nitrogen fixation takes place. Drought stress generally impairs this symbiosis. Consequently legume plants under soil moisture stress can develop nitrogen deficiency.

The first effect of drought is the reduction in rhizobium inoculation of root in a dry soil and subsequent reduction in nodule number and individual nodule size or weight (Zahran 1999). In bean subjected to soil drying (Ramos et al. 1999) the number and weight of nodules was reduced. Nitrogenase activity, as determined by the acetylene reduction assay, was also reduced and was almost stopped at 30% soil field capacity. At this level of soil moisture there was a marked increase in nodule oxygen diffusion resistance and nodule senescence. Abscisic acid which accumulates in roots under drought stress may induce a decline in nitrogen fixation

(Gonzalez et al. 2001). Exogenous ABA treatment reduced N_2 fixation via reduction in Leghaemoglobin content of nodules.

Resistance of the Symbiosis to Drought The analysis of drought resistance in the legumes often does not differentiate between plant resistance and the resistance of the symbiosis.

Selected drought resistant soybean lines in terms of yield under drought stress expressed better nitrogen fixation under drought stress to lower soil water potential than the more susceptible lines (Sinclair et al. 2007). It was concluded (Serraj and Sinclair 1998) that large variation in nodulation sensitivity to water deficit exists among soybean cultivars and that the response of N_2 fixation rates to drought is related in part to nodule formation and growth. Dehydration avoidance of the plant and maintenance of high plant water status was an important factor in maintaining soybean N_2 fixation under drought stress (Patterson and Hudak 1996).

Sensitivity to drought stress varies with the variety of rhizobial strains; hence the selection of drought tolerant strains might carry a benefit (Athar and Johnson 1996; Rehman and Nautiyal 2002). In groundnuts, the selection for improved nitrogen fixation under drought stress was an effective approach to improve drought resistance in terms of pod weight (Pimratch et al. 2009).

Reactive oxygen species are a ubiquitous danger for aerobic organisms (section "Antioxidants"). This risk is especially elevated in legume root nodules due to the strongly reducing conditions, the high rates of respiration, the tendency of leghemoglobin to autoxidize, the abundance of non-protein iron and the presence of several redox proteins that leak electrons to O_2. Consequently, nodules are particularly rich in both quantity and diversity of antioxidant defenses. These include enzymes such as superoxide dismutase and ascorbate peroxidase and metabolites such as ascorbate and thiol tripeptides. Nodule antioxidants have been the subject of intensive molecular, biochemical and functional studies. The emerging view is that antioxidants are critical for the protection and optimal functioning of N_2 fixation (Becana et al. 2001).

Legume species that transport ureides from the nodules were found to be much more drought-sensitive than those that transport amides. It was concluded (Vadez and Sinclair 2001) that a feedback mechanism involving ureide level may control N_2 fixation under drought. Consistent with this observation, the drought tolerance of N_2 fixation in soybean was associated with low concentrations of ureides in plant tissues. Experimental evidence for a direct inhibition of nitrogenase activity by ureides application supported the feedback hypothesis. The basis for ureide accumulation is hypothesized to result from decreased ureide catabolism in the leaf. Asparagine (Asn) is known to chelate Mn ions, which are the co-factor of the enzyme allantoate amidohydrolase, and as a result may cause an inhibition of ureide breakdown in the shoot. These data are consistent with the hypothesis that Asn would be a signal for the feedback inhibition of N_2 fixation (Serraj et al. 2001). However, because the relationship between ureides abundance in the plant and N_2 fixation was poor under drought stress it was concluded that ureides assay in plant tissues as a measure of N_2 fixation is unreliable under drought stress, pending further refinements (Purcell et al. 2004).

The Inheritance of Root Traits

Considering the above discussion of root interaction with the shoot on one hand, and with soil conditions on the other, studies of root inheritance and genetics are difficult and often result in inconsistent results. Observed root development, architecture and various associated traits will vary immensely according to the experiment growing conditions and methods of measurement. Undoubtedly genotype × root medium interaction is possible. The root phenotype might interact with the water regime and even with the irrigation method used in the experiment, especially in cereals and grasses. Most genetic studies of roots require large plant populations which dictate some form of easily-managed root culture system and often for a short growing duration. The most common approach to plant culture is long vessels containing soil or sand amenable to washing and root recovery. In situ root observations in genetic studies are rare. Some past research attempted to conclude about root inheritance from the study of seedlings, for the obvious reason that seedlings are easy to work with. This line of research led nowhere.

Genetic variation for root traits has long been in evidence (O'Toole and Bland 1987). The heritability of root traits has been published for various crops and studies without any common denominator to all. Interested readers should search the literature for the specific case of interest. In certain cases root depth or maximum root length were found to have high heritability (e.g., Sayar et al. 2007). Of all major root architectural traits maximum root depth or length in a non-limiting medium appears to be relatively highly heritable.

Obviously, molecular mapping for root traits has become a popular and important effort in view of the difficulty in performing phenotypic selection for roots. The mapping exercise is susceptible to all the problems mentioned above regarding root phenotyping. Meaningful molecular mapping of root traits with respect to genetic improvement of dehydration avoidance should be performed under perfect phenotyping conditions in both drought stress and non-stress conditions as done by Price et al. (2002), Venuprasad et al. (2002) and others. It goes without saying that MAS is probably the most cost-effective approach to root trait improvement. This is seen for example from the successful development via marker-assisted backcross program to develop near-isogenic rice lines for root depth (Shen et al. 2001).

Price and his associates invested many years of work in studying rice roots in relations to dehydration avoidance with main consideration for root QTL identification and subsequent MAS program. They achieved progress in the release of 'Birsa Vikas Dhan 111' rice cultivar. At the same time their concluding paper on the subject (Price et al. 2002) is probably the best and most candid "take-home" message on roots, MAS and dehydration avoidance research. One does not have to agree with all the conclusions made in the paper, but the point is that these conclusions are important because they are derived from very extensive and long-term experience. What they bring forward is the fact that at the end of the day QTLs for dehydration avoidance expression in the field (in delayed leaf rolling or reduced leaf firing) did not co-locate in a consistent manner across studies with QTLs for major root traits such as deep roots. They suggest that this seems to cast a doubt on the notion that dehydration avoidance can be improved by improving root traits. They accept past evidence that deep and thick roots in rice were shown to contribute to drought resistance in more than few experiments. Certainly this conclusion receives wide support from other crops and growing conditions. Price et al. therefore sought an explanation why rice root QTLs did not co-locate well with dehydration avoidance expression in the field. Their following conclusions are candid and therefore very important:

1. Root studies were done in an artificial environment typical of root work in large populations. Due to the various soil environmental effects and root interaction with shoot development artfactual results were expected. This may have affected root QTLs and their reputability, especially for those with smaller effect.
2. When QTLs for root traits are correlated with "drought resistance," the definition of drought resistance in the field is crucial. If drought resistance is defined for example by yield under stress there may be additional factors affecting yield under stress beyond the effect of root depth on dehydration avoidance.
3. Price et al. mention the problem of field drought screening during the dry off-season, as used by many. It is characterized also by high temperatures which can develop specific interaction with dehydration avoidance and thus impair the expected correlation with root QTLs. Furthermore, there is a question whether there was sufficient soil moisture at depth in the trials so that the advantage of deep roots would be expressed.
4. Finally, field variability can destroy any field phenotyping study whether this variability is driven by soil physical, chemical or biotic factors. As shown by Price et al. this could be a factor in at least one of their field experiments.

All of the above is true also for any other drought phenotyping work towards molecular mapping or phenotypic selection. This is a worthy lesson which Price et al. teach us and despite their disappointment in some of their results, the lesson itself as published is a valuable document.

Roots and Dehydration Avoidance – Conclusion

Three major root traits determine soil moisture capture and thus impact dehydration avoidance: maximum root depth, root penetration capacity and soil moisture extraction capacity.

Maximum root depth is as obvious as a long rope in a deep well. Maximum root depth depends on genotype as well as on soil conditions, plant ontogeny, and interaction with shoot traits such as tillering. Basically, maximum root length is a constitutive trait and the resulting root depth will depend on the interaction with the environment.

Root penetration is important in soils of high bulk density and where a hardpan is present in the soil horizon. The capacity of roots to penetrate dense soil is different from the capacity to penetrate a hardpan. The physiological basis of superior root penetration capacity is not clear. Root turgor, hormones and root thickness may be important in this respect. Genetic variation for the trait has been recorded. It should be remembered that the genetic capacity for coping with high soil impedance is not only a mechanical or a hydraulic issue but also a hormonal one. Breeders have been attempting to select for root penetration by the use of the wax layer technique, with variable result.

Finally, three major facts should be considered when roots are candidates for improving dehydration avoidance:

1. Maximum root length and depth are the most common root traits in genetic manipulation of roots in breeding for drought resistance.
2. Whether genetic root manipulation should be considered in breeding for dehydration avoidance depends on the drought profile in the target environment.
3. Effective phenotypic selection for root function as a component of dehydration avoidance can be performed by assessing shoot water status under conditions which ascribe an advantage to the given root trait, be it depth, penetration or capacity for moisture extraction.

Stem Xylem Cavitation

Xylem cavitation and the resultant embolism can be an important reason for reduced stem conductance which results in reduced leaf water status under drought stress. Most likely the problem occurs mainly in long stems and tree trunks rather than in roots and it is generally associated with plant height. While certain herbaceous crop plants such as rice might be prone to xylem cavitation (Stiller et al. 2003) the issue is more serious with trees.

It is therefore quite understandable that trees prone to cavitation vulnerability are less capable of dehydration avoidance under drought stress. This has been verified in more than a few cases. For example, when various poplar (Populus spp.) and willow (Salix spp.) clones were tested for cavitation vulnerability (Cochard et al. 2007) it was found that variation in vulnerability to cavitation across clones was poorly correlated with anatomical traits such as vessel diameter, vessel wall strength, wood density and fiber wall thickness; however, a striking correlation was established between cavitation resistance and aboveground biomass production. Using the "cavitron" technique it was found that *Prunus* species resistant to drought stress were less prone to cavitation and that this could be phenotyped by certain anatomical characteristics. This is in agreement with studies in other tree species (Harvey and van den Driessche 1997).

This author is unaware of any successful breeding program for dehydration avoidance by the reduction of xylem cavitation vulnerability.

Osmotic Adjustment

The net cellular accumulation of solutes under water deficit is defined as osmotic adjustment (OA) and it serves to maintain cellular hydration, turgor and high leaf relative water content (RWC). For this reason OA is considered as a component of dehydration avoidance. The corollary is that for the same leaf water potential a genotype with OA capacity will maintain higher RWC and higher turgor than a genotype with low OA (Fig. 2.3). Deeper root growth and deeper soil moisture extraction were found to be associated with OA capacity across diverse breeding materials (e.g., Tangpremsri et al. 1991; Chimenti et al. 2006). Barley cultivars with higher OA had relatively higher stomatal conductance and better grain yield under drought stress (Gonzalez et al. 1999). The capacity for osmotic adjustment was the main physiological attribute associated with wheat resistance under cyclic water stress which enabled plants to recover from water deficit (Izanloo et al. 2008).

Apart from sustaining plant hydration the positive effects for OA include also the protection of cellular membrane stability (Bohnert and Shen 1999) and enzymes (Paleg et al. 1985) and probably protection against oxidative damage (Shen et al. 1997). The protective functions of OA may depend on the specific solutes, whether these are ions or various organic solutes such as proline, glycinebetaine or sucrose.

Proline amino acid has for long (Singh et al. 1972) been associated with drought resistance. It is synthesized *de novo* upon tissue dehydration and metabolized upon rehydration. Opinion still differ on the exact role of proline with the realization that in most cases its accumulation under dehydration stress does not amount to an effective increase in OA. It is thought to have a protective role in membranes, cellular organelles and proteins as well as a possible role in the return to normal metabolism upon rehydration. Recently with the increase in interest in antioxidants in abiotic stress resistance proline has been found to have a role also in this respect. Despite the fact that few transgenic plants over-expressing proline accumulation were claimed to be somewhat more resistant to drought stress, there is no known successful use of proline accumulation in breeding for drought resistance. Therefore proline accumulation *per se* is not considered here as a candidate trait in drought resistance, at this time.

The major function of OA to be considered here in terms of breeding is in its hydraulic role. There is wide and ample genetic variation for OA under drought stress in many crops (Table 3.1). Not all data for OA are comparable since some estimates were derived as RWC at given osmotic potential rather than MPa of OA (Chap. 4, section "Osmotic Adjustment"). Still it is notable that sorghum, the millets, rice and chickpeas have a relatively high capacity and a wide range of genetic diversity for OA. Wheat, barley, maize, sunflower and pigeon-pea have a relatively moderate capacity for OA but often still display a wide range of genotypic diversity.

The measurement of OA is difficult and subject to many pitfalls. A major difficulty is the fact that that OA is a direct function of the rate of tissue water deficit (e.g., Basnayake et al. 1996). Therefore all materials tested for OA must be analyzed when all are at a reasonably same low level of RWC. Data should not be adjusted or normalized for RWC but actually derived from tissues subjected to a similar level of dehydration as measured by RWC. This guideline is not always observed.

When genetic differences are solid the measured values of OA may change a little from one study to another but the ranking of genotypes is generally consistent (Blum et al. 1999). Wheat OA ranking in 6 greenhouse experiments were well correlated with cultivar performance under stress in the field (Moinuddin et al. 2005).

Table 3.1 Observed genetic variation for osmotic adjustment (OA) in different crops and its association with yield under drought stress where measured

Crop	OA range across genotypes (MPa)	Relations of OA to yield under stress	Reference
Barley	RWC[a]	Positive correlation	Gonzalez et al. (1999)
Chickpeas	0.40–1.00	Increase by up to 20%	Morgan et al. (1991)
Chickpeas	0.37–0.71	Increase by 48%	Moinuddin and Khanna-Chopra (2004)
Chickpeas	0.40–1.20	No effect under terminal drought	Turner et al. (2007a, b)
Maize	0.06–0.49	Increase	Chimenti et al. (2006)
Maize populations	0.06–0.47	Increase	Chimenti et al. (2006)
Pea	RWC	Positive correlation with yield	Rodriguezmaribona et al. (1992)
Peanut	0.90–1.37	Increase by 164% in 1 of 2 years	Erickson and Ketring (1985)
Sorghum	0.20–0.52	Increase by 24%	Ludlow et al. (1990)
Sorghum	0.20–0.45	Increase by 15–34%	Santamaria et al. (1990)
Sorghum	L & H[b]	Increase by 11%	Tangpremsri et al. (1995)
Sunflower	0.04–0.23	Increase by 30%	Chimenti et al. (2002)
Wheat	0.21–0.75	Positive contribution	Blum and Pnuel (1990)
Wheat	RWC[a]	Positive correlation with yield	Morgan (1995)
Wheat	0.39–0.63	Correlation with yield nearly significant ($p = 0.09$)	Blum et al. (1999)
Wheat	0.80–0.99	Positive correlation with yield	Moinuddin et al. (2005)
Wheat	0.39–0.57	Positive correlation with yield	Moinuddin et al. (2005)
Wheat	RWC[b]	Increase by 7–17% over 56 trials	Morgan et al. (1986)

[a]Recorded genotypic variation in OA was expressed as RWC at given osmotic potential
[b]Low and high OA groups of genotypes

On the other hand, OA of different chickpea genotypes did not correspond between two studies suggesting that OA capacity in chickpeas was not a stable trait (Basu et al. 2007; Turner et al. 2007a, b). However, the method used for OA in these studies was not strict. Osmotic potential at full turgor of stressed plants (OP100) should be compared with OP100 of the same cultivar never exposed to any stress, such as well-watered plants (Chap. 4, section "Osmotic Adjustment"). In both studies, OP100 of stressed plants was compared with that of plants that were not free of any stress (e.g., 80% RWC in Turner et al. 2007a, b). This can introduce an error into the estimate of OA. Furthermore chickpea leaves are covered with glandular trichomes containing malic acid. Malic acid is a known osmolyte. When such leaves are processes for osmotic potential measurement leaf surface malic acid can perhaps bias results for OA. There is no data on this issue and for now it is a speculation.

A positive association between OA and yield under stress (Table 3.1) was seen in at least 17 studies including wheat, barley, sorghum, maize, sunflowers, chickpeas, pea, pigeon-pea and peanuts. No relationship to yield under stress was indicated in two studies (chickpeas and maize). This table does not include numerous cases of proven genetic diversity for OA without the study of yield.

Where genetic variation for OA exists, a lack of association between OA and yield under stress across diverse germplasm should not be taken at face value. Other drought resistance mechanisms may become dominant in a certain drought scenarios and certain breeding populations and thus obscure or even eliminate the contribution of OA to yield. For example variation in root depth and deep soil moisture extraction among rice genotypes was more important in affecting yield than variation in OA when deep soil moisture was available (Blum unpublished data). In fact, this could be the reason why poor root development was associated with high capacity for OA in diverse rice germplasm (Lilley et al. 1996; Babu et al. 2001). In chickpeas, OA at the grain filling growth stage was not associated with yield under drought stress (Basu et al. 2007) probably because under such conditions stem reserve utilization for grain filling was relatively more important towards yield in this species (Leport et al. 1999). Genetic variation in OA capacity at an earlier growth stage was positively associated with chickpea yield under drought stress (Morgan et al. 1991).

Finally, the agronomic value of OA in stabilizing yield under stress is supported by the release in Australia of 'Mulgara', a dryland wheat cultivar which is characterized as having high OA capacity (Munns and Richards 2007). It is important to point out that in the course of time OA as a yield promoting component of drought resistance was challenged by voiced and published opinions that OA was irrelevant, that it involved a "cost" in terms of yield or that it was just supporting survival. This debate was very important in that it promoted new ideas and further research. Two published opinionated reviews expressed those past negative views. Munns (1988) published her opinion before most of the evidence in Table 3.1 has been published, while Serraj and Sinclair (2002) presented their negative view when most of that evidence was available. There is also additional support for OA received from more recent transgenic work (notwithstanding yield data) but I chose to rely mainly on evidence from crop physiology and breeding research.

Genetic Variation and Inheritance of Osmotic Adjustment

The capacity for OA is an inherited trait. OA in wheat was found to be conditioned by alternative alleles at a single locus, with high response being recessive (Morgan 1991) (Fig. 3.7). The gene appeared to condition primarily potassium accumulation, with amino acid accumulation as a possible secondary and dependent response. Analysis of a single chromosome substitution series of Chinese Spring × Red Egyptian wheat cross indicated that genes coding potassium accumulation and OA are located on chromosome 7A (Morgan and Tan 1996). The heritability of OA and turgor potential of wheat as computed by pooled analysis of variance across six greenhouse experiments was 0.74, 0.73, and 0.79, respectively indicating high expected progress in selection for OA (Moinuddin et al. 2005).

Fig. 3.7 The effect of one gene for OA (Morgan 1991) under drought stress in high OA wheat line ('H.Osm-134') as compared to a low OA line ('L.Osm-136') in a controlled pot experiment performed by the author. The difference between the two lines is expressed in the relative number of desiccated and rolled leaves

In sorghum, the inheritance of OA was investigated in populations derived from three bi-parental crosses between high OA lines (Tx2813 and TAM422) one low OA line (QL27) (Basnayake et al. 1995). Narrow sense heritability was high in the three crosses (0.76, 0.65 and 0.54, respectively). Two independent major genes for high OA were identified. Tx2813 possessed a recessive gene named oa1, whereas the line TAM422 possessed an additive gene named OA2. Few minor genes that influence the expression of OA in these crosses were not overruled. A divergent selection experiment for OA in maize provided solid evidence that the trait was highly heritable and that selection was effective in improving OA and grain yield under drought stress (Chimenti et al. 2006).

Glycinebetain is an important osmolyte in certain plants. Yang et al. (1995) developed in maize a series of near-isogenic F_8 pairs of glycinebetaine-containing and glycinebetaine deficient lines. The pairs of lines differed for alternative alleles at a single locus; the wild-type allele conferring glycinebetaine accumulation was designated Bet1. This gene affected glycinebetaine biosynthesis at the level of choline oxidation and was located near the centromere on the short arm of chromosome 3.

Mapping QTLs Controlling Osmotic Adjustment

Since the accurate measurement of OA in large populations is cumbersome, the option for using MAS for this trait should be developed. The first molecular mapping of this trait under drought stress was reported for rice by Lilley et al. (1996). The study was based on 52 rice recombinant inbred lines, a randomly

Table 3.2 Summary of results for molecular mapping of osmotic adjustment. Only studies with correct protocol for OA are included (see text)

Species	Population	Population used[a]	QTL number	Reference
Rice	CO39 × Moroberekan	RIL	1 major	Lilley et al. (1996)
Rice	CT9993 × IR62266	DHL	5	Zhang et al. (2001)
Wheat	Songlen/Condor4 × 3Ag14	RIL	1 major	Morgan and Tan(1996)
Sunflower	PAC2 × RHA266	RIL	1 major + 4 minor	Poormohammad et al. (2007)

[a] *RIL* – Recombinant inbred lines, *DHL* – Doubled haploid lines

sampled subset of a population originally developed to study the genetics of resistance to rice blast. Water deficit was induced by withholding water starting 24 days after sowing. A major QTL explaining one-third of the phenotypic variation for OA was identified at the RG1 region of chromosome 8.

While not many mapping studies were performed for OA (Table 3.2) it is encouraging to see the relatively frequent number of cases where a major QTL for OA has been reported. Phenotyping a mapping population for OA is not a simple task and the exact protocols must be followed while avoiding known pitfalls. The papers by Teulat et al. (1997, 1998) are an educational case in question. The authors claimed to have mapped several QTLs for OA and OA-related traits in barley. However, the protocol used for estimating OA was not correct. Firstly, plants were not stressed to similar plant water status, which was reflected in the reported variation in RWC data (Fig. 1 in Teulat et al. 1998). Furthermore, most plants were not stressed enough for full OA expression as RWC was higher than 80% and closer to 90% (Teulat et al. 1997). Secondly, the OA values for both parents (Tadmor and Er/Apm) were close to zero (Table 3 in Teulat et al. 1998). Finally, the differences in OA between the two parents and among the RILs were not significant statistically, which suggests that this population could not present any useful segregation for the genetic study of OA.

Saranga et al. (2001) measured leaf osmotic potential at full turgor (OP100) at dawn in the field after drought was relieved by irrigation. OP100 in itself is not an estimate of OA and they were careful not to indicate that it was. However, they argued that most of the variation in OP100 was due to osmotic adjustment. In this claim they neglected to consider the likely possibility that genotypes could differ in field measured OP100 also due to a probable variation in their level of water deficit experienced during the drying cycle prior to the recovery by irrigation. For that reason estimating genotypic variation in OA is practically impossible in field experiments.

Stay-Green (Non-senescence)

The normal process of leaf senescence can be visually recognized as the loss of leaf greenness due to chlorophyll breakdown followed by leaf desiccation in most cases. Genotypes where senescence is delayed or slowed are defined in practice as

Fig. 3.8 An outstanding example of stay-green (*left*) as compared with a normal senescent (*right*) sorghum line under post-flowering drought stress

"non-senescent" or "stay-green." Thomas and Howarth (2000) defined five types of stay-green. The three most common ones are delayed onset of senescence, low rate of senescence and the combination of the two. Here all types will be defined as Stay-green (SG) as most experiments did not follow the definitions of Thomas and Howarth. SG crop cultivars are generally recognized as important for plant production under post-flowering drought stress. SG is largely a constitutive trait which can be expressed under well watered conditions but it becomes more prominent as compared to normal senescent genotypes when subjected to post flowering drought stress (Fig. 3.8).

Since the initial discovery of SG sorghum in the Texas sorghum conversion program (Dahlberg 2000) it has become an important component of their drought resistance breeding program, defined also as "post-flowering drought resistance." Consequently much of the breeding research on SG was done with sorghum.

SG and sweet sorghums generally tend to accumulate more stem TNC during grain filling than non-sweet or senescent grain sorghums. Non-senescent sorghums also store more assimilates in leaves. This led to the conclusion that the recombination of both high yield potential and the "stay-green" trait might be complex (Tuinstra et al. 1997).

Work with sorghum F_7 RILs of B35 × Tx7000 cross which segregated for SG indicated a positive correlation ($r = 0.75^{**}$) between SG and leaf RWC under drought stress (H.T. Nguyen, personal communication). This is supported by an experiment with SG transgenic cotton where SG gene reduced plant wilting under moderate drought stress (Yan et al. 2004). Therefore, besides its effect towards chlorophyll maintenance SG can also sustain high leaf water status. This is why SG is discussed here under "dehydration avoidance."

Borrell et al. (2000) produced conclusive evidence showing that SG in sorghum, whether by way of delayed onset or reduced rate of senescence was very important for supporting biomass and grain yield under post-flowering drought stress. Borrell and Hammer (2000) went on to show that leaf nitrogen and nitrogen uptake during grain filling were the main physiological cause behind genotypic variations in delayed onset and reduced rate of leaf senescence in sorghum. Plant nitrogen regimen was important also in SG of maize (Ma and Dwyer 1998), *Festuca* sp. (Hauck et al. 1997) and *Lolium* sp. (Bakken et al. 1997). SG was identified as an important component of high yield in maize, whether under well-watered or stress conditions (Zaidi et al. 2004).

The SG phenotype of wheat cultivar Seri-M82 was simply ascribed to its superior ability to extract deep soil moisture where available during grain filling (Christopher et al. 2008). Where deep soil moisture was not available, this cultivar did not present the trait. This may have also been the case where drought resistance of wheat was highly associated with persistence of green flag leaf (Foulkes et al. 2007). It must therefore be recognized that a senescent phenotype can also be a function of poor resource capture (water and nitrogen) by roots or a function of its inherent programmed senescence.

It has long been suggested that plant hormones are tightly linked to senescence control, where generally ABA promote senescence and kinetin delays it. Subsequently it was also established that kinetin enhance SG and leaf chlorophyll retention in plants (e.g., Gan and Amasino 1995). Over-expression of kinetin production in transgenic tobacco during its reproductive stage dramatically improved SG under drought stress and resulted in better leaf water status, improved recovery from stress and a better yield of capsules as compared with the wild type (Rivero et al. 2009).

In contrast to Kinetin, ABA appears to enhance senescence and the export of carbon and nitrogen from leaves of a cereal crop plant such as rice (Yang et al. 2003). It has long been established that the effect of ABA on senescence could be mediated by its enhancement of ethylene production. Ethylene is a known senescence enhancing plant hormone (e.g., Buchanan-Wollaston 1997). More recently salicylic acid has also been implicated in signaling leaf senescence (Abreu and Munné-Bosch 2008).

SG has become a noted trait in breeding programs for water-limited environments, partly because it is easily recognized visually, especially under post flowering drought stress. It has often been observed by maize breeders that when maize hybrids were selected for improved drought resistance in terms of yield the better hybrids tended to have delayed leaf senescence and noticeably greener leaves besides other traits. Thus, whatever might be the mechanism behind it, the retention of greener viable leaves at grain filling is an impressive phenotype which seems to be preferred by breeders.

However it should already be mentioned at this point that the SG trait for supporting yield under post-anthesis drought stress might be in opposition to use of stem reserves for grain filling under stress. This relationship and its implications will be discussed below (section "Stem Reserve Utilization for Grain Filling").

The Inheritance of Stay-Green

The most extensive studies of SG inheritance were performed in sorghum. Initial studies (Tuinstra et al. 1997) identified two QTLs with major effects on yield and SG under post-flowering drought. These loci were also associated with yield under fully irrigated conditions. Four major QTLs for SG were identified (Kebede et al. 2001; Sanchez et al. 2002; Harris et al. 2007), designated Stg1 through Stg4. Each was found to be effective independently. These QTLs showed consistency across genetic background and different environments. Two of the QTLs corresponded to SG QTL regions in maize. The Stg1 and Stg2 regions also contain the genes for key photosynthetic enzymes, heat shock proteins, and an ABA responsive gene (Xu et al. 2000). Work done with an Ethiopian donor of SG and two Indian sorghum cultivars identified three SG QTLs that were consistent across genetic materials and environments (Haussmann et al. 2003).

A stay-green rice mutant was isolated via chemical mutagenesis and it was found to be controlled by a single recessive nuclear gene, tentatively symbolized as sgr(t) (Cha et al. 2002).

In wheat the coincidence of QTLs for senescence on chromosomes 2B and 2D under drought-stress and optimal environments, respectively, indicated a complex genetic control of stay green with respect to the environment (Verma et al. 2004).

3.4.2.2 Dehydration Tolerance

Dehydration tolerance is the ability of the plant as a whole or in any of its components to function under low plant water status. At the onset it must be very clear that a difference in dehydration tolerance between two genotypes cannot be resolved, deduced or declared unless the difference is measured and established when the two genotypes are at the *same plant water status*. Neglect to observe this simple requirement is a surprisingly common oversight in published research, especially when tests of transgenic plants are reported.

Desiccation tolerance is often discussed and indicated as a component of drought resistance. There is a difference between *dehydration tolerance and desiccation tolerance*. Dehydration tolerance basically expresses a capacity of the plant to function at low plant water status. Functions can be growth of certain organs or tissues, fertility processes or translocation processes. Desiccation tolerance expresses the capacity of the whole plant (such as the case of resurrection plants) or certain organs or tissues to survive extreme desiccation as evidenced by their recovery upon rehydration. Desiccation tolerance is discussed here under dehydration tolerance as well as under "resurrection plants" in Sect. 5.5.

Plant Survival and Lethal Water Status

The maintenance of life in a desiccated state or survival at low plant water status is a long standing criterion of dehydration and desiccation tolerance which trace back to the early studies of ecological botany and xerophytic plants. Ecological survival is a time related phenomenon whereas it is traditionally defined as the length of time the plant remains alive under drought stress. Various plant functions, including appearance, can be used to define plant longevity and survival under stress, where plant death is the end point. Plant death by desiccation is determined by its failure to recover upon rehydration.

Survival can also be driven by dehydration avoidance. Where dehydration and the reduction of plant water status under drought stress are delayed, survival is extended. Plant recovery after severe dehydration is a popular and practical measure of survival. It has been shown that the rate of recovery from sever desiccation of various genotypes may depend on their RWC at peak stress just before water was applied to induce recovery (Blum 2005). Dehydration avoidance was also found to be behind the better field recovery after stress of certain rice (Lilley and Fukai 1994) and tall fescue (Huang et al. 1998) genotypes and certain Angiosperm species (Farrant et al. 1999). Therefore, in order to account for the effect of dehydration avoidance on dehydration tolerance, true dehydration tolerance as well as desiccation tolerance must be normalized for plant water status.

This is the idea behind the "lethal water status" (LWS) parameter as an accurate estimate of dehydration tolerance in terms of absolute survival (Flower and Ludlow 1986; Basnayake et al. 1993). LWS is the leaf water potential or RWC at which a standard leaf dies. In diverse sorghum lines lethal RWC and leaf water potential ranged from 58 to 68% and −3.1 to −3.9 MPa, respectively. In other cases (Lilley et al. 1996) leaf osmotic potential just before the leaf died was taken as an estimate of LWS. At such low water status turgor is null so that osmotic potential is equal to water potential. This is a reasonable approach because osmotic potential is easier to measure than the leaf water potential in very desiccated leaves. While LWS is a standardized measure for leaves, it should also be recognized that LWS might be different in different organs in the plant, such as exposed leaves on one hand and meristems on the other. The exact point of whole plant death is difficult to determine, if such a point exist at all. However, where plant selection is concerned large differences in survival are sought and plant death is determined on a rather broad phenotypic expression, depending on the case.

Use of soil water potential (Volaire and Lelievre 2001) or soil moisture content (Likoswe and Lawn 2008; Xiong et al. 2007) at plant death is another approach. Lethal soil water status or content at plant death can include a component of dehydration avoidance as the case might have been in the results of Volaire et al. (2005).

The physiological and molecular basis of the survival of extreme desiccation is not well understood. Research into this area is too often hindered by overlooking the need to differentiate between avoidance and tolerance in the experiment. Furthermore, as seen above, whole plant physiology during dehydration may strongly influence tolerance in terms of survival and recovery. The translation of the available knowledge into application in breeding is also hindered by the fact that most of this knowledge has been developed for natural vegetation, resurrection plants or transgenic model plants. There is need to bridge this gap in order to apply this knowledge to breeding. For example, extensive and exciting research into the physiology and molecular biology of desiccation tolerance in resurrection plants has been performed since the early 1990s (Sect. 5.5). Regretfully there is still no progress towards application of this information in plant breeding at this time.

The embryo in the dry seed is alive at a typical moisture content of about 10%. This capacity for desiccation tolerance depends on the embryo being in an absolutely dormant state. As soon as imbibition is affected and germination commences, the recovering embryo which develops into a heterotrophic seedling loses its tolerance (e.g., Blum et al. 1980). A high level of desiccation tolerance cannot be sustained together with growth. This is also apparent from the drought response of resurrection plants which can maintain life only in a dormant state. Alpert (2000) discussed this issue and underlined the apparent negative relationship between desiccation tolerance and growth. In more than few respects desiccation tolerance is similar to freezing tolerance whereas both require plant dormancy and cannot occur in a growing plant. Consequently, the value of desiccation tolerance in crop plants and agriculture might be limited to much defined situations that will be discussed further below.

The normal progress towards a dormant state in crop plants begins with the death of all leaves and ends with the survival of meristems. Younger plants are better at survival and recovery whereas they still maintain viable meristems such as the apical meristem, tiller meristems or root meristems. The conservation of life in a dormant desiccated state requires adjustments and adjustments require time. It has already been discussed above that slow dehydration is prerequisite for hardening and adaptation. As much as we know, time is used for the accumulation of certain specific protective proteins (e.g., LEA proteins, dehydrins), sugars, amino acids (e.g., proline), antioxidants and the enhancement of cell membrane stability, all of which serve to conserve cell life. The preservation of the native structure of macromolecules and membranes and the conservation of hydration of macromolecules is an essential part of desiccation tolerance. Storage of carbon and nitrogen in cells may also be important for recovery and regrowth. Certain sugars may have a role in repairing damage after sever desiccation (Peters et al. 2007). ABA accumulation under dehydration stress may have a significant role in signaling genes required for the accumulation of protective agents. ABA is also responsible for reducing metabolism and initiating a dormant state (Sreedhar et al. 2002). While the role of various cellular protectants has been researched extensively mainly by use of model transgenic plants, the problem remains in quantifying the effect on survival. Such quantification should allow assessment and ranking of metabolites for their value towards plant breeding. Since different experiments use different criteria of survival, comparisons across studies are difficult.

It has already been mentioned above that OA has a role in desiccation tolerance, due to the protective role of certain osmolytes. The results obtained by Basnayake et al. (1993) in sorghum are therefore intriguing. In their study, maximum osmotic adjustment was inversely related to desiccation tolerance. Lethal leaf relative water content and lethal leaf water potential increased linearly as maximum osmotic adjustment increased. Despite their reduced desiccation tolerance, lines with high osmotic adjustment survived 10 days longer. Similarly, high OA capacity and thus turgor maintenance was important for soybean survival (James et al. 2008).

Within the context of these studies, dehydration avoidance seems to be negatively associated with desiccation tolerance in terms of LWS. It is however clear that OA delays (in time) leaf and plant death during the course of a drying cycle. However, the plant might then die at a higher plant water status as compared with a plant that lacks OA. Therefore measurement of plant death in terms of lethal soil water status should be accompanied also by the measurement of time to plant death.

It is an interesting question why most semi-arid region plants did not evolve the capacity for desiccation tolerance since most plants are subjected from time to time to extreme desiccation. The trait seems to be more common in perennial grasses than in annuals. Annuals perhaps evolved dehydration avoidance and escape mechanisms since they naturally terminate life with reproduction while perennials remain to survive through a dry season into the next year. In that sense most of our annual crop plants seem to be devoid of such natural capacity for desiccation tolerance.

The value of desiccation tolerance and survival in crop plants in agriculture is limited to much defined situations. The negative association between plant production and desiccation tolerance has been often pointed out as the main impediment to using desiccation tolerant genetic resources in crop plant breeding (e.g., Toldi et al. 2009). Occasionally the opinion has been expressed that survival traits are not important at all for crop plants since survival is in the way of crop productivity under limited water supply. Direct link between drought resistance and crop production is certainly a prime consideration. However, the variety of drought scenarios and the ecological and social context of the various crop production systems indicate that desiccation tolerance and survival can sometimes have an important role in farmers' livelihood.

Dehydration tolerance and survival mechanisms underlying delayed plant death allows continued soil moisture extraction (Volaire and Lelievre 2001) and the extension of plant life towards an oncoming rainfall event and subsequent recovery. While this may not be necessarily expressed in a conspicuous economic benefit in terms of yield, the trait is extremely important in subsistence agriculture and at times it can even impact human survival.

In modern field crop farming in certain ecological zones such as the Mediterranean, seedling survival can carry an important economic impact. Farmers often plant early into dry soil and the first winter rain germinate the seed. The next rain event can be delayed and seedlings begin to wilt. If seedling survival capacity is high then the next rain will revive the field. Else, a second sowing will be required which involves a cost and a delay of the crop growing cycle. A delay in crop cycle can involve hazards such as diseases and pests or a sever terminal drought stress.

Cell Membrane Stability (CMS)

The cellular (plasma) membranes are central to cell life and function. Cellular membranes are composed of proteins and phospholipids arranged in a fluid bilayer, where some proteins "float" in the phospholipid bilayer while others are embedded in the membrane. Phospholipid molecules are lined side by side with their polar surfaces facing the aqueous phase on either side of the membrane and their hydrocarbon chains forming a central hydrophobic region. The significant characteristic of this arrangement is in the maintenance of its functional fluidity. The loss of fluidity is detrimental to function. Water is required for the maintenance of the membrane fluidity and functional structure. The loss of water is associated with significant structural changes in the phospholipid bilayer, which are largely physicochemical, but may also result from lipid peroxidation. As membranes are "protected" from desiccation-induced damage by the prevalence of membrane-compatible solutes, such as sugars or proline, a link may exist between the capacity for osmotic adjustment and the degree of membrane protection against the effect of dehydration. This proposition is highly attractive also in view of the effect of pretreatment by dehydration ("hardening") toward an increase in membrane stability in a subsequent stress cycle. The movement of metabolites and other life supporting compounds across the plasma membrane is regulated by various specific transporters and ion channels. Their specific function in response to cellular dehydration and dehydration tolerance is not well resolved.

Membrane structural or functional disorders are often expressed and measured by the leakage of solutes from cells. Leaked substances include various electrolytes, amino acids, sugars, organic acids, hormones, phenolics, and fluorescent materials. When leakage from water-stressed tissue samples is taken as an index of injury to cellular membranes, it may be argued that a given rate of leakage does not necessarily represent a given level of injury in all cells, but rather it may represent the proportion of dead cells in the tissue. This would be supported by Fellows and Boyer (1978) who suggested that desiccation-induced cellular structural changes occur on a cell-by-cell basis rather than in all cells at once. However, the reversibility of the response, in terms of membrane repair and the associated reduced leakage upon recovery from water stress (Bewley 1979; Wang and Huang 2004), does not support the gradual cell-by-cell death concept.

Drought stress-induced ABA accumulation was associated with membrane leakage in various Kentucky Bluegrass lines (Wang et al. 2004). This however could be a simple expression of a non-causative correlation of events occurring during plant desiccation. On the other hand pretreatment with ABA increased CMS in the moss *Atrichum androgynum* (Beckett 2001) and in orchid protocorms (Wang et al. 2002) subjected to desiccation and barley subjected to osmotic stress (Bandurska 1998). The exact role of plant hormones in CMS remains to be established.

CMS as expressed by leakage of electrolytes which is measured conductometrically (Sect. 4.2.3.3) has been used in the evaluation of and screening for freezing, chilling and heat tolerance since the late 1960s (Blum 1988). Solid evidence has been provided for the positive association between CMS for heat tolerance and yield in hot environments across diverse materials of wheat (Saadalla et al. 1990; Blum et al. 2001; Reynolds et al. 1994), Soybean (Martineau et al. 1979), cotton (Rahman et al. 2004), and cowpea (Thiaw and Hall 2004). CMS assay for dehydration tolerance as performed in vitro with leaves submerged in PEG (polyethylene glycol) solution was found to represent well the relative advantage of durum wheat cultivars in seedling growth under drought stress and in their field performance

(Bajji et al. 2002). The relative drought resistance of various turfgrass accessions was well represented by electrolyte leakage (Abraham et al. 2004).

The significance of CMS towards dehydration tolerance is derived from the fact that leaf desiccation cause cellular leakage which can be measured electroconductometrically (e.g., Blum and Ebercon 1981; Riga and Vartanian 1999). A positive association has been found between certain membrane lipid composition and CMS under desiccation stress. Riga and Vartanian (1999) made a detailed account of the physiological events taking place during the dehydration of tobacco plants. They concluded that high CMS was crucial for desiccation tolerance and prolonged survival after turgor was completely lost. Similarly the summer survival of *Festuca arundinacea* and *Lolium perenne* strains was linked to high CMS (Jiang and Huang 2001) and it was ascribed to the accumulation of low molecular weight dehydrins during drought stress in *Dactylis glomerata* (Volaire 2002).

Transgenic rice lines over-expressing barley HVA1 gene for LEA protein accumulation performed better than wild types under prolonged drought stress (Babu et al. 2004). These transgenic lines had higher CMS which was not associated with OA. Transgenic HVA1 mulberry (*Morus indica*) also had higher CMS and better drought resistance than the wild type (Lal et al. 2008). Transgenic tobacco expressing a novel LEA gene had improved CMS under drought stress as compared with the wild type (Wang et al. 2006). It is quite possible that antioxidants also promote CMS under drought stress as seen in a transgenic rice plant expressing high manganese superoxide dismutase activity (Wang et al. 2005a, b).

On the other hand, dehydration tolerance, in terms of reduced leakage from leaf disks stressed in vitro in a solution of PEG, was not related to seedling recovery from stress in several sorghum varieties (Blum and Ebercon 1981) or to seedling growth rate under stress in various wild emmers (*T. dicoccoides*) or wild emmer wheat derivatives (Blum 1988). A certain disagreement about the value of CMS testing as an absolute physiological marker for dehydration tolerance emerges from the various results and it should be clarified here. CMS as a method for dehydration tolerance has been tested by two approaches: (1) leakage from leaf samples desiccated in vitro by immersion in PEG solution and (2) leakage in dehydrated leaves as sampled in vivo from plants subjected to natural drought stress. The in vitro method is repeatable and well standardized but stress is un-natural being applied by PEG in a test tube. On the other hand, in vivo sampling is subjected to artifacts if all samples are not at the same leaf water status when sampled. Since CMS decreases with leaf dehydration, the comparison of different genotypes for CMS must be performed when all leaves are sampled at the same low leaf water status, within reason. This was done by Tripathy et al. (2000).

The inheritance of CMS has been studied for heat and cold tolerance where the method has been used more extensively, but hardly for dehydration tolerance. CMS as measured in rice leaves sampled in the field at 60–70% RWC presented a broad-sense heritability of 0.34 (Tripathy et al. 2000). Phenotypic variation in CMS was explained by nine individual QTLs, accounting for 13.4–42.1% of the variation.

In conclusion, the above review indicates an important role for CMS in dehydration tolerance. A large number of studies indicated it can predict genetic variation for heat and dehydration tolerance. The assay is detailed in Sect. 4.2.3.3.

Stem Reserve Utilization for Grain Filling

Grain filling in the cereals depends on the supply of assimilates at the time when the grain is growing exponentially. In addition to the demand for grain growth, grain and whole plant respiration is an additional sink for carbon. The source of assimilates to provide for these sinks is current photosynthesis. While the demand for carbon by these sinks is increasing as grain filling proceeds, the source is diminishing due to natural leaf senescence. When drought or any other stress occurs during grain filling the photosynthetic source is further inhibited and grain growth is slowed or arrested. In many grain crops stem reserve stored during pre-anthesis and anthesis growth stages is remobilized into the grain to provide the balance of carbon needed for grain/fruit growth. Stem reserve utilization can be a major source for grain filling, the importance of which increases as current photosynthesis is diminished by stress during grain filling. Even during three atypically dry years in the UK it was found that stem reserve utilization for grain filling was a most important trait for conserving yield (Foulkes et al. 2002). Estimates of the relative contributions of stem reserves to total grain mass per ear or to grain yield of wheat vary among the different reports according to the experimental conditions and cultivars used. These contributions were estimated to be anywhere between 6 and 100%, as reviewed by Blum (1981, 1998), Schnyder (1993), and Yang and Zhang (2006).

Hsiao (1973) and Boyer (1976) already determined that translocation is one of the more dehydration tolerant processes in plants. It would proceed at levels of water deficit sufficient to inhibit photosynthesis. Therefore, the capacity to mobilize stem reserves in a dehydrated or desiccated plant is taken as a unique expression of whole plant dehydration tolerance which is tightly linked to grain yield.

Storage of reserves in stems up to anthesis and even a few days later depends on the availability of assimilates beyond the immediate demand by growth processes. Total wheat stem non-structural carbohydrates (TNC) at anthesis can vary from 50 to 350 g kg^{-1} dry mass. TNC is stored mainly as starch or fructan depending on the crop species. Any conditions that would enhance assimilation during stem elongation are expected to enhance storage. For example, under dryland field conditions only half the TNC was available for remobilization during grain filling, as compared with irrigated conditions. When ambient CO_2 concentration was raised to increase assimilation, more carbon was stored in the stems (Winzeler et al. 1989).

Storage and its availability for remobilization may vary along the stem. The peduncle and penultimate internode (and leaf sheath) contain the most storage in barley and wheat. Various aspects of stem anatomy with respect to storage are of interest. An early study did not find any consistent difference in total stem reserves accumulation between solid and hollow-stemmed wheat (Lopatecki et al. 1962). A more recent study (Saint Pierre et al. 2010) found that total amount of pith-fill in the upper stem internode was highly correlated with the total content of water-soluble carbohydrates (WSC) per stem across diverse wheat materials. A positive correlation was also found between pith-fill and grain yield under drought stress which was explained by the positive contribution of WSC to grain yield. More data is needed on stem solidness and stem pith fill as important factors in stem reserve utilization before this trait can be confidently used in selection.

Stem length, as affected by height genes can be important in affecting stem reserve storage. The Rht1 and Rht2 dwarfing genes of wheat were found to reduce reserve storage by 35 and 39%, respectively, as a consequence of a 21% reduction in stem length (Borrell et al. 1993). However, under favorable conditions the advantage of the tall genotype in reserve storage was not expressed in greater mobilization to the ear probably because the tall genotype had a smaller sink demand than the dwarf. Longer stems presented better reserve storage availability for use also in sunflower (Sadras et al. 1993) and sorghum (Blum et al. 1997). Within species the variation in stem length density (dry weight per unit stem length) expresses reasonably well the relative variation in TNC storage.

With the increasing recognition that stem reserves are an important component of breeding for improved yield under drought stress (Blum 1998), more attention is now given to selection for TNC storage in stems (e.g., Ruuska et al. 2006). However, storage is only part of the requirement. It is quite possible to have ample reserve storage with insufficient remobilization (e.g., Blum et al. 1994).

Reserve utilization is not dependant only on the amount stored. Stem reserve mobilization is affected by sink demand, by the environment and by cultivar. The demand by the growing grain is a primary factor in determining stem reserve mobilization. When sink size was reduced by degraining, more reserves remained in the stem, as compared with intact ears (Kuhbauch and Thome 1989). The most prominent environmental effect on remobilization is that which reduces current assimilation, namely abiotic and biotic stress during grain filling. Thus, for example, shading of barley or wheat plants after anthesis promoted the use of stem reserves for grain filling (Bonnett and Incoll 1992; Kiniry 1993). What seems to be important are the reduction in assimilation and not the nature of stress causing the reduction. Tolerance to *Septoria tritici* leaf blotch in wheat is expressed and measured by sustained grain filling under severe disease epiphytotics. It has been demonstrated that mobilized stem reserve was a major component of *Septoria* tolerance (Zilberstein et al. 1985) as well as other leaf diseases of wheat (Bancal et al. 2007).

Grain filling under heat stress is also supported by stem reserve mobilization (Blum et al. 1994). Heat as well as drought during grain filling reduces the duration of grain filling. There is normally an increase in the rate of grain dry matter accumulation under high temperatures, but it is not sufficient to compensate for the decrease in duration. When grain filling under such stress depends on remobilized stem reserves, the rate at which these reserves are metabolized and transported to the grain is crucial. There is a limit to this rate. Thus, a genetically longer grain filling duration seems to be an advantage in this respect. Short grain filling duration may allow the escape of post-flowering stress while longer duration may allow greater utilization of stem reserves for grain filling under stress provided that storage is available.

Stem reserve mobilization to the grain requires the breakdown of storage into transportable forms of carbon towards its assimilation in the grain. Drought stress during grain filling in rice promoted remobilization in tandem with enhanced sucrose synthase and starch branching enzyme activities in grains (Yang et al. 2001). In wheat drought stress promoted activity in stems of fructan exohydrolase and acid invertase and conversion of fructans to fructose (Wardlaw and Willenbrink

2000). ABA but not zeatin or zeatin riboside in stems was positively correlated with remobilization of pre-stored carbon and grain filling rate. Exogenous ABA spray application to rice plants reduced chlorophyll in the flag leaves, enhanced stem reserve remobilization, and increased grain filling rate. Spraying with kinetin had the opposite effect. An enhanced carbon remobilization and accelerated grain filling rate were attributed to an elevated ABA level in wheat plants when subjected to drought stress (Yang et al. 2003). Stress-enhanced ABA in wheat stems was suggested to regulate fructan exohydrolase and sucrose phosphate synthase activities (Yang et al. 2004). Stem TNC storage as measured in two wheat varieties was not, on its own, a reliable criterion to identify potential grain yield in wheat exposed to water deficits during grain filling. Mobilization of reserve was important and that was conditioned by stem fructan exohydrolase activity under drought stress (Zhang et al. 2009). In conclusion, reserve mobilization and grain filling under drought stress can differ among genotypes of the same stem reserve storage capacity. This has an important bearing on the method to select for stem reserve utilization for grain filling under drought stress (Sect. 4.2.3.4).

It has already been mentioned above in the discussion of the "stay green" trait that effective stem reserve utilization is strongly linked to enhanced plant senescence, for obvious reasons (Fokar et al. 1998; Yang et al. 2003; Kumar et al. 2006). Extensive remobilization as a result of stress or demand by a large sink predisposed maize (Dodd 1979) and sorghum (Tenkouano et al. 1993) plants to stalk rots. The "stay green" trait reduced stalk rots and lodging.

It is therefore very doubtful if the stay-green trait can be recombined with superior stem reserve utilization for grain filling as the two traits appear to be mutually exclusive. Where breeding for stress resistance during grain filling is concerned the breeder will have to choose one of the two resistance ideotypes. While many considerations are involved in this choice, an important consideration is that stem reserve utilization for grain filling is an effective solution for drought and heat tolerance as well as certain abiotic tolerances. On the other hand stay-green can enhance resistance to stem rots and reduce lodging.

Conclusive data on the inheritance of this trait is not available and it is suspected to be complex in view of the physiology and the metabolism involved. A study with three wheat populations revealed seven to 16 QTLs controlling the storage of TNC (Rebetzke et al. 2008a, b). Due to the small effects of QTLs it was concluded that MAS for storage was not possible and phenotypic selection will be required. As discussed above, storage is only a partial component of the trait and the capacity for re-mobilization must also to be considered.

Antioxidants

Reactive oxygen species (ROS) are continuously produced in plants as byproducts of aerobic metabolism. Depending on the nature of the ROS species, some are highly toxic and are rapidly detoxified by various cellular enzymatic and nonenzymatic agents (*syn.* antioxidants). ROS levels increase during abiotic, biotic or wounding stresses. Plants appear to generate ROS also as signaling molecules to control various plant processes. ROS are therefore assumed to play a complex role in plants (Apel and Hirt 2004; Miller et al. 2010).

Oxidative stress as part of the abiotic stress syndrome in plants has recently become a very popular theme in plant stress research, especially at the laboratory level. There is even a claim suggesting that superior performance of certain given wheat cultivars under dry conditions could be ascribed to their better antioxidant activity (Fan et al. 2009).

Transgenic technology should now permit a critical evaluation of enhanced antioxidant activity as an agent of plant resistance to various stresses, including drought. While the development of such transgenic plants appears relatively straightforward, the phenotypic and functional studies done by using them does not clarify the role of enhanced antioxidant activity in dehydration tolerance. This is especially of concern considering the fact that ROS production and antioxidant activity is a normal phenomenon in plants and that in more than a few cases the increase in activity in the transgenic plant was not outstanding to the extent that it corresponds with the observed stress resistance. In certain cases (e.g., Selote and Khanna-Chopra 2004) drought susceptibility was ascribed to relatively lower antioxidant activity while at the same time other possible explanations for susceptibility were in evidence, such as a low plant water status in the susceptible genotype.

Apart from the lack of a solid causative link between antioxidants and dehydration tolerance there is the question of quantifying their effect. Whereas antioxidant production is a normal process in non-stressed plants, how much of an increase is required to enhance stress tolerance? Even in the medical science there is still a debate on the level of antioxidant enhancement required for the protection against certain stress driven pathologies.

Following are few brief examples of research into the role of antioxidants in drought resistance which should raise doubts about antioxidants as serious candidates in breeding for drought resistance at this time.

Two wheat cultivars "known to differ in their drought resistance" also differed respectively in their antioxidant activity when detached leaves were subjected to osmotic stress in vitro (Lascano et al. 2001). However they did not differ in their antioxidant system behavior in the field. In the field these cultivars differed in their dehydration avoidance and not in their antioxidant activity.

Transgenic tobacco plants over-expressing ascorbate peroxidase3 had greater photosynthesis because of higher stomatal conductance (Yan et al. 2003). It is not clear how antioxidants improved stomatal conductance unless the transgenic plants simply also had a higher leaf water status (not measured). Over-expression of glutamate dehydrogenase in tobacco plants helped maintain high plant water potential under drought stress (Mungur et al. 2006). It is not clear how antioxidants influenced plant water potential, unless an unaccounted hydraulic reason was involved. Transgenic tobacco plants expressing antiquitin-like ALDH7 had relatively higher antioxidative activity (Rodrigues et al. 2006) when grown in pots. At the same time they also had higher turgidity under drought stress as compared with wild type plants. It is not clearly explained how this gene helped maintain turgidity under stress or whether higher turgidity was sustained by way of alternative means, such as a smaller size of the transgenic plant or osmotic adjustment (not measured).

Alfalfa over-expressing Mn-containing superoxide dismutase (SOD) was tested for 3 years in the field under unspecified drought and cold conditions (Mckersie et al. 1996). Dry matter production by the transgenic plants was generally significantly greater than the wild type. In the growth chamber transgenic line "5" had higher antioxidant activity than transgenic line "30" within a range even greater than the difference between line "30" and the wild type. However, in these field experiments transgenic line "30" yielded better than "5" which does not correspond with the respective levels of assumed protection by SOD, unless the level of line "30" was already optimal. In a different study (Rubio et al. 2002) alfalfa over-expressing Mn-containing SOD in the mitochondria and chloroplasts of leaves and in the mitochondria of root nodules performed slightly better in net photosynthesis under mild drought but they did not differ from wild type in their resistance under moderate and severe drought stress. If SOD has a role in drought resistance one would expect its contribution to increase as stress increases from mild to moderate to severe.

When antioxidant over-expression was introduced into chloroplasts (see below), the improvement of net photosynthesis under drought stress seemed to be a reasonable outcome of photosystem protection. However, the reader should be reminded that when drought stress occurs net photosynthesis is initially reduced primarily by stomatal closure while advanced leaf desiccation is required before chloroplast and photosystem biochemistry are affected. Therefore the observed improvement of net photosynthesis by chloroplast antioxidants under mild stress (Rubio et al. 2002; Badawi et al. 2004) and its ineffectiveness under more severe stress (Mckersie et al. 1996; Rubio et al. 2002) cannot be reconciled with present knowledge on photosynthesis under drought stress.

When all of the above is taken together the unavoidable conclusion is that more research is needed to explain the observed and claimed advantages of enhanced antioxidant activity towards drought resistance and especially the consequence for yield under stress, if any at all.

At present antioxidants cannot be considered as serious candidates in plant breeding for drought resistance.

Dehydration Tolerant Reproduction

Drought stress at flowering is a major cause of reproductive failure in crop and fruit production. The early reproductive stages which involve the development of the reproductive organs, fertilization and early embryo development are the most stress sensitive plant developmental stages with respect to yield (Fig. 2.9).

Plant reproduction under drought stress is determined first and foremost by dehydration avoidance. It is only when plant water status is reduced when the question of dehydration tolerance in reproduction comes up. Therefore, while the subject of reproduction is discussed under dehydration tolerance, there is not always a clear delineation between avoidance and tolerance in affecting reproductive success under drought stress. For example Liu et al. (2006) found in rice that *cv.* Moroberekan had better anther dehiscence under drought stress as compared with IR-64. However it was not clearly established if the difference was in the inherent anther response to low water status (tolerance) or whether the difference between the two cultivars was only in the capacity of Moroberekan to maintain higher plant water status (dehydration avoidance) due to its deeper roots.

During drought stress the developing ovary and pollen may or may not reduce their water status, depending on how protected they are from the atmosphere and how close they are to the main leaf-to-root water potential gradient path. Rice spikelets are especially sensitive to water loss and they can be readily desiccated when plant water potential is reduced (O'Toole et al. 1984). On the other hand wheat spikelets are relatively more dehydration avoidant when leaf water potential is reduced under drought stress (Saini and Aspinall 1981). When wheat leaf water potential declined under the effect of a drying soil, ovaries and pollen water potential also declined but their turgor remained as high as in well-watered plants most likely due to OA.

Irregular carbohydrate metabolism in the developing pollen was suggested to be involved with pollen sterility in drought stressed wheat. Selective transcriptional down-regulation of anther invertases during pollen meiosis was the first stage in the development dysfunctional pollen under stress (Koonjul et al. 2005) which seemed to be irreversible when plants were re-watered. Down regulation of ovary invertases is also involved in ovary abortion (Andersen et al. 2002; Mclaughlin and Boyer 2004). Ovary abortion is an important cause of reproductive failure under drought stress, especially in maize (Boyer and Westgate 2004), and sugar starvation has been suggested as a possible cause. However, while sucrose feeding to drought stressed maize improved ovary survival, it was not clear to what extent whole plant carbohydrate status was linked to ovary carbohydrate status. Sugar responsive genes could be involved in directly signaling ovary metabolism in response to plant dehydration. It has already been suggested for wheat that carbohydrate starvation per se may not be the reason for pollen dysfunction (Sheoran and Saini 1996). The inability of wheat ovaries to metabolize incoming sucrose to hexoses was the preferred explanation.

On the other hand it has been suggested that rice sterility under drought stress at flowering was specifically caused by poor anther dehiscence (Liu et al. 2006). Two causes were suggested for the relatively better dehiscence of 'Moroberekan' cultivar: (1) constitutively superior development of fibrous structures in the endothecium at the anther apex and base and (2) better maintenance of pollen size.

Reproduction under drought stress in the pulses has been less investigated than in the cereals. Pod abortion has been indicated as the most important reason for yield reduction in chickpeas when stress develops after fertilization, with differences in this respect between genotypes (Leport et al. 2006). Pod abortion in drought stressed soybean was associated with pod water potential while pod growth appeared to be controlled by a hormonal root signal (Liu et al. 2004a). However in another study (Liu et al. 2004b) pod abortion in drought stressed soybean was linked to impaired capacity to metabolize incoming sucrose to hexoses together with reduced carbohydrate flux from leaves to pods. In this respect the reason for ovary abortion in maize and pod abortion in soybean seems to have the same basis. Seed abortion in drought stressed pea occurred only if stress took place before the initiation of the linear phase of seed growth (Ney et al. 1994). Once the linear phase of growth was initiated stem reserve mobilization to the seed prevented abortion. Abortion of set fruit in various fruit trees is a well recognized consequence of drought stress.

A conceptual model of the plant being able to sense its total available carbon pool at flowering and subsequently adjust the number of set grain/fruit is an attractive speculation. Sugar responsive genes could be important in this respect. However, from the above discussion it can be realized that neither water status nor carbon status can satisfactory explain all cases of reproductive failure under drought stress.

The role of high ABA content in affecting sterility under drought stress has been discussed in Sect. 2.4.2. Plant reproduction can be impacted by in situ production of ABA in the dehydrating shoot or by incoming root-sourced ABA. The role of ABA and sugar in controlling successful reproduction is given support by the indicated interaction between the glucose and abscisic acid signaling pathway in the growing rice seed embryo (Toyofuku et al. 2000). Therefore, in addition to the above model of a carbohydrate pool sensing plant, there seems to be an early warning system of soil water status mediated by ABA which allows the plant to adjust its reproduction accordingly.

It is therefore quite reasonable that lower plant ABA content has been linked to improved plant reproduction in stress resistant genotypes (e.g., Oliver et al. 2007; Sanguineti et al. 1999; Tang et al. 2007; Yang et al. 2007). If the impaired plant reproduction under drought stress is not an immediate function of carbon starvation and if the hormonal early warning produced by the roots is all too early according to agronomic considerations, then a genetic reduction of ABA signaling or the reduced sensitivity to ABA might carry a potential for yield improvement in water-limited environments. The manipulation of ABA production (Giuliani et al. 2005) or ABA sensitivity (Blum and Sinmena 1995) in crop plants under stress offers a potentially important avenue for breeding or at least the acquisition of important information towards ABA and plant breeding.

In the cereals, arrested exertion of the inflorescence during drought stress at the boot stage is tightly linked to sterility and reduced yield. Inflorescence enclosed within the boot fail to carry out normal reproduction (O'Toole and Namuco 1983). Poor inflorescence exertion is caused by inhibited peduncle growth. It is not clear if the cause for inhibited peduncle growth is also the cause for both poor exertion and sterility or that the development of the inflorescence within the flag leaf sheath is in itself the cause of sterility. Poor inflorescence exertion under drought stress is a reliable symptom of drought stress and resistance (Fig. 3.9).

In conclusion, there are four potential drought resistance mechanisms with respect to the reproductive process:

1. Plant dehydration avoidance to conserve reproductive organ water status. This is the most effective way to protect the reproductive system from failure.

Fig. 3.9 Poorly exerted panicles (*left*) and well exerted panicles (*right*) in two different sorghum lines subjected to pre-flowering drought stress. The two genotypes have the same flowering date under well-watered conditions. Notice the greater leaf senescence in the fully headed genotype as a consequence of the developing "sink load" (Sect. 2.7)

2. Dehydration tolerance by maintaining carbohydrate status of the plant.
3. Case specific constitutive traits of the reproductive system which might enhance their resistance to desiccation.
4. Low plant ABA status or low ABA sensitivity.

The genetic control of specific plant reproductive processes under drought stress is practically unknown while relatively better information exists for reproduction under extreme temperature stress.

Anthesis-to-silking interval (*ASI*) is a crucial developmental trait that has a very large impact on maize reproductive success and yield. Drought stress during the reproductive stage delays silking and thus pollen shed does not correspond with silk receptivity. Maize pollen is shed for several days so that ASI can extend for a few days. Longer ASI causes fertility failure.

ASI-based drought resistant maize sustains short ASI despite stress (e.g., Bolanos and Edmeades 1996). Genetic correlation between ASI and yield became higher with drought stress and selection for ASI under drought stress was very effective in enhancing yield. Slow or delayed silking is a distinct expression of reduced silk growth. The allele for leaf growth maintenance under drought stress was found to be that also for maintained silk elongation rate (Welcker et al. 2007). The cause for reduced silk growth under drought stress is not quite clear especially since silk water status apparently changes with its growth (Schoper et al. 1987). As ABA inhibits growth it would be expected indeed that maize lines selected for high leaf ABA concentration would have delayed silking under drought stress as compared with low ABA lines (Landi et al. 2001). Thus, ABA produced in the plant or imported from the root could have a role in extending ASI. This is in support of the negative role of high ABA content on plant growth and reproduction.

The association between ASI and yield under stress is not robust under all conditions. For example, variation in silk receptivity due to silk morphology and rate of aging can affect fertilization (Anderson et al. 2004) and consequently yield. However, short ASI, together with yield under drought and nitrogen stress, has become a major selection index in maize breeding in Africa with outstanding progress (Bänziger et al. 2006).

The inheritance of ASI in maize is understandably complicated by the fact that it involves a time span determined by two developmental events: male and female flowering times. It may also be complicated by the fact that different QTLs are responsible for ASI under stress and non-stress conditions. Extensive work by Ribaut et al. (1996) identified six putative QTLs for ASI under drought on chromosomes 1, 2, 5, 6, 8 and 10, which together accounted for approximately 47% of the phenotypic variance. Under drought stress, four QTLs were common for the expression of male flowering and female flowering, one for the expression of ASI and male flowering and four for the expression of ASI and female flowering. The number of common QTLs for two traits was related to the linear correlation between these two traits. Segregation for ASI was found to be transgressive with the drought-susceptible parent contributing alleles for reduced ASI at two QTL positions. Alleles contributed by the resistant line at the other four QTLs were responsible for a 7-day reduction of ASI. These four QTLs represented around 9% of the linkage map, and were stable over years and stress levels. Welcker et al. (2007) described a much simpler situation where alleles for leaf elongation under drought stress were the same as these responsible for silk growth, namely short ASI.

A Dehydration Tolerant Photosystem and Respiration?

Photosystem Function Photosynthesis is probably one of the most investigated plant process with respect to performance under drought stress. Despite the numerous reports on the subject it is impossible to conclude that there is genetic variation for tolerance in any photosystem functions under low photosynthetic tissue water status. When real genetic variation in this respect is sought in the literature, the reported results can be divided into three groups:

1. Comparative tests under some forms of drought stress do not show any differences in photosynthesis between genotypes.
2. Where differences were found between genotypes in photosynthesis it is not clear if the difference can be ascribed to photosystem tolerance or tissue dehydration avoidance.
3. Real evidence for conserved photosystem function during desiccation was found in lower plant forms such as certain moss (e.g., Proctor et al. 2007). Even in resurrection plants the recovery of photosynthesis after extreme desiccation is not due to its conservation but most likely due to *de novo* transcription of the genome upon the recovery of leaf water status (Farrant and Kruger 2001).

It can therefore be concluded that hard evidence for dehydration tolerance in photosystem function of crop plants is still impending. In most cases of crop plants if not all of them, it is the maintenance of leaf water status by dehydration avoidance that sustain leaf photosynthesis under drought stress. It is also reminded here that injury to the photosystem by leaf dehydration basically takes place after full stomatal closure.

Respiration A critical examination of respiration at low water status in different crop plant genotypes is limited. Atkin and Macherel (2009) reviewed the literature on respiration under drought stress. They conclude that respiration may increase or decrease or remain unchanged under drought stress as compared with well watered conditions. Understandably, this does not allow a simple explanation of the role for respiration in dehydration tolerance. They point out that changes in respiration under drought stress are relatively small when compared with photosynthesis. From their own review and discussion of the literature they propose a model by which mitochondrial respiration enables survival and rapid recovery of productivity under water stress conditions.

A study with isolated wheat mitochondria showed that mitochondria from drought-sensitive genotypes had low oxidative phosphorylation efficiency after dehydration and rewatering, whereas the drought resistant genotype mitochondria had higher phosphorylation rates (Vassileva et al. 2009). Pastore et al. (2007) suggested two possible roles for mitochondria function under drought stress, including involvement in ROS production and inhibition.

In conclusion, at this time neither photosystem function nor respiration is a practical candidate towards breeding for dehydration tolerance.

3.4.2.3 Drought Escape

Flowering time and the duration of crop growth is a cornerstone of plant breeding towards environmental adaptation. It is commonly referred to also as plant phenology. Its inheritance is often relatively simple. Its control by photoperiod, temperature and vernalization is well understood for most crops and environments. The establishment of an optimal flowering time in the target environment involves extensive field experiments which evaluate major environmental variables and their effect on yield. Crop growth models and geographical information systems can also aid in optimizing plant phenology towards a given environment.

"Drought escape" is a trait derived from sufficiently early flowering and maturity or harvest time to allow the escape in full or in part of a drought stress occurring during the grain filling or fruit development growth stage. Such stress is often called also "terminal stress," "post-flowering stress" or "late-season stress." Often, this stress also involves high temperatures. In some geographical regions it can also involve chilling temperatures. Exposure to terminal stress cause a reduction in yield due to poor grain filling and sometimes failure to set grain or fruit. In certain crops it can also cause poor product quality.

Traditional crop varieties or landraces were evolved through a very long process of selection by farmers to achieve optimal fitness to their environment in terms of phenology. In most cases landraces are photoperiod sensitive. Their photoperiod sensitivity usually determines a fixed flowering and maturity date so that the harvest will be securely dated at the termination of the rainy season, irrespective of planting date. Planting date in most dryland farming systems must be flexible in tune with the beginning of the rains, which rarely occur on a fixed date. Thus, photoperiod sensitivity provides the farmer some flexibility with respect to sowing time while maintaining security with respect to harvest conditions. However, landraces of the tropics and semi-arid tropics tend to have a very long growing period in correspondence with their adaptation to long rainy seasons. They are therefore highly sensitive to photoperiod which is expressed in lush vegetation, relatively smaller inflorescence and a low harvest index.

The choice of early flowering as a solution to escape terminal stress should be treated with care. It can involve drawbacks. For example, in some parts of Australia early flowering of wheat involves greater risk of frost injury during flowering. Late flowering is associated with reduced risk of frost and a greater risk of terminal stress. Late flowering genotypes offer a potential for regrowth and productivity upon recovery from stress (e.g., Villnlobos-Kodruigez and Shibles 1985). On the same calendar day, late-maturing genotypes are younger than early ones and therefore carry more viable meristems from which recovery can take place.

However the major drawback of earliness is the widely known positive correlation between growth duration and yield. Yield potential increases with growth duration, to a limit. The correlation holds well on well watered conditions but under conditions where earliness is an effective escape solution, the advantage of late flowering in yield turns into a disadvantage. Figure 3.10 demonstrates that under a favorable water regime the correlation across different wheat genotypes between growth duration and yield was positive. As seasonal rainfall reduced this correlation became negative.

There is therefore a balance between growth duration, yield potential and the severity and frequency of terminal drought experienced by the crop. Sufficient data must be acquired with the crop in the target environment to establish the optimal growth duration which would result in a yield advantage in most years.

Derouw and Winkel (1998) saw an advantage in variability of flowering date within a millet field under conditions of unpredictable drought occurrences (note that they define escape as "avoidance"). In the case of drought escape an optimal growth duration and flowering date is based on a relatively predictable drought scenario with respect to terminal stress. Where diversity in flowering date is opted for, modern farming practice chooses for obvious reasons to plant several varieties each with a slightly different flowering date rather than one variety diverse in flowering date.

Early maturity involves reduced total seasonal crop evapotranspiration (ET) because of the short growth duration. However, as growth duration is genetically linked with leaf number and often with leaf size, early genotypes have a small

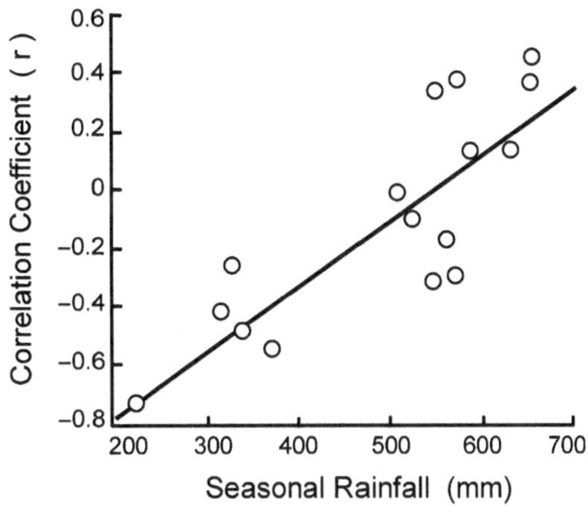

Fig. 3.10 The linear relationship between the coefficient of correlation between grain yield and days to heading across 12 common wheat cultivars within any given trial and trial seasonal precipitation, over 16 trials. Reduced precipitation involved increased terminal drought stress. $y = -1.20 + 0.025x$; $R^2 = 0.62$

leaf-area index than late genotypes, on the same date. Thus, early genotypes express reduced ET during most growth stages, up to the point where a full ground cover is achieved (Blum 1970; Lafitte and Courtois 2002; Bidinger et al. 2005).

In a study of early and late flowering isogenic lines of sorghum grown in hydroponics (Blum et al. 1997) it was found that the earlier genotypes had a faster rate of adventitious root initiation and growth which corresponded with their faster shoot development. However, the later genotype has a relatively larger root volume due to more root branching. When these isogenic lines were grown in soil filled glass-paneled root boxes (Blum and Arkin 1984) it was found that indeed total plant root length corresponded to total leaf area of the shoot. The early maturity gene ascribed relatively smaller total root length density and smaller total leaf area as compared with the late maturity gene, at almost all growth stages. Since the root boxes used for these studies were not deep enough for sorghum, it could not be determined if the maturity genotypes differed in maximum root depth. In barley however it was found that late maturing cultivars had deeper roots at flowering and could extract deep soil moisture as compare with earlier flowering cultivars (Mitchell et al. 1996). This is reasonable since extended growth duration involve also extended root growth duration.

The potential growth duration and flowering time are well expressed under potential growing conditions. Deviation in flowering time can take place when plants are subjected to drought stress before flowering. Cereal crops generally delay flowering under severe drought stress. Similar response was observed in common bean (Muñoz-Perea et al. 2006). The general delay in flowering under drought stress is consistent with the finding that maize lines with high ABA content were later in flowering than low ABA lines (Sanguineti et al. 1996). In the cereals the delay is expressed in late or even incomplete exertion of the inflorescence (Fig. 3.9). A minimal delay in flowering (or heading) under drought stress as compared with well-watered control serve as a phenotypic indication of relative drought resistance (Sect. 4.2.1.3).

3.4.2.4 Summer Dormancy

In regions with a dry summer, such as the Mediterranean the survival of native cool-season perennial grasses depends on their capacity to persist through the dry and often very hot summer. Surviving the summer in these grasses is achieved by way of summer dormancy. Summer dormancy is expressed by complete cessation of growth and senescence of foliage. In the case of summer dormant range grasses, the dry dead foliage is often grazed and the remaining basal part of the dormant grass will recover and grow upon the relief of drought in the autumn.

Summer-dormant and summer-active cultivars of tall fescue [*Lolium arundinaceum* (Schreb.) Darbysh.], orchardgrass (*Dactylis glomerata* L.), perennial ryegrass (*Lolium perenne* L.), and hardinggrass (*Phalaris aquatica* L.) were tested for several years under the dry summer conditions at Vernon, Texas (Malinowski et al. 2009). Summer-active types of these grasses were not adapted to this environment and could not survive the first summer after planting. Summer-dormant types were fully adapted and persisted for at least 5–8 years, depending on species. The most adapted were tall fescue, hardinggrass, and to some extent orchardgrass [subsp. *hispanica* (Roth) Nyman].

Florence Volaire and co-workers did extensive research on summer dormancy and the survival of native grasses under drought stress (Volaire and Norton 2006; Volaire 2008; Norton et al. 2009; Volaire et al. 2009a, b). According to that research summer dormancy can be defined by four criteria: (1) reduction or cessation of leaf production and expansion; (2) senescence of mature foliage; (3) dehydration of surviving organs; and (4, optional) formation of resting organs. Two levels of summer dormancy are recognized: (a) complete dormancy, when cessation of growth is associated with full senescence of foliage and induced dehydration of leaf bases; and (b) incomplete dormancy, when leaf growth is partially inhibited and is associated with moderate levels of foliage senescence. Summer dormancy is expressed under increasing photoperiod and temperature. It is under hormonal control and usually associated with flowering followed by reduction in metabolic activity in meristematic tissues. Dehydration tolerance and dormancy are independent phenomena and differ from the adaptations of resurrection plants.

For example, Volaire et al. (2009a) found out that even under irrigation, cessation of leaf elongation, senescence of lamina and relative dehydration of basal tissues were triggered only by a day length longer than 13 h 30 min in plants of *Poa bulbosa* and *Dactylis glomerata* cv. 'Kasbah' which exhibited complete dormancy. Dormancy was not triggered by dehydration stress. Drought stress alone cannot induce summer dormancy under early-spring short days, although water deficit under late-spring long days reinforces it and could enhance plant survival through the summer. The only way by which summer dormancy can be identified is by lack of any growth recovery after irrigation during the summer.

The open and most interesting question is how the plant recovers from dormancy and proceeds to re-grow at the end of summer. How the return to short days breaks dormancy and how the meristems which were survived extreme desiccation and heat throughout the summer months are able to initiated growth in the autumn upon rehydration. Summer dormancy is an important feature in breeding perennial range grasses.

3.5 Water-use Efficiency (WUE)

WUE was originally developed by agriculture engineers as a ratio between yield and irrigation water in order to assess returns for irrigation input and cost. WUE is an important yardstick to measure irrigation efficiency. The WUE term was later adopted by soil scientists and agronomists for a wider use in agronomy, including dryland farming. Physiologists found the term useful also at the leaf level in studies of gas exchange where WUE (or "transpiration ratio") is defined as the ratio of carbon fixation to transpiration. WUE can therefore be used at various levels of the crop, from the single leaf to the field.

Studies of water use efficiency at the whole plant and field level were cumbersome due to the work load and costs involved in assessing whole plant or crop water use, especially when large plant populations in plant breeding were considered. The breakthrough came with the development of better understanding of stomatal dynamics, gas exchange and photosystem function, leading to the carbon isotope discrimination (delta) assay as a heritable marker for WUE at the whole plant level (Farquhar et al. 1989; Hall et al. 1994). The reader is referred to these papers for details on the theory and the method. In the majority of cases low carbon isotope discrimination (low delta) as measured in the grain or the leaves was found to be well correlated with high WUE and vice versa, with few exceptions (e.g., Turner et al. 2007a, b) where delta was not associated with WUE.

With the increasing global concern for water scarcity, WUE has become a contemporary topic, including rainfed conditions. An important contribution of the carbon isotope discrimination method was that it enhanced research on WUE and provided extensive data on the subject especially in the context of breeding and genetic diversity. At the same time the large volume of published information on delta, WUE and their implications towards selection for water limited environments created some confusion in the plant breeding community. Confusion was largely created by the fact that the relations between delta (WUE) and yield were sometimes positive and sometimes negative, depending on the crop growing conditions. The following excerpt from a highly qualified review of the subject (Hall et al. 1994) serves to demonstrate the point:

> …For cowpea and wheat, selecting for high delta may increase productivity in well-watered environments. For some environments and species an intermediate level of delta

may be most adaptive. It is possible that a broadly adapted cultivar would exhibit high delta in well-watered environments, but low delta when grown on limited moisture stored in the soil. Selection for changes in delta with drought would be difficult to achieve by measurement of delta.

It therefore appears that WUE as a target in breeding for water-limited environments is obscure if not constantly moving. Plant breeders discussing carbon isotope discrimination and WUE expressed confusion on two primary questions: (1) under what environmental conditions selection for carbon isotope discrimination is expected to result in yield gain, and (2) which direction should selection be made, high (low delta) or low (high delta) WUE.

WUE is often equated in a simplistic manner with drought resistance without considering the fact that it is a ratio between two physiological (photosynthesis and transpiration) or agronomic (yield and crop water use) variables. As a ratio it is often susceptible to misinterpretation, especially when the dynamics of the nominator and the denominator are ignored. A discussion of WUE in the context of plant breeding for water limited environments is presented by Blum (2005). A second paper (Blum 2009) provides further insight into WUE in breeding and explains the source of the confusion about WUE in breeding and why it is an ambiguous selection criterion for yield in most water limited environments. The paper argues that "effective use of water" (EUW) rather than WUE is the target of breeding for water-limited environments. That paper is reproduced here in full with permission.

3.5.1 Effective Use of Water (EUW) and Not WUE Is the Important Driver of Yield Under Drought Stress

The full paper by A. Blum is reprinted here from Field Crops Res 112:119–123 with permission. Format compatible with Field Crops Research requirement.

Abstract Water-use efficiency (WUE) is often considered as an important determinant of yield under stress and even as a component of crop drought resistance. It has been used to imply that rainfed plant production can be increased per unit water used, namely resulting in "more crop per drop."

This opinionated review argues that selection for high WUE in breeding for water limited conditions will most likely lead under most conditions to reduced yield and reduced drought resistance. As long as the biochemistry of photosynthesis can not be improved genetically, greater genotypic transpiration efficiency (TE) and WUE are driven mainly by plant traits that reduce transpiration and crop water-use, processes which are crucially important for plant production. Since biomass production is tightly linked to transpiration, breeding for maximized soil moisture capture for transpiration is a most important target for yield improvement under drought stress. Effective use of water (EUW) implies maximized soil moisture capture for transpiration which also involves reduced non-stomatal transpiration and minimized water loss by soil evaporation. Even osmotic adjustment which is a major stress adaptive trait in crop plants is recognized to enhance soil moisture capture and transpiration. High harvest index (HI) expresses successful plant reproduction and yield in terms of the reproductive functions and assimilate partitioning towards reproduction. In most rainfed environments crop water deficit develops during the reproductive growth stage thus reducing HI. EUW by way of improving plant water status helps sustain assimilate partitions and reproductive success. It is concluded that EUW is a major target for yield improvement in water limited environments. It is not a coincidence that EUW is an inverse acronym of WUE because very often high WUE is achieves at the expense of reduced EUW.

Keywords Water-use efficiency • Drought resistance • Transpiration • Stomata • Grain yield • Harrvest index • Plant breeding • Soil moisture • Roots.

The coined slogan "more crop per drop" (Kijne et al. 2003) as a target for crop improvement in water-limited environments emerged in recent years in the press and among research administrators and sponsors. It is a very catchy slogan indeed, but also a misleading one. It does not serve well the cause of breeding for water limited environments and especially rainfed conditions. It led breeders to believe that crop production under water-limited conditions can be genetically improved by increasing plant production per given amount of water used by the crop. A misconception also developed that improved water-use efficiency (WUE) is synonymous with drought resistance and high yield under drought stress. It is possible to achieve "more crop per drop" by certain crop and soil management practices used in dryland farming. However, this review concentrates on genetic improvement.

This paper is therefore designed to clarify the meaning and consequences of WUE if used in breeding either in practice or in concept. It raises the argument that the important determinant of plant production under most conditions of limited water supply is the effective use of water (EUW) and not high WUE.
Passioura (1996) proposed to view grain yield as a partial function of WUE:

$$Y = WU \times WUE \times HI \quad (1)$$

Where Y is grain yield, WU is water-use, and HI is harvest index.

This expression is still quite popular among breeders and agronomists since it is simple and has some educational merit. However, it should not be taken simplistically at face value. It should be fully understood in terms of the underlying crop physiology as well as its mathematics. It implies that WUE is an independent variable in affecting grain yield. In this expression WUE equals B/WU, where B is biomass, therefore:

$$Y = WU \times (B/WU) \times HI \quad (2)$$

In the above expression WU can be cancelled out and then we return to the basics (Donald and Hamblin, 1976), namely:

$$Y = B \times HI \quad (3)$$

De Wit (1958) set the cornerstone for relating plant production to water-use:

$$B = mT/E_0 \quad (4)$$

where B is crop biomass, m is a crop constant, T is crop transpiration and E_0 is free water (potential) evaporation. This relationship is very solid and it stood in the test of time. For genotypic comparisons E_0 is common and can be removed from this expression. Thus for genotypic comparisons, biomass of a genotype is a function of crop transpiration and a crop constant which is independent of T. A comparison of m constants received in various experiments (Hanks 1983) confirmed their fair consistency from one species to the other. It was, however, noted that the m constant may vary with varieties within a species. It was pointed out by Fischer and Turner (1978) that m diverged significantly between C_3 and C_4 species, indicating the evident importance of the photosystem in determining m. Since the basic biochemistry of photosynthesis has not yet been genetically improved within a crop species (Horton, 2000; Parry et al. 2007; Sheehy et al. 2000), m offers limited scope for significant genetic improvement in biomass. The remaining practical and quantitatively effective option for biomass improvement under drought stress is greater T or its seasonal manipulation with respect to plant development and a predictable water supply (see below).

Transpiration efficiency (TE) which is WUE at the leaf level is determined by the delicate interplay between transient photosystem activity, substomatal cavity CO_2 concentration and stomatal activity (Farquhar et al. 1989). Carbon isotope discrimination (delta) measurement in stover or grain has been developed as a method for estimating seasonal plant TE as affected by these transients. Low delta is reasonably correlated with high TE (e.g. Hall et al. 1994). Delta was therefore taken to represent crop WUE (e.g. Condon et al. 2002).

The amount of water transpired per given unit of CO_2 fixation (TE) is an interesting physiological yardstick but our interest here is in maximizing CO_2 fixation under drought stress per plant and per unit land area as an engine of plant production. Stomatal closure, which is a generally negative response in this respect (not considering survival under severe drought), can be driven by a variety of determinants such as internal leaf CO_2 concentration, cellular solutes, specific ions, pH and ABA produced in the leaf or imported from the root. Higher yielding genotypes of cotton, wheat and rice have greater stomatal conductance and transpiration under drought stress (Blum et al. 1982; Izanloo et al. 2008; Sanguineti et al. 1999) or well-watered conditions (Fischer et al. 1998; Horie et al. 2006; Lu et al. 1994; Lu and Zeiger 1994; Reynolds et al. 1994; Shimshi and Ephrat 1975) as indicated by leaf gas exchange measurements or canopy temperature measurements. Therefore high plant production requires high stomatal conductance over time, to allow greater CO2 fixation per unit land area, under different conditions. This translates into maximized soil water use for transpiration. In most cases as will be seen further on, higher T and high stomatal conductance under drought stress will express lower TE or WUE.

The delicate interplay between transient photosystem activity, substomatal cavity CO_2 concentration and stomatal activity is not the only determinant of plant transpiration and WU at the whole plant and crop level. Since drought stress is the issue here, interest is in plant water use for stomatal transpiration at given soil moisture content. Various plant constitutive and stress-adaptive traits have a role in this respect, namely in enhancing an effective use of water (EUW) for transpiration.

Synchronizing growth duration with the expected or the predicted seasonal soil moisture supply is often the first and foremost step in breeding for water limited environments. Two major considerations are important in this respect: (a) short growth duration dictates moderate water-use and the escape of terminal (reproductive stage) drought stress; and (b) long duration genotypes generally have a greater water use and larger and deeper root system that allow deep soil moisture extraction – if indeed deep soil moisture is available (Mitchell et al. 1996).

Leaf permeability is crucial. Leaves can loose water through the cuticle. This non-stomatal "leak" increases crop transpiration without an associated benefit in CO_2 fixation. For example, rice has a highly conductive cuticle while the drought resistant sorghum has relatively impermeable cuticle. Eliminating this leak by higher epicuticular wax deposition will increase stomatal transpiration (Kerstiens 1997, 2006). Stomata leakiness at night can also add to water loss without any advantage in CO2 assimilation. Nocturnal transpiration can be significant (Caird et al. 2007).

Accelerated leaf desiccation and death is a mean by which plants reduce water requirement under drought stress. Plant foliage desiccates in a progressive manner from lower (older) leaves to upper (younger) leaves, despite the fact that lower leaves are generally at a better water status than upper leaves. It has been found in sorghum that this strategy reduce transpiration in older and less productive leaves while deviating water use to younger and fully productive ones (Blum and Arkin 1984).

An important part of the available soil moisture for transpiration is evaporated directly from soil to the atmosphere without taking part in transpiration. For example, up to 40% of the total available soil water was found to be lost by soil evaporation in wheat in Australia (French and Schultz 1984; Siddique et al. 1990). Soil surface shading by the crop canopy is crucial for reducing this water loss. Reduced soil evaporation by fast vigorous seedling growth is therefore a target in the Australian wheat breeding program (Rebetzke and Richards 1999) which is actually directed at increasing T.

The major plant adaptive response to drought at the cellular level which has a proven effect on yield under drought stress is osmotic adjustment (OA) (Blum 2005). It has been repeatedly demonstrated that OA has two major functions towards plant production under drought stress: (a) It enables leaf turgor maintenance for the same leaf water potential thus supporting stomatal conductance (e.g. Ali et al. 1999; Sellin 2001), and (b) it improves root capacity for water uptake (e.g. Chimenti et al. 2006; Tangpremsri et al. 1991).

Therefore the enhancement of biomass production under drought stress can be achieved primarily by maximizing soil water capture while diverting the largest part of the available soil moisture towards stomatal transpiration. This is defined as effective use of water (EUW) and it is the major engine for agronomic or genetic enhancement of crop production under limited water regime. The consequence of EUW towards WUE is practically irrelevant.

There is a wide consensus that the reproductive growth stage is the most sensitive to water deficit. This is well depicted by the classic and widely used example in rice (O'Toole 1982; Fig. 6). It is also recognized that drought stress at the reproductive stage is the most prevalent problem in rainfed drought prone agriculture, at least simply because in most rainfed ecosystem the crop season's rains diminish towards flowering and harvest time. Therefore, irrespective of biomass production up to flowering, sustained WU and T into the reproductive growth stage is crucial for reproductive success (e.g. Merah 2001; Kato et al. 2008). An effective mean to achieve reproductive success under drought stress is soil moisture capture by deep root system where deep soil moisture is available (e.g. Kirkegaard et al. 2007). ABA production in the shoot or the root under stress may also impede reproductive processes (e.g. Davies and Jones 1991), but ABA accumulation might be at least partly repressed by higher WU and the resultant improvement of plant water status (Westgate et al. 1996).

Ample information has been developed by carbon isotope discrimination analysis of plants in order to understand the relationship between TE, WUE and yield under different water-regimes (e.g. Hall et al. 1994). It should be noted here that delta estimates TE at the leaf level which is not crop WUE. However in most studies when delta (TE) is found to be related to yield then the term WUE is used in the report. Following are some major conclusions which serve to reject the notion that high WUE can be equated with drought resistance or the improvement of yield under water limited conditions.

Conflicting results were obtained in various crops under different growing conditions on the association between delta and yield ((Hall et al. 1994; Matus et al. 1996; Monneveux et al. 2007; Morgan et al. 1993; Munoz et al. 1998; Ngugi et al. 1994; Ngugi et al. 1996; Read et al. 1991; Saranga et al. 2004; Sayre et al. 1995; Specht et al. 2001). These range from no relationship between delta and yield to negative or positive relationships, depending on the crop and the environment. Sometimes the relationship was biased by phenology or plant height (or perhaps HI) (e.g. Rebetzke et al. 2008; Sayre et al. 1995). It appears that a solid association between WUE and plant production as a biologically legitimate variable across species and environments is impossible. If solid relationships were found they were limited to narrow environmental conditions within a given crop phenology. In this sense WUE in rainfed agriculture is an elusive index. On the other hand its value in economizing the use of expensive irrigation water is well established.

Deep or dense root system which would promote soil moisture capture and WU is correlated across genotypes with low WUE (Pinheiro et al. 2005). Kobata et al. (1996) concluded that "the high dry matter production of those rice cultivars known to be drought resistant under field conditions is caused not by high WUE, but by high ability to maintain transpiration, which is supported by deep root systems."

Thus, it is not surprising that favorable genotypic plant water status under drought stress as reflected in measurements of relative water content or canopy temperature is correlated with low WUE across genotypes (Araus et al. 1993; Frank et al. 1997; Read et al. 1991; Zong et al. 2008).

Genotypic variation in WUE under limited water regime is affected more by variation in the denominator (WU) rather than by variation in the nominator (biomass) (Blum 2005). This has also been determined for TE and stomatal conductance at the single leaf level (e.g., Juenger et al. 2005; Monclus et al. 2006, Monneveux et al. 2006). Hence, selection for high WUE under limited water supply tend to result in a genetic shift towards plant traits that limit WU, such as early flowering and smaller leaf area

(Martin et al. 1999; Menendez and Hall 1995; Ngugi et al. 1994; Sayre et al. 1995; White et al. 1990). The successful and widely cited case for dryland wheat grain yield improvement with selection for high WUE (low carbon isotope discrimination) in NSW Australia (Condon et al. 2002) can be explained by the fact that wheat is grown there mainly on stored soil moisture. A major avenue for yield improvement is the control of WU during the earlier part of the growing season in order to avoid lack of soil moisture during reproduction. This was earlier attempted by selection for reduced root xylem diameter (Richards and Passioura 1989) and it can also be achieved by reduced leaf area and growth duration as done in the past in sorghum grown under stored soil moisture conditions (Blum 1970; 1972; Blum and Naveh 1976). Such plants that allow optimized seasonal distribution of soil moisture use express high WUE for grain yield due to their relative moderate WU and high HI. The same genetic materials, selected for high WUE were not successful in Western Australia where rainfed wheat does not grow on stored soil moisture (Fig. 6 in Condon et al. 2002).

It is therefore not surprising that drought resistance was found to be associated with low WUE when analyzed by delta under limited water supply (e.g. Araus et al. 2003; Morgan et al. 1993; Ngugi et al. 1994; Solomon and Labuschagne 2004). A drought resistant Coffea canephora clone had relatively lower WUE than a drought susceptible one, where resistance was associated with deeper roots and presumably greater WU (Pinheiro et al. 2005).

Finally, crop WUE has long been known to increase with increasing drought stress and reduced water supply (e.g. Meyers et al. 1984). This has been more recently confirmed with delta analysis (Craufurd et al. 1999; Ismail et al. 1994; Li et al. 2000; Peuke et al. 2006). It corresponds well with the fact that plant water deficit result in high WUE (#3 above). Assume therefore that two different cultivars are planted side by side and exposed to drought stress. If the one with higher WUE is selected it will most likely to be the one relatively more stressed and at a lower plant water status, namely the drought susceptible one.

Therefore, for all practical purposes plant breeders targeting water-limited environments should consider skipping the use and reference to WUE and consider plant constitutive and adaptive traits which drive the effective use of water (EUW) and the resultant dehydration avoidance as major traits for yield improvement in drought prone environments. This discussion does not refer to very shallow soils with very limited soil water holding capacity. These extremely difficult conditions require another discussion on plant survival and recovery and not plant production.

In conclusion, crop WUE as estimated under rainfed conditions by delta or any other method is an elusive ratio. Reynolds and Tuberosa (2008) concluded in tune with the original expression of Passioura (1996) that "water uptake (WU), water-use efficiency (WUE), and harvest index (HI) are drivers of yield." Indeed WU and HI are drivers of yield but I suggest that WUE is just a passenger. Whereas HI (in terms of assimilate partitioning and reproductive success under drought stress) is also largely influenced by WU and plant water status, it can be concluded that WU alone is the main (not the exclusive) driver of yield under drought stress.

Therefore, it is not a coincidence that EUW is an inverse acronym of WUE whereas very often high WUE is achieved at the expense of reduced EUW, and vice versa.

References

Ali, M., Jensen, C.R., Mogensen, V.O., Andersen, M.N., and Henson, I.E., 1999. Root signaling and osmotic adjustment during intermittent soil drying sustain grain yield of field grown wheat. Field Crops Res. 62, 35–52.

Araus, J.L., Reynolds, M.P., Acevedo, E., 1993. Leaf posture, grain yield, growth, leaf structure, and carbon isotope discrimination in wheat. Crop Sci. 33, 1273–1279.

Araus, J.L., Villegas, D., Aparicio, N., García Del Moral, L.F., El Hani, S., Rharrabti, Y., Ferrio, J.P., Royo, C., 2003. Environmental factors determining carbon isotope discrimination and yield in durum wheat under mediterranean conditions. Crop Sci. 43, 170–180.

Blum, A. 1970. Effects of plant density and growth duration on sorghum yield under limited water supply. Agron. J. 62, 333–336.

Blum, A., 1972. Effect of planting date on water-use and its efficiency in dryland grain sorghum. Agron. J. 64, 775–778.

Blum, A., 2005. Drought resistance, water-use efficiency, and yield potential – are they compatible, dissonant, or mutually exclusive? Aust. J. Agric. Res. 56, 1159–1168.

Blum, A., Arkin, G.F., 1984 Sorghum root growth and water-use as affected by water supply and growth duration. Field Crops Res. 9, 131–142.

Blum, A., Mayer, J., Gozlan, G., 1982. Infrared thermal sensing of plant canopies as a screening technique for dehydration avoidance in wheat. Field Crops Res. 5, 137–146.

Blum, A., Naveh, M., 1976. Improved water-use efficiency by promoted plant competition in dryland sorghum. Agron. J. 68, 111–116.

Caird, M.A., Richards, J.H., Hsiao, T.C., 2007. Significant transpirational water loss occurs throughout the night in field-grown tomato. Funct. Plant Biol. 34, 172–177.

Chimenti, C.A., Marcantonio, M., Hall, A.J., 2006. Divergent selection for osmotic adjustment results in improved drought tolerance in maize (Zea mays L.) in both early growth and flowering phases. Field Crops Res. 95, 305–315.

Condon, A.G., Richards, R.A., Rebetzke, G.J., Farquhar, G.D., 2002. Improving intrinsic water-use efficiency and crop yield. Crop Sci. 42, 122–131.

Craufurd, P.Q., Wheeler, T.R., Ellis, R.H., Summerfield, R.J., Williams, J.H., 1999. Effect of temperature and water deficit on water-use efficiency, carbon isotope discrimination and specific leaf area in peanut. Crop Sci. 39, 136–142.

Davies, W.J., Jones, H.G., 1991. Abscisic acid: Physiology and Biochemistry. Bios Scientific Publishers, London, pp. 266.

De Wit, C.T., 1958. Transpiration and crop yields, Versl. Landbouwk. Onderz., Institute of Biological and Chemical Research on Field Crops and Herbage, Wageningen, The Netherlands, 64:6.

Donald, C.M., Hamblin, J., 1976. The biological yield and harvest index of cereals as agronomic and plant breeding criteria. Adv. Agron. 28, 361–405.

Farquhar, G.D., Ehleringer, J.R., Hubick K., 1989. Carbon isotope discrimination and photosynthesis. Ann. Rev. Plant Physiol. Plant Mol. Biol. 40, 503–537.

Fischer, R.A., Turner, N.C., 1978. Plant productivity in the arid and semiarid zones, Ann. Rev. Plant Physiol., 29, 277–317).

Fischer, R.A., Rees, D., Sayre, K.D., Lu, Z.M., Condon, A.G., Saavedra, A.L., 1998. Wheat yield progress associated with higher stomatal conductance and photosynthetic rate, and cooler canopies. Crop Sci. 38, 1467–1475.

Frank, A.B., Ray, I.M., Berdahl, J.D., Karn, J.F., 1997. Carbon isotope discrimination, ash, and canopy temperature in three wheatgrass species. Crop Sci. 7, 1573–1576.

French, R.J., Schultz, J.E., 1984. Water use efficiency of wheat in a Mediterranean-type environment. I. The relation between yield water use and climate. Aust. J. Agric. Res. 35, 743–764.

Hall, A.E., Richards, R.A., Condon, A.G., Wright, G.C., Farquhar, G.D. 1994. Carbon isotope discrimination and plant breeding. Plant Breed. Rev. 12, 81–113.

Hanks, K.J., 1983. Yield and water-use relationships, an overview. In: Limitations to Efficient Water Use in Crop Production. Taylor, H.M., Jordan, W.R., Sinclair, T.R., (Eds), American Society of Agronomy, Madison, Wisconsin (USA), pp. 393–410.

Horie T., Matsuura S., Takai T., Kuwasaki K., Ohsumi A., Shiraiwa T. 2006. Genotypic difference in canopy diffusive conductance measured by a new remote-sensing method and its association with the difference in rice yield potential. Plant Cell Environ. 29, 653–660.

Horton, P., 2000. Prospects for crop improvement through the genetic manipulation of photosynthesis, morphological and biochemical aspects of light capture. J. Exp. Bot. 51, 475–485.

Ismail, A.M., Hall, A.E., Bray, E.A., 1994. Drought and pot size effects on transpiration efficiency and carbon isotope discrimination of cowpea accessions and hybrids Aust. J. Plant Physiol., 21, 23–35.

Izanloo, A., Condon, A.G., Langridge, P., Tester, M., Schnurbusch, T., 2008. Different mechanisms of adaptation to cyclic water stress in two South Australian bread wheat cultivars J. Exp. Bot. 59, 3327–3346.

Juenger, T.E., Mckay, J.K., Hausmann, N., Keurentjes, J. J. B., Sen, S., Stowe, K.A., Dawson, T.E., Simms, E. L., Richards, J.H., 2005. Identification and characterization of QTL underlying whole-plant physiology in Arabidopsis thaliana, 13C, stomatal conductance and transpiration efficiency. Plant Cell Environ. 28, 697–708.

Kato, Y., Kamoshita, A., Yamagishi, J., 2008. Preflowering abortion reduces spikelet number in upland rice (Oryza sativa L.) under water stress. Crop Sci. 48, 2389–2395.

Kerstiens, G., 1997. In vivo manipulation of cuticular water permeance and its effect on stomatal response to air humidity. New Phytol. 137, 473–480.

Kerstiens, G., 2006. Water transport in plant cuticles, an update. J. Exp. Bot. 57, 2493–2499.

Kijne, J. W., Barker, Randolph, Molden, D. J. (Eds), 2003. Water Productivity in Agriculture: Limits and Opportunities for Improvement, CABI, UK, 332 pp.

Kirkegaard, J.A., Lilley, J.M., Howe, G.N., Graham, J.M., 2007. Impact of subsoil water use on wheat yield. Aust. J. Agric. Res. 58, 303–315.

Kobata, T., Okuno, T., Yamamoto, T. 1996. Contributions of capacity for soil water extraction and water use efficiency to maintenance of dry matter production in rice subjected to drought. Japanese J. Crop Sci. 65, 652–662.

Li, C.Y., Berninger, F., Koskela, J., Sonninen, E., 2000. Drought responses of Eucalyptus microtheca provenances depend on seasonality of rainfall in their place of origin. Aust. J. Plant Physiol. 27, 231–238.

Lu, Z.M., Zeiger, E., 1994. Selection for higher yields and heat resistance in pima cotton has caused genetically determined changes in stomatal conductances. Physiol. Plant. 92, 273–278.

Lu, Z.M., Radin, J.W., Turcotte, E.L., Percy, R., Zeiger, E., 1994. High yields in advanced lines of pima cotton are associated with higher stomatal conductance, reduced leaf area and lower leaf temperature. Physiol. Plant. 92, 266–272.

Martin, B., Tauer, C.G., Lin, R.K., 1999. Carbon isotope discrimination as a tool to improve water-use efficiency in tomato. Crop Sci. 39, 1775–1783.

Matus, A., Slinkard, A.E., Vankessel, C., 1996. Carbon isotope discrimination and indirect selection for transpiration efficiency at flowering in lentil (Lens culinaris medikus), spring bread wheat (Triticum aestivum L.) durum wheat (T.turgidum l), and canola (Brassica napus L.). Euphytica 87, 141–151.

Menendez C.M. Hall W.E., 1995 Heritability of carbon isotope discrimination and correlations with earliness in cowpea. Crop Sci. 35, 673–678, 3032.

Merah, O., 2001. Potential importance of water status traits for durum wheat improvement under Mediterranean conditions. J. Agric. Sci. 137, 139–145.

Meyers, R.J.K., Foale, M.A., Done, A.A., 1984. Response of grain sorghum to varying irrigation frequency in the Ord irrigation area. II. Evapotranspiration water-use efficiency. Aust. J. Agric. Res. 35, 31–42.

Mitchell, J.H., Fukai, S., Cooper, M., 1996. Influence of phenology on grain yield variation among barley cultivars grown under terminal drought. Aust. J. Agric. Res. 47, 757–774.

Monclus, R., Dreyer, E., Villar, M., Delmotte, F.M., Delay D., Petit, J.M., Barbaroux, C., Thiec, D., Bréchet, C., Brignolas, F., 2006. Impact of drought on productivity and water use efficiency in 29 genotypes of Populus deltoides × Populus nigra. New Phytol. 169, 765–777.

Monneveux, P., Rekika, D., Acevedo, E., Merah, O., 2006. Effect of drought on leaf gas exchange, carbon isotope discrimination, transpiration efficiency and productivity in field grown durum wheat genotypes. Plant Sci. 170, 867–872.

Monneveux, P., Sheshshayee, M.S., Akhter, J., Ribaut, J.M., 2007. Using carbon isotope discrimination to select maize (Zea mays L.) inbred lines and hybrids for drought tolerance. Plant Sci. 173, 390–396.

Morgan, J.A., Lecain, D.R., Mccaig, T.N., Quick, J.S., 1993. Gas exchange, carbon isotope discrimination, and productivity in winter wheat. Crop Sci. 33, 178–186.

Munoz, P., Voltas, J., Araus, J.L., Igartua, E., Romagosa, I., 1998. Changes over time in the adaptation of barley releases in north-eastern Spain. Plant Breed. 117, 531–535.

Ngugi, E.C.K., Austin, R.B., Galwey, N.W., Hall, M.A., 1996. Associations between grain yield and carbon isotope discrimination in cowpea. Europ. J. Agron. 5, 9–17.

Ngugi, E.C.K., Galwey, N.W., Austin, R.B., 1994. Genotype x environment interaction in carbon isotope discrimination and seed yield in cowpea (Vigna unguiculata l. walp.). Euphytica 73, 213–224.

O'Toole, J.C., 1982. Adaptation of rice to drought-prone environments. In: Drought Resistance in Crops With Emphasis on Rice. IRRI (Eds), International Rice Research Institute, Los Banos, Phillippines, pp. 195–213.

Parry, M. A. J., Madgwick, P. J., Carvahlo, J. F. C., and Andralojc, P. J., 2007. Prospects for increasing photosynthesis by overcoming the limitations of Rubisco. J. Agric. Sci. 145, 31–43.

Passioura, J.B., 1996. Drought and drought tolerance. Plant Growth Reg.20, 79–83.

Peuke, A.D., Gessler, A., Rennenberg, H., 2006. The effect of drought on C and N stable isotopes in different fractions of leaves, stems and roots of sensitive and tolerant beech ecotypes. Plant Cell Environ. 29, 823–835.

Pinheiro, H.A., Damatta, F.M., Chaves, A.R.M., Loureiro, M.E., 2005. Drought tolerance is associated with rooting depth and stomatal control of water use in clones of Coffea canephora. Ann. Bot. 96, 101–108.

Read, J.J., Johnson, D.A., Asay, K.H., Tieszen, L.L., 1991. Carbon Isotope Discrimination, Gas Exchange, and Water-Use Efficiency in Crested Wheatgrass Clones. Crop Sci. 31, 1203–1208.

Rebetzke, G.J., Condon, A.G., Farquhar, G.D., Appels, R., Richards, R.A., 2008. Quantitative trait loci for carbon isotope discrimination are repeatable across environments and wheat mapping populations. Theor. Appl. Gen. 118, 123–137.

Rebetzke, G.J., Richards, R.A., 1999. Genetic improvement of early vigour in wheat. Aust. J. Agric. Res. 50, 291–301.

Reynolds, M.P., Balota, M., Delgado, M.I.B., Amani, I., Fischer, R.A., 1994. Physiological and morphological traits associated with spring wheat yield under hot, irrigated conditions. Aust. J. Plant Physiol. 21, 717–730.

Reynolds, M., Tuberosa, R. 2008. Translational research impacting on crop productivity in drought-prone environments Curr. Opin. Plant Biol. 11, 171–179.

Richards, R.A., Passioura, J.B., 1989. A breeding program to reduce the diameter of the major xylem vessel in the seminal roots of wheat and its effect on grain yield in rain-fed environments Aust. J. Agric. Res. 40, 943–950.

Sanguineti, M.C., Tuberosa, R., Landi, P., Salvi, S., Maccaferri, M., Casarini, E., Conti, S., 1999. QTL analysis of drought related traits and grain yield in relation to genetic variation for leaf abscisic acid concentration in field-grown maize. J. Exp. Bot. 50, 1289–1297.

Saranga, Y., Jiang, C.X., Wright, R.J., Yakir, D., Paterson, A.H., 2004. Genetic dissection of cotton physiological responses to arid conditions and their inter-relationships with productivity. Plant Cell Environ. 27, 263–277.

Sayre, K.D., Acevedo, E., Austin, R.B., 1995. Carbon isotope discrimination and grain yield for three bread wheat germplasm groups grown at different levels of water stress. Field Crops Res. 41, 45–54.

Sellin, A., 2001. Hydraulic and stomatal adjustment of Norway spruce trees to environmental stress. Tree Physiol. 21, 879–888.

Sheehy, J.E., Mitchell, P.L., Hardy, B, (eds.) 2000. Redesigning Rice Photosynthesis to Increase Yield. Elsevier Science, Amsterdam, (The Netherlands), pp. 300.

Shimshi, D., Ephrat, J., 1975. Stomatal behavior of wheat cultivars in relation to their transpiration, photosynthesis and yield. Agron. J. 67, 326–331.

Siddique, K.H.M., Tennan, t D., Perry, M.W., Belford, R.K., 1990. Water use and water use efficiency of old and modern wheat cultivars in a Mediterranean-type environment Aust. J. Agric. Res. 41, 431–447.

Solomon, K.F., Labuschagne, M.T., 2004. Variation in water use and transpiration efficiency among durum wheat genotypes grown under moisture stress and non-stress conditions. J. Agric. Sci. 141, 31–41.

Specht, J.E., Chase, K., Macrander, M., Graef, G.L., Chung, J., Markwell, J.P., Germann, M., Orf, J.H., Lark, K.G., 2001. Soybean response to water; a QTL analysis of drought tolerance. Crop Sci.41, 493–509.

Tangpremsri, T., Fukai, S., Fischer, K.S., Henzell, R.G., 1991. Genotypic variation in osmotic adjustment in grain sorghum.2. Relation with some growth attributes. Aust. J. Agric. Res. 42, 759–767.

Westgate, M.E., Passioura, J.B., Munns, R., 1996. Water status and ABA content of floral organs in drought-stressed wheat. Aust. J. Plant Physiol. 23, 763–772.

White, J.W., Castillo, J.A., Ehleringer, J., 1990. Associations between productivity, root growth and carbon isotope discrimination in Phaseolus-vulgaris under water deficit. Aust. J. Plant Physiol.17, 189–198.

Zong, Lin Zhu, Liang, Suo, Xu, Xing, Li, Shu Hua, Jing, Ji Hai, Monneveux P., 2008. Relationships between carbon isotope discrimination and leaf morpho-physiological traits in spring-planted spring wheat under drought and salinity stress in Northern China. Aust. J. Agric. Res. 59, 941–949.

After the paper by Blum (2009) was published, Condon et al. (2009) explained why WUE (as measured by carbon isotope discrimination) often had no relations or negative relations to the yield of different wheat genotypes under drought stress. In the same conference Rebetzke et al. (2009) demonstrated very well for wheat that low canopy temperature which is a physiological marker for dehydration avoidance and plant production was negatively associated with WUE. These arguments lend further support to the conclusions made by Blum (2009) that high WUE is a poor selection index for plant production in most rainfed water limited environments.

In this context it is also very interesting to note that C_4 species (e.g., maize, sorghum) are known to have high WUE than C_3 species (e.g., wheat, rice), by the token of their different CO_2 assimilation metabolism. A study of C_3 and C_4 *Panicoid* grasses indicated that the C_3 species were generally more drought resistant than the C_4 species in terms of maintaining photosynthesis under stress and in terms of recovery after stress (Ripley et al. 2010). Still, this interesting case is not necessarily the rule.

Barbour et al. (2010) found that several barley cultivars differed significantly in mesophyll conductance, where low conductance was linked to high transpiration ratio as measured by carbon isotope discrimination. According to their results and interpretation increased mesophyll conductance will increase photosynthesis with no effect on transpiration. This is therefore a unique case for an increase in WUE at the leaf level based on increased photosynthesis rather than on decreased transpiration. They caution that mesophyll conductance is a dynamic trait. Differences between genotypes could result from variation in a number of leaf traits, including the surface area of chloroplasts exposed to the intercellular spaces, the mesophyll cell wall thickness, and the permeability of plasma and chloroplastic membranes. However their convincing study points at an important potential avenue for the genetic improvement of plant production. Ramifications at the crop and field level as well as under drought stress condition are a challenge for future research.

The heritability of delta is generally high (Hall et al. 1994; Condon et al. 2004) especially when measured at late plant growth stages. It was noted for wheat that heritability was reduced when the analysis was delayed into the final growth stages. QTLs for delta were repeatable across environments and wheat mapping populations (Rebetzke et al. 2008a, b). However, it was also stated in the conclusion of that study that care must be taken to avoid confounding genotypic differences in delta with plant height and phenology in environments experiencing terminal drought. Polygenic control and small size of individual QTLs for delta could reduce the potential for delta QTLs in marker-assisted selection for improved yield of wheat.

Since the cost of carbon isotope discrimination assay is relatively high, attempts were made to correlate delta with simpler and cheaper plant measurements. Delta has been correlated with specific leaf weight (SLW) (Rajabi et al. 2008) and leaf ash content (Masle et al. 1993; Mian et al. 1996). However, the underlying reasons for these relationships and their consistency across environments and populations are not well resolved to the extent that these associations are solid enough for use in breeding.

3.5.1.1 Oxygen Isotope Enrichment as a Physiological Marker for EUW

As explained in the benchmark paper by Barbour (2007), the absolute isotope composition of a material is difficult to measure directly, so isotope ratios are generally compared with that of a standard. In the case of $^{18}O/^{16}O$ ratio, the standard is commonly Vienna-Standard Mean Oceanic Water (VSMOW), with an isotope ratio of 2.0052×10^{-3}.

Plant isotope compositions are expressed as relative deviations from VSMOW, and denoted $\Delta\delta^{18}O_p$ where $\delta^{18}O_p = R_p/R_{st} - 1$, and R_p and R_{st} are the isotope ratios of the plant tissues and the standard, respectively. Variation in the isotope composition of source (soil water in this case) may be removed from $\Delta\delta^{18}O_p$ by presenting the composition as an enrichment above source water: $\Delta^{18}Op = R_p/R_s - 1$, where R_s is the $^{18}O/^{16}O$ ratio of source water. $\Delta^{18}O_p$ may be calculated from $\delta^{18}O_p$ by:

$$^{18}\Delta_p = \frac{\delta^{18}\Delta_s - \delta^{18}\Delta_p}{1 + \delta^{18}\Delta_s}.$$

Oxygen isotope enrichment ($\Delta^{18}O_p$) is associated with plant transpiration and stomatal conductance especially under well watered conditions. According to theory, enrichment is proportional to the amount of water passing through the plant via transpiration, as controlled by stomatal conductance. Since stomatal conductance is correlated with crop yield (Sect. 2.7), it is expected that oxygen isotope enrichment as measured towards the end of plant growth in leaves or grain will be associated with yield. This was generally verified for irrigated conditions (Barbour 2007). This association and the fact that the assay of oxygen isotope enrichment in plant samples is not as expensive as the assay of carbon isotope discrimination make this analysis a very attractive proposition for estimating EUW as a major integrated indication of dehydration avoidance. Oxygen isotope enrichment might replace short-term measurements of stomatal conductance or canopy temperature as estimates of transpiration.

However, Barbour (2007) cautioned that the association between oxygen isotope enrichment and transpiration or stomatal conductance may not be as pronounced under drought stress as it is under well watered conditions and that the association with yield may even be less significant at times. There is some evidence that these associations might depend on the specific crop and the specific nature of the stress environment and the response of the crop to stress. For example, in case of effective substantial use of stem reserves for grain filling under drought stress, $\Delta^{18}O_p$ is not likely to be in good association with yield. Ferrio et al. (2007) did not find a significant association between $\Delta^{18}O_p$ and yield of wheat cultivars under water limited conditions. They concluded that $\Delta^{18}O_p$ of grains is not a proper physiological trait to breed for suboptimal water conditions, as its variability is almost entirely determined by crop phenology. In maize (Cabrera-Bosquet et al. 2009a) kernel $\Delta^{18}O_p$ correlated negatively with grain yield under well-watered and intermediate water stress conditions, while it correlated positively under severe water stress conditions. In another study with maize, Cabrera-Bosquet et al. (2009b) found that genotypic variation in grain yield was mainly explained by the combination of ash content and $\Delta^{18}O_p$. It seems that more research is needed verify the potential of $\Delta^{18}O_p$ as a physiological marker for EUW and yield under drought stress.

In conclusion, selection for high or low WUE (by use of carbon isotope discrimination analysis of plant material) should be considered with great care and understanding, always remembering that WUE is not a physiological determinant but a result of several plant processes. Never select for WUE if you do not understand the consequences for yield in the selection environment and the target environment. While in irrigation farming saving water is an important consideration and WUE is a common yardstick for that purpose, dryland rainfed farming is about maximizing the effective use of all the water available to the crop via stomatal transpiration. In most of not all cases of dryland/rainfed conditions this is likely to result in low WUE. *Therefore, when all published data on the subject are taken together, even the controversial ones, it can be concluded that high WUE (as expressed by low carbon isotope discrimination in plant materials or seed) is generally positively related to yield under well-watered conditions while it is negatively related to yield under most drought conditions, with exceptions.* Therefore, irrespective of the high cost of carbon isotope discrimination analysis or any correlations which might exist between delta and other plant traits, carbon isotope discrimination analysis might be considered as useful for phenotyping drought resistance if it is used as a surrogate for *low* WUE under pre-determined conditions. However, at this time it is impossible to assure a global definition of the water-limited conditions where this method will be effective. Each breeder will probably have to research the method with his own crop, his own selection environment and the target environment of his program before deciding on its adoption or rejection. This is a transient conclusion depending on further and future research.

A method for estimating EUW by analyzing leaf or grain for oxygen isotope enrichment is emerging. It might take the same long path of carbon isotope discrimination method evaluation before conclusive opinions on its value towards breeding will be reached.

3.6 Summary of Plant Constitutive Traits Controlling Drought Resistance

The recent literature dealing with the molecular biology of drought resistance is very much concerned with stress responsive genes, stress signaling and genes dynamically involved with adaptation to the developing stress. However, at the same time the approach to drought resistance from the field and whole plant physiology as discussed above indicates that plant constitutive traits which do not require stress responsive genes for their expression under drought stress have a major impact on drought resistance and sustained plant production under drought stress. Certainly these traits are responsive to stress, but they are controlled by genes which do not require stress for their expression. For example, OA is a classical drought responsive and adaptive trait, unless it has been genetically engineered to be expressed constitutively. On the other hand, large storage of stem reserves is a constitutive trait that can support dehydration tolerance in terms of grain filling. At the same time reserve mobilization to the grain can be partially constitutive and partially stress responsive.

A brief summary of these constitutive traits is therefore warranted because breeding for these traits does not require selection under drought stress, except sometimes for reasons of validization.

1. *Drought escape* as conditioned by early flowering is a constitutive trait which can and most likely should be selected for under potential conditions.
2. *Plant size* which exercise control over plant and crop water use is an important trait under certain drought conditions. Traits which control plant water use and may serve to moderate use involve *tillering* in the cereals, *leaf area* and *growth duration*, all of which can be selected for under normal growing conditions.
3. *Basic root architecture* is a constitutive trait. While root development and architecture is highly plastic in response to the soil environment and its water status, the major features of the root such as its potential size and *maximum length* can

be selected without requiring exposure to drought. This is especially true when selection is based on comparisons among phenotypes within a given population.

4. Certain morphological features of plants were found to be relevant towards plant resistance to drought or heat and these can be selected for under any conditions. *Awns* in the cereals can contribute significantly to ear assimilation under stress (Tambussi et al. 2007) and this contribution becomes more significant when leaves are dysfunctional under stress. Certain constitutive *leaf forms* (Stiller et al. 2004) and leaf anatomical features (Bacelar et al. 2004) can affect plant water use and performances under drought. Even *leaf color* as determined by the potential content of leaf chlorophyll have been suggested as important for leaf performance under drought and heat stress (Watanabe et al. 1995; Sanchez et al. 2001).
5. *Seedling* and early growth vigor in wheat and barley might be important for plant water status under drought stress since faster ground cover reduces soil surface evaporation and increase the amount of soil moisture available for transpiration (Rebetzke and Richards 1999).

Most constitutive plant traits which are important for drought resistance are mainly involved with dehydration avoidance and EUW.

3.7 The Drought Resistant Ideotype

The term "ideotype" is sometimes considered as a novel concept in plant breeding. The ideotype describes the desirable plant phenotype which determines its adaptation, productivity and consumer demand. However, this is a very old concept. The first human being who selected a non-shattering wild emmer as a first step towards agriculture already selected an ideotype. Even the local rice farmer in Mali has a defined ideotype of a rice plant which he seems to prefer (Efisue et al. 2008). In some respects the theory and practice of farmers participation in modern breeding programs (Ceccarelli and Grando 2007) is based on the farmer's preferred ideotype.

Donald (1968) was the first to introduce the physiological "ideotype" concept into plant breeding based on the understanding of how yield is being formed.

Drought resistant ideotypes were already published as a matter of concept and information, such as the case for CIMMYT wheat (e.g., Fig. 3.11). These however were not working ideotypes directed at specific target environments but rather a compilation of the most important traits towards a conceptual ideotype. This early stage of ideotype formulation is important since it concentrates on the important plant traits out of a large maze of published information on what different authors consider important or not. It helps to plan the breeding program. In maize, the short ASI under drought stress has become a specific dominant ideotype in breeding for stress conditions during flowering especially in CIMMYT work in Africa (Bänziger et al. 2006).

The conceptual ideotype cannot be applied to breeding without linking it to the specific target environment. These environmentally-linked ideotypes are a result of

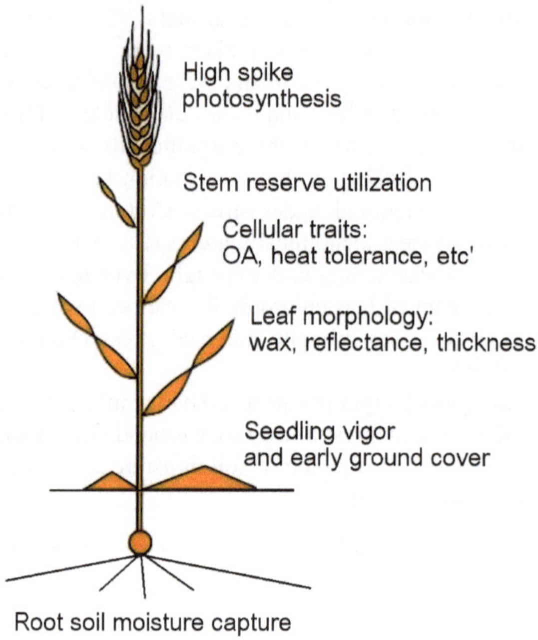

Fig. 3.11 The CIMMYT drought resistant wheat ideotype (After Reynolds et al. (2006))

models – not necessarily crop simulation models but also conceptual models based on the collective experience and formal knowledge in plant breeding. Common sense is no less important in the design than data and knowledge. Crop simulation models, statistical approach, and geographical information systems are beginning to help in designing a desirable ideotype. It is reassuring that most of these tools seem to verify what experienced breeders already know.

Four major realistic drought stress scenarios with respect to the design of a drought resistant ideotype can be considered:

1. Soil moisture is always available at depth.
2. Water is available at depth but soil is hard and offers large resistance to root penetration.
3. Soil moisture is stored in the profile from a previous rainy season, without or with very little effective seasonal rainfall.
4. Water is not available at depth and only the shallow soil layer undergoes a seasonal wetting-drying cycles typical of arid zone precipitation regime.

Superimposed over these drought scenarios there are four major plant growth stages when drought stress can occur:

Stage A: Seedling development stage.
Stage B: Vegetative growth stage.
Stage C: Flowering and fruit/seed set stage.
Stage D: Fruit/grain development stage.

All four by four combinations will not be discussed here. Only the principles involved with each will be discussed here and the reader should be able to form his own strategy based on the relevant situation he faces.

3.7.1 The Ideotype with Respect to Drought Stress Scenarios

3.7.1.1 Soil Moisture Is Always Available at Depth

Deep soil moisture with or without a fluctuating water table is common to many cropping environments. Deep roots are therefore essential in capturing this moisture as discussed above. For obvious reasons much research has been done with rice root development since in many rainfed rice environments deep soil moisture is available while rice is inherently limited in deep root growth.

Deep root growth is achieved towards the end of the growing season and deep soil moisture (or subsoil moisture) can be used mainly towards flowering and maturity. At the same time this ideotype might be subjected to drought stress at the seedling or the vegetative growth stages, so that additional pertinent design of protection might be required. As already discussed, in the cereals tillering and root development interact. A major cause of this interaction is developmental while part of it might be genetic. Considering this developmental aspect, limited tillering at the vegetative stage can enhance deeper root development especially when the top soil dries out.

This drought stress profile should remind the breeder of a common pitfall concerning roots and their role in field selection work. It is crucial to ascertain that the cropping sequence and resultant soil conditions in the breeder's field would not produce a drought stress scenario different from what one would expect in the target environment. For example, let us assume that the target environment is mainly of type #4 above. Assume that in the breeder's selection field a previously irrigated crop or a high water-table result in available deep soil moisture (type#1 above). Such a situation would provide a selective advantage to deep rooted genotypes which have no value in the target environment.

3.7.1.2 Water Is Available at Depth but Soil Is Hard and Offers Large Resistance to Root Penetration

The capacity for root growth into deep soil is not the same as the capacity for root penetration of hard soil or a soil barrier. As already discussed above these are different traits. There is very limited experience in selection for root penetration. Although selection for root penetration can be done with simulated barriers such as wax-petrolatum layer it is not quite clear to what extent superior penetration of wax-petrolatum layer is universally related to real capacity in the field.

As with other abiotic stress problems it should be remembered that sometimes breeding can offer only limited solutions. Problem solving might require a combination of breeding and management approach as typical of salinity problems for example. In the case of root penetration it has long been known that the problem can be amended by deep tillage (ripping), tractor traffic control and deep placement of suitable organic or inorganic materials.

Hard soil resistance to root penetration is only one example of a host of factors which can constrain root growth into deep soils despite the fact the genotype has the capacity to grow long roots. Biotic factors such as soil born diseases, nematodes, etc. should be considered as well as soil mineral toxicities such as aluminum toxicity. Thus, when deep rooted genotypes which were selected in the greenhouse do not express their advantage under deep soil moisture conditions in the field, a soil problem might be suspected.

3.7.1.3 Stored Soil Moisture with Limited or No Seasonal Rainfall

This scenario is found in parts of the Mediterranean region with summer crops such as sunflower, sorghum, maize and dryland cotton. It is found in parts of Eastern Australia where winter crops such as wheat are grown. It is found in parts of India rabi (winter) season where various crops are grown. An important part of the global dryland farming land is under this drought scenario.

There is a certain advantage to this drought scenario as compared with other severe drought environments. Whereas the total amount of soil moisture available for the crop season is basically known at planting time, the expected drought stress is more or less predictable, pending the seasonal evapotranspirational demand which is often also predictable. Therefore management practices can be planned ahead and the appropriate cultivar can be chosen to fit the expected moisture regime. In sorghum under these conditions, for example, planting date, plant density, planting geometrical arrangement and cultivar phenology could be developed, adjusted and selected in order to maximize yield and effectively use the given amount of soil moisture (Blum 1970, 1972; Blum and Naveh 1974). A major consideration in crop management and breeding under these conditions is to reduce crop soil moisture use during the first half of the season in order to conserve sufficient moisture for the reproductive growth stage. Reduced plant size and early flowering provide an appropriate ideotype. Measurement of the residual soil moisture at depth after harvest can indicate if breeding for deeper roots is required in the specific case.

3.7.1.4 Water Is Not Available at Depth

This is the most difficult scenario to consider in breeding and crop management since it is unpredictable in terms of when drought will occur and at what intensity. Long-range weather data analysis might help to develop probabilities of certain drought scenarios to occur. Total seasonal rainfall and the amount per rainfall event are important data. Residual water in the soil from previous rainy season should also be considered in relations to crop management, such as the case is for fallow systems. Soil type in terms of water storage capacity is important. When all is considered it may be possible to foresee a certain pattern of water supply and probabilities for drought stress during the crop season. However, this environment can only be approached in terms of probabilities. The ideotype will then comprise of resistance at the most probable growth stages to be affected, from seedling to grain/fruit growth. It can be seen that survival and recovery from one rainfall event to the other in this environment can be important if this happens during the early vegetative stage when the plant still has active meristems. As the case was discussed above, maintenance of RWC at any given leaf water potential during stress is the main condition for continued growth or successful recovery from stress. Osmotic adjustment (OA) is therefore a crucial adaptive trait that will allow maintenance of RWC and superior dehydration avoidance. While survival as a drought resistance trait is often rejected as irrelevant in crop production it should be pointed out that seedling survival under extreme desiccation (Chap. 4, section "The Seedling Wilting Test") is linked to cowpea yield performance under drought and it is part of the cowpea breeding program in the Sahel region (Agbicodo et al. 2009).

Recovery is an agronomic option as long as the delay caused by drought does not place the crop maturity beyond the normal season into cold or other abiotic or biotic problems. The skeptics should be aware of the fact that recovery has already been put to use in normal agronomic practice as exemplified in winter wheat. In parts of the USA Midwest region, dryland winter wheat is sometimes grazed by cattle just after tillering. The crop is then allowed to recover and produce grain yield which is not seriously below the un-grazed crop. Grazing damage at this stage is not very different from drought stress in terms of the required recovery processes. The main difference is in drought stress being prolonged. However in wheat (Blum et al. 1990) (and most other crops) recovery is always better if stress develops earlier during the vegetative growth stage.

The rate of soil moisture extraction to a given low soil moisture content or soil water potential is not a set value. Variation exists in this respect among crops and probably among genotypes. Root length density in a given soil volume is one factor controlling extraction while root and shoot osmotic properties could be an additional factor. The breeder selecting plants in a drying soil should realize that greater soil moisture extraction by roots to lower soil water potential can be an important asset for this drought scenario.

3.7.2 *The Ideotype with Respect to Timing of Stress*

In a global survey performed among plant breeders engaged in breeding for water limited conditions, the frequency of drought stress at different plant growth stages of various crops was explored. As can be seen from Fig. 3.12 drought stress at the seedling and early vegetative growth stage is relatively the least frequent in these target environments. The frequency and impact of drought increases as plants enter the reproductive growth stages.

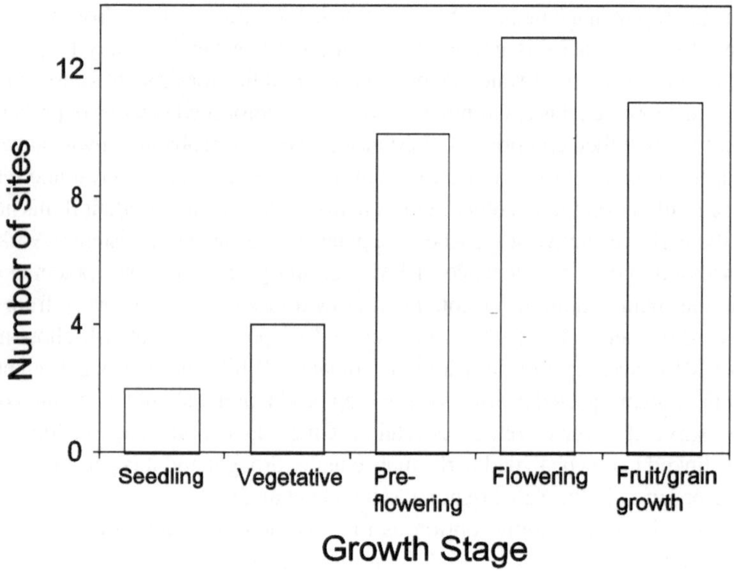

Fig. 3.12 The frequency and importance of drought stress at different plant growth stages as estimated by breeders working in various geographical sites where breeding programs are concerned with drought (Mahalakshmi and Blum 2008)

3.7.2.1 Stress at Seedling Development Stage

This is a relatively less common stress period since in most cases the season begins with sufficient rainfall. However, in certain cases emerged seedlings are subjected to a lag in rains. Seedling survival is then an important attribute. The capacity for seed germination under moisture stress is not related to the survival capacity of the autotrophic seedling (Blum et al. 1980). Significant variation in seedling survival has been recorded in crops that are often subjected to stress at this stage, such as wheat (Tomar and Kumar 2004), cowpea (Okosun et al. 1998) or range grasses (Johnson and Asay 1993). The most apparent basis for seedling survival is OA which allows retaining some hydration in plant tissues to lower plant water potential. Knowledge derived from work with resurrection plants might be useful in engineering better seedling tolerance (Sect. 5.5). It is considered that seedling survival is an opportune trait for genetic engineering without being subjected to some of the complexities involve with developed plants and reproductive processes. Success in engineering seedling survival is relatively easier to assess and prove. Constitutive accumulation of osmotically active solutes and other protective agents in seedlings seems to be a real possibility for designing better survival.

3.7.2.2 Stress at the Vegetative (Pre-flowering) Growth Stage

A superior ideotype at this growth stage should be able to continue growing and developing during drought stress and/or be capable of recovery after sever stress in order to continue growth upon rehydration.

At this growth stage the plant is growing actively while various vegetative and reproductive organs are being differentiated. The plant still carries active meristems and dormant primordia. This plant is therefore at a complex growth stage to analyze with respect to its various drought responses, but on the other hand it has the maximum developmental options for coping with stress and the recovery from stress.

The huge amount of publications on the subject covering many diverse crops plants and drought conditions can be confusing. At the end of the day it can be concluded that the main characteristic of importance is the plant capacity to maintain high plant water status during the drying cycle. Dehydration avoidance by way of maintaining plant water potential or by way of osmotic adjustment as reflected in RWC is the main attribute for this ideotype. How these traits can be incorporated into the ideotype has been discussed above under "dehydration avoidance" and in relations to the specific drought environment. Constitutive traits in support of dehydration avoidance are important. Variations in drought resistance at this stage can sometimes be expressed also at later plant growth stages. For example, delayed heading and delayed flowering in many crops subjected to drought stress express relative susceptibility to stress during the pre-flowering growth stage.

The sorghum breeding program at Lubbock Texas was extensively involved with breeding for drought resistance for a long time. Darrel Rosenow and his colleagues (e.g., Rosenow et al. 1996) selected in the field for post flowering drought resistance (expressed mainly in stay-green) and pre flowering drought resistance (expressed in minimal leaf rolling and less leaf desiccation). Understandably the two traits were independent of each other. Pre-flowering drought resistance was found to be associated with three QTLs in one study (Kebede et al. 2001). It is most interesting to note the statement made by this knowledgeable group that "More than 80% of commercial sorghum hybrids in the United States are grown under non-irrigated conditions and most of them have pre-flowering drought resistance" (Sanchez et al. 2002). This must be the result of continuous selection of hybrids and hybrid parent materials for drought resistance at this growth stage under dryland field conditions. Their work attends to the fact that vegetative growth stage drought resistance is achievable.

3.7.2.3 Stress at Flowering and at the Stage of Fruit/Seed Set

Flowering is widely recognized as the most drought sensitive plant growth stage. A drought resistant ideotype for this stage is therefore extremely important. In many dryland or rainfed cropping systems water normally becomes scarce towards the latter part of the season owing to reduced rainfall and/or depleted soil moisture. Breeding for upland rainfed rice for example is strongly directed at an ideotype resistant to flowering stage drought; similarly to winter cereals in regions of the Mediterranean and many temperate climates. Drought resistance at the reproductive stage has been discussed above. It can be a function of both dehydration avoidance and dehydration tolerance affecting the reproductive proficiency of ovaries and pollen as well as minimizing early abortion of flowers, zygotes or fruits. Drought escape is a common solution, placing this stage under conditions of more favorable plant water supply.

If the exact cause of reproductive failure under drought stress is well established, say high ABA accumulation causing distinct sterility symptoms, perhaps a laboratory bioassay for pollen function in response to ABA can be used as a screen (Frascaroli and Tuberosa 2006).

Selection for this stage resistance can be done by using a managed stress environment and simply observing various phenotypic expressions of fertility dysfunctions which are normally very obvious. The problem with this technique is that the breeding population must flower within a very short time span of few days. Wider span of flowering time would result in many escapes from the designated drought stress. Because of the difficulties in such field phenotyping work, marker assisted selection is a sought after as a desirable solution. Success stories in this respect were discussed above, as the case was for rice (section "Dehydration Tolerant Reproduction").

3.7.2.4 Stress at Fruit or Grain Development Stage

Being the last stage of plant development, fruit and grain formation is very often subjected to drought stress. Besides drought escape an ideotype suitable for coping with this stress might be dehydration avoidant by the token of deep soil moisture extraction where deep soil moisture is available. Alternatively, a constitutively limited crop water use prior to flowering would provide some remaining moisture for the fruit/grain developmental stage as seen in certain drought resistant millets in India (Kholov et al. 2010). The ideotype might also be designed by one of two mutually exclusive traits: non-senescence (stay-green) and stem reserve utilization for grain filling. The two options and the considerations for deploying one or the other depend on the crop, on other stresses such as stem diseases, etc, as discussed above (section "Stem Reserve Utilization for Grain Filling"). It should be considered that sufficient storage of stem reserves for grain filling requires reasonably favorable growing conditions during the pre-flowering stages.

Irrespective of non-senescence and the capacity for grain filling from stem reserves, wheat and barley were shown to have an additional source of assimilates for grain filling which is the inflorescence and the awns. The cereal inflorescence as a whole and especially the awns may remain photosynthetically active under drought and heat stress when leaves (and especially the flag leaf) are rendered inactive by dehydration. It has also been found that CO_2 released by ear and grain respiration is re-fixed by the green parts of the inflorescence and in quantitative terms it can have an effect on yield (Tanbussi et al. 2007). Thus, awned small-grain cultivars are very typical of dryland ecosystems.

Beyond the above-described framework of ideotypes specific to environments and plant growth stages the breeder might recognize additional specific or general drought adaptive ideotypes. One example is the drought resistant legume as based on drought resistant rhizobium symbiosis, which is important throughout the plant's life and under most drought stress scenarios. The improvement of root nodule function in a drying soil can have a huge impact on pulses and legumes productivity in rainfed farming. There is ample knowledge on the physiology of rhizobium-plant interaction but very little use of this knowledge in breeding for water limited environments, to date. Genetic variation for the trait is constantly reported but approach to and design of breeding programs for the trait is not adequately developed.

Finally one should always remember that the drought resistant ideotype is only one feature of the general and complete crop ideotype. The final ideotype for release requires all the additional traits which are required by the grower and consumer.

References

Abraham EM, Huang B, Bonos SA et al (2004) Evaluation of drought resistance for Texas bluegrass, Kentucky bluegrass, and their hybrids. Crop Sci 44:1746–1753

Abreu ME, Munné-Bosch S (2008) Salicylic acid may be involved in the regulation of drought-induced leaf senescence in perennials: a case study in field-grown *Salvia officinalis* L. *plants*. Environ Exp Bot 64:105–112

Alpert P (2000) The discovery, scope, and puzzle of desiccation tolerance in plants. Plant Ecol 151:5–17

Al-Yassin A, Grando S, Kafawin O et al (2005) Heritability estimates in contrasting environments as influenced by the adaptation level of barley germplasm. Ann Appl Biol 147:235–244

Andersen MN, Asch F, Wu Y et al (2002) Soluble invertase expression is an early target of drought stress during the critical, abortion-sensitive phase of young ovary development in maize. Plant Physiol 130:591–604

Anderson SR, Lauer MJ, Schoper JB et al (2004) Pollination timing effects on kernel set and silk receptivity in four maize hybrids. Crop Sci 44:464–473

Angadi SV, Entz MH (2002) Root system and water use patterns of different height sunflower cultivars. Agron J 94:136–145

Apel K, Hirt H (2004) Reactive oxygen species: metabolism, oxidative stress, and signal transduction. Ann Rev Plant Biol 55:373–399

Araus JL, Sánchez C, Cabrera-Bosquet L (2010) Is heterosis in maize mediated through better water use? New Phytol 187:392–406

Athar M, Johnson DA (1996) Nodulation biomass production and nitrogen fixation in alfalfa under drought. J Plant Nutr 19:185–199

Atkin OK, Macherel D (2009) The crucial role of plant mitochondria in orchestrating drought tolerance. Ann Bot 103:581–597

Auge RM (2004) Arbuscular mycorrhizae and soil/plant water relations. Can J Soil Sci 84:373–381

Babu RC, Shashidhar HE, Lilley JM et al (2001) Variation in root penetration ability, osmotic adjustment and dehydration toleance among accessions of rice adapted to rainfed lowland and upland ecosystem. Plant Breed 120:233–238

Babu RC, Zhang J, Blum A et al (2004) HVA1, a LEA gene from barley confers dehydration tolerance in transgenic rice (*Oryza sativa* L) via cell membrane protection. Plant Sci 166:855–862

Bacelar EA, Correia CM, Moutinho-Pereira JM (2004) Sclerophylly and leaf anatomical traits of five field-grown olive cultivars growing under drought conditions. Tree Physiol 24:2 33–239

Badawi GH, Yamauchi Y, Shimada E et al (2004) Enhanced tolerance to salt stress and water deficit by overexpressing superoxide dismutase in tobacco (*Nicotiana tabacum*) chloroplasts. Plant Sci 166:919–928

Bajji M, Kinet JM, Lutts S (2002) The use of the electrolyte leakage method for assessing cell membrane stability as a water stress tolerance test in durum wheat. Plant Growth Regul 36:61–70

Bakken AK, Macduff J, Humphreys M (1997) A stay-green mutation of *Lolium perenne* affects NO_3-uptake and translocation of N during prolonged n starvation. New Phytol 135:41–50

Baldocchi OO, Verma SB, Kosenberg NJ et al (1983) Leaf pubescence effects on the mass and energy exchange between soybean canopies and the atmosphere. Agron J 75:537–541

Bancal MO, Robert C, Ney B (2007) Modelling wheat growth and yield losses from late epidemics of foliar diseases using loss of green leaf area per layer and pre-anthesis reserves. Ann Bot 100:777–789

Bandurska H (1998) Implication of ABA and proline on cell membrane injury of water deficit stressed barley seedlings. Acta Physiol Plant 20:375–381

Bänziger M, Setimela PS, Hodson D et al (2006) Breeding for improved drought tolerance in maize adapted to Southern Africa. Agric Water Manag 80:212–224

Barbour MM (2007) Stable oxygen isotope composition of plant tissue: a review. Funct Plant Biol 34:83–94

Barbour MM, Warren CR, Farquhar GD et al (2010) Variability in mesophyll conductance between barley genotypes, and effects on transpiration efficiency and carbon isotope discrimination. Plant Cell Environ 33:1176–1185

Barker T, Campos H, Cooper M et al (2005) Improving drought tolerance in maize. Plant Breed Rev 25:173–253

Basnayake J, Ludlow M, Cooper M et al (1993) Genotypic variation of osmotic adjustment and desiccation tolerance in contrasting sorghum inbred lines. Field Crops Res 35:51–62

Basnayake J, Cooper M, Ludlow MM et al (1995) Inheritance of osmotic adjustment to water stress in three grain sorghum crosses. Theor Appl Genet 90:675–682

Basnayake J, Cooper M, Henzell RG et al (1996) Influence of rate of development of water deficit on the expression of maximum osmotic adjustment and desiccation tolerance in three grain sorghum lines. Field Crops Res 49:65–76

Basu PS, Berger JD, Turner NC et al (2007) Osmotic adjustment of chickpea (*Cicer arietinum*) is not associated with changes in carbohydrate composition or leaf gas exchange under drought. Ann Appl Biol 150:217–225

Becana M, Dalton DA, Moran JF et al (2001) Reactive oxygen species and antioxidants in legume nodules. Physiol Plant 109:372–381

Beckett RP (2001) ABA-induced tolerance to ion leakage during rehydration following desiccation in the moss *Atrichum androgynum*. Plant Growth Regul 35:131–135

Benjamin JG, Nielsen DC (2006) Water deficit effects on root distribution of soybean, field pea and chickpea. Field Crops Res 97:248–253

Bernier J, Kumar A, Ramaiah V et al (2007) A large-effect QTL for grain yield under reproductive-stage drought stress in upland rice. Crop Sci 47:507–516

Betrán FJ, Beck D, Bänziger M et al (2003a) Secondary traits in parental inbreds and hybrids under stress and non-stress environments in tropical maize. Field Crops Res 83:51–65

Betrán FJ, Beck D, Bänziger M et al (2003b) Genetic analysis of inbred and hybrid grain yield under stress and nonstress environments in tropical maize. Crop Sci 43:807–817

Bewley JO (1979) Physiological aspects of desiccation tolerance. Ann Rev Plant Physiol 30:195–205

Bidinger FR, Serraj R, Rizvi SMH et al (2005) Field evaluation of drought tolerance QTL effects on phenotype and adaptation in pearl millet [*Pennisetum glaucum* (L) R Br] topcross hybrids. Field Crops Res 94:14–32

Bidinger FR, Nepolean T, Hash CT et al (2007) Quantitative trait loci for grain yield in pearl millet under variable postflowering moisture conditions. Crop Sci 47:969–980

Blum A (1970) Effects of plant density and growth duration on sorghum yield under limited water supply Agron J 62:333–336

Blum A (1972) Effect of planting date on water-use and its efficiency in dryland grain sorghum. Agron J 64:775–778

Blum A (1973) Components analysis of yield responses to drought of sorghum hybrids. Exp Agric 9:159–170

Blum A (1988) Plant breeding for stress environments. CRC Press, Boca Raton

Blum A (1997) Constitutive traits affecting plant performance under drought stress. In: Edmeades GO, Banziger M, Mickelson HR et al (eds) Developing drought and low N tolerant maize. CIMMYT, El Batan

Blum A (1998) Improving wheat grain filling under stress by stem reserve mobilization. Euphytica 100:77–83
Blum A (2004) Sorghum physiology. In: Nguyen HT, Blum A (eds) Physiology and biotechnology integration for plant breeding. CRC Press, Boca Raton
Blum A (2005) Drought resistance, water-use efficiency, and yield potential – are they compatible, dissonant, or mutually exclusive? Aust J Agric Res 56:1159–1168
Blum A (2009) Effective use of water (EUW) and not water-use efficiency (WUE) is the target of crop yield improvement under drought stress. Field Crops Res 112:119–123
Blum A, Arkin GF (1984) Sorghum root growth and water-use as affected by water supply and growth duration. Field Crops Res 9:131–142
Blum A, Ebercon A (1981) Cell membrane stability as a measure of drought and heat tolerance in wheat. Crop Sci 21:43–47
Blum A, Naveh M (1976) Improved water-use efficiency by promoted plant competition in dryland sorghum. Agron J 68:111–116
Blum A, Pnuel Y (1990) Physiological attributes associated with drought resistance of wheat cultivars in a Mediterranean environment. Aust J Agric Res 41:799–810
Blum A, Sinmena B (1995) Isolation and characterization of variant wheat cultivars for ABA sensitivity. Plant Cell Environ 18:77–78
Blum A, Sullivan CY (1986) The comparative drought resistance of landraces of sorghum and millet from dry and humid regions. Ann Bot 57:835–846
Blum A, Arkin GF, Jordan WR (1977a) Sorghum root morphogenesis and growth. I. Effect of maturity genes. Crop Sci 17:149–153
Blum A, Jordan WR, Arkin GF (1977b) Sorghum root morpho-genesis and growth. II. Manifestation of heterosis. Crop Sci 17:153–157
Blum A, Sinmena B, Ziv O (1980) An evaluation of seed and seedling drought tolerance screening tests in wheat. Euphytica 29:727–736
Blum A, Gozlan G, Mayer J (1981) The manifestation of dehydration avoidance in wheat breeding germplasm. Crop Sci 21:495–499
Blum A, Mayer J, Gozlan G (1982) Infrared thermal sensing of plant canopies as a screening technique for dehydration avoidance in wheat. Field Crops Res 5:137–146
Blum A, Golan G, Mayer J et al (1989) The drought response of landraces of wheat from the Northern Negev desert in Israel. Euphytica 43:87–96
Blum A, Ramaiah S, Kanemasu ET et al (1990) Recovery of wheat from drought stress at the tillering developmental stage. Field Crops Res 24:67–85
Blum A, Sinmena B, Mayer J et al (1994) Stem reserve mobilisation supports wheat grain filling under heat stress. Aust J Plant Physiol 21:771–781
Blum A, Munns R, Passioura JB et al (1996) Genetically engineered plants resistant to soil drying and salt stress: how to interpret osmotic relations? Plant Physiol 110:1051
Blum A, Golan G, Mayer J et al (1997) The effect of dwarfing genes on sorghum grain filling from remobilized stem reserves, under stress. Field Crops Res 52:43–54
Blum A, Zhang JX, Nguyen HT (1999) Consistent differences among wheat cultivars in osmotic adjustment and their relationship to plant production. Field Crops Res 64:287–291
Blum A, Klueva N, Nguyen HT (2001) Wheat cellular thermotolerance is related to yield under heat stress. Euphytica 117:117–123
Bohnert HJ, Shen B (1999) Transformation and compatible solutes. Sci Hort 78:237–260
Bolanos J, Edmeades GO (1996) The importance of the anthesis-silking interval in breeding for drought tolerance in tropical maize. Field Crops Res 48:65–80
Bonnett GD, Incoll LD (1992) Effects on the stem of winter barley of manipulating the source and sink during grain-filling. 1. Changes in accumulation and loss of mass from internodes. J Exp Bot 44:75–82
Borrell AK, Hammer GL (2000) Nitrogen dynamics and the physiological basis of stay-green in sorghum. Crop Sci 40:1295–1307
Borrell AK, Incoll LD, Dalling MJ (1993) The influence of the rht1 and rht2 alleles on the deposition and use of stem reserves in wheat Ann Bot 71:317–326

Borrell AK, Hammer GL, Henzell RG (2000) Does maintaining green leaf area in sorghum improve yield under drought? II. Dry matter production and yield. Crop Sci 40:1037–1048
Boyer JS (1976) Photosynthesis at low potentials. Philos Trans R Soc Lond Ser B 273:501–511
Boyer JS, Westgate ME (2004) Grain yields with limited water. J Exp Bot 55:2385–2394
Boyer JS, Johnson RR, Saupe SG (1980) Afternoon water deficits and grain yields in old and new soybean cultivars. Agron J 72:981–986
Buchanan-Wollaston V (1997) The molecular biology of leaf senescence. J Exp Bot 48:181–199
Busscher WJ, Lipiec J, Bauer PJ et al (2000) Improved root penetration of soil hard layers by a selected genotype. Commun Soil Sci Plant Anal 31:3089–3101
Cabrera-Bosquet L, Sánchez C, Araus JL (2009a) Oxygen isotope enrichment reflects yield potential and drought resistance in maize. Plant Cell Environ 32:1487–1499
Cabrera-Bosquet L, Sanchez C, Araus JL (2009b) How yield relates to ash content, ^{13}C and ^{18}O in maize grown under different water regimes. Ann Bot 104:1207–1216
Campos H, Cooper M, Habben JE et al (2004) Improving drought tolerance in maize: a view from industry. Field Crops Res 90:19–34
Carleton AH, Foote WH (1968) Heterosis for grain yield and leaf area and their components in two six-rowed barley crosses. Crop Sci 8:554–560
Castiglioni P, Warner D, Bensen RJ et al (2008) Bacterial RNA chaperones confer abiotic stress tolerance in plants and improved grain yield in maize under water-limited conditions. Plant Physiol 147:446–455
Castleberry CW, Crum CW, Krull CF (1984) Genetic yield improvement of US maize cultivars under varying fertility and climatic environments. Crop Sci 24:33–37
Ceccarelli S (1987) Yield potential and drought tolerance of segregating populations of barley in contrasting environments. Euphytica 36:265–273
Ceccarelli S (1989) Wide adaptation: how wide? Euphytica 40:197–205
Ceccarelli S, Grando S (1991) Environment of selection and type of germplasm in barley breeding for low-yielding conditions. Euphytica 57:207–219
Ceccarelli S, Grando S (2007) Decentralized-participatory plant breeding: an example of demand driven research. Euphytica 155:349–360
Ceccarelli S, Grando S, Impiglia A (1998) Choice of selection strategy in breeding barley for stress environments. Euphytica 10:307–318
Cha KW, Lee YJ, Koh HJ et al (2002) Isolation, characterization, and mapping of the stay green mutant in rice. Theor Appl Genet 104:526–532
Chapman SC, Edmeades GO (1999) Selection improves drought tolerance in tropical maize populations. II. Direct and correlated responses among secondary traits. Crop Sci 39:1315–1324
Chimenti CA, Marcantonio M, Hall AJ (2006) Divergent selection for osmotic adjustment results in improved drought tolerance in maize (*Zea mays* L) in both early growth and flowering phases. Field Crop Res 95:305–315
Christopher JT, Manschadi AM, Hammer GL et al (2008) Developmental and physiological traits associated with high yield and stay-green phenotype in wheat. Aust J Agric Res 59:354–364
Clark LJ, Whalley WR, Barraclough PB (2003) How do roots penetrate strong soil? Plant Soil 255:93–104
Clarke JM, McCaig TN, DePauw RM (1994) Inheritance of glaucousness and epicuticular wax in durum wheat. Crop Sci 34:327–331
Cochard H, Casella E, Mencuccini M (2007) Xylem vulnerability to cavitation varies among poplar and willow clones and correlates with yield. Tree Physiol 27:1761–1767
Collins NC, Tardieu F, Tuberosa R (2008) Quantitative trait loci and crop performance under abiotic stress: where do we stand? Plant Physiol 147:469–486
Condon AG, Richards RA, Rebetzke GJ (2004) Breeding for high water-use efficiency. J Exp Bot 55:2447–2460
Condon AG, Kirkegaard JA, Rebetzke GJ (2009) Wheat yield and water use: what do they have to do with carbon isotope discrimination? In: Interdrought-III, Shanghai (in press)
Cooper M, Hammer GL (eds) (1996) Plant adaptation and crop improvement. CABI, Oxon
Cox TS, Shroyer JP, Liu B-H et al (1988) Genetic improvement in agronomic traits of hard red winter wheat cultivars from 1919 to 1987. Crop Sci 28:756–760

Dahlberg JA (2000) Collection, conversion, and utilization of sorghum. In: Smith CW, Frederiksen RA (eds) Sorghum: origin, history, technology and production. Wiley, New York
Degenkolbe T, Do P, Zuther et al (2008) Expression profiling of rice cultivars differing in their tolerance to long-term drought stress. Plant Mol Biol 69:133–153
Derouw A, Winkel T (1998) Drought avoidance by asynchronous flowering in pearl millet stands cultivated on-farm and on-station in Niger. Exp Agric 34:19–39
Dodd JL (1979) Grain sink size and predisposition of *Zea mays* to stalk rot. Phytopathology 70:534–535
Du WJ, Fu SX, Yu DY (2009) Genetic analysis for the leaf pubescence density and water status traits in soybean [*Glycine max* (L) Merr] Plant Breed 128:259–265
Duvick DN (1997) What is yield In: Edmeades GO, Banziger M, Mickelson HR et al (eds) Developing drought and low-N tolerant maize. CIMMYT, El Batan
Eapen D, Barroso ML, Ponce G et al (2005) Hydrotropism: root growth responses to water. Trends Plant Sci 10:44–50
Edmeades GO, Bolanos J, Chapman SC et al (1999) Selection improves drought tolerance in tropical maize populations. I. Gains in biomass, grain yield, and harvest index. Crop Sci 39:1306–1315
Efisue A, Tongoona P, Derera J et al (2008) Farmers' perceptions on rice varieties in sikasso region of mali and their implications for rice breeding. J Agron Crop Sci 194:393–400
Erickson PI, Ketring DL (1985) Evaluation of peanut genotypes for resistance to water stress in situ. Crop Sci 25:870–876
Fan X-W, Li F-M, Song L et al (2009) Defense strategy of old and modern spring wheat varieties during soil drying. Physiol Plant 136:310–323
Farquhar GD, Ehleringer JR, Hubick K (1989) Carbon isotope discrimination and photosynthesis. Ann Rev Plant Physiol Plant Mol Biol 40:503–537
Farrant JM, Kruger LA (2001) Longevity of dry *Myrothamnus flabellifolius* in simulated field conditions. Plant Growth Regul 35:109–120
Farrant JM, Cooper K, Kruger et al (1999) The effect of drying rate on the survival of three desiccation-tolerant angiosperm species. Ann Bot 84:371–379
Febrero A, Fernandez SM, Cano JL et al (1998) Yield, carbon isotope discrimination, canopy reflectance and cuticular conductance of barley isolines of differing glaucousness. J Exp Bot 49:1575–158
Fellows KJ, Boyer JS (1978) Altered ultrastructure of cells of sunflower leaves having low water potentials. Protoplasma 93:381–386
Ferrio JP, Mateo MA, Bort J et al (2007) Relationships of grain δ13C and δ18O with wheat phenology and yield under water-limited conditions. Ann Bot 150:207–215
Fischer RA, Wood JT (1979) Drought resistance in spring wheat cultivars. III. Yield associations with morpho-physiological traits. Aust J Agric Res 30:1001–1010
Fischer RA, Rees D, Sayre KD et al (1998) Wheat yield progress associated with higher stomatal conductance and photosynthetic rate, and cooler canopies. Crop Sci 38:1467–1475
Fischer KS, Lafitte R, Fukai S (eds) (2003) Breeding rice for drought-prone environments. International Rice Research Institute, Los Baños
Flexas J, Medrano H (2002) Drought-inhibition of photosynthesis in C_3 plants: stomatal and non-stomatal limitations revisited. Ann Bot 89:183–189
Flexas J, Bota J, Loreto F et al (2004) Diffusive and metabolic limitations to photosynthesis under drought and salinity in C_3 plants. Plant Biol (Stuttg) 6:269–279
Flower DJ, Ludlow MM (1986) contribution of osmotic adjustment to the dehydration tolerance of water-stressed pigeonpea (*Cajanus cajan* [L] millsp) leaves. Plant Cell Environ 9:33–40
Fokar M, Blum A, Nguyen HT (1998) Heat tolerance in spring wheat. II. Grain filling. Euphytica 104:9–15
Foulkes MJ, Scott RK, Sylvester-Bradley R (2002) The ability of wheat cultivars to withstand drought in UK conditions: formation of grain yield. J Agric Sci 138:153–169
Foulkes MJ, Sylvester-Bradley R, Weightman R et al (2007) Identifying physiological traits associated with improved drought resistance in winter wheat. Field Crops Res 103:11–24
Frahm MA, Rosas JC, Mayek-Perez N et al (2004) Breeding beans for resistance to terminal drought in the Lowland tropics. Euphytica 136:223–232

Fukai S, Pantuwan G, Jongdee B et al (1999) Screening for drought resistance in rainfed lowland rice. Field Crops Res 64:61–74

Galle A, Feller U (2007) Changes of photosynthetic traits in beech saplings (*Fagus sylvatica*) under severe drought stress and during recovery. Physiol Plant 131:412–421

Gan S, Amasino RM (1995) Inhibition of leaf senescence by autoregulated production of cytokinin. Science 270:1986–1988

Giese BN (1976) Roles of the cer-j and cer-p loci in determining the epicuticular wax composition on barley seedling leaves. Hereditas 82:137–148

Giuliani S, Sanguineti MC, Tuberosa R et al (2005) Root-ABA1, a major constitutive QTL, affects maize root architecture and leaf ABA concentration at different water regimes. J Exp Bot 56:3061–3070

Gonzalez A, Ayerbe L (2010) Effect of terminal water stress on leaf epicuticular wax load, residual transpiration and grain yield in barley. Euphytica 172:341–349

Gonzalez A, Martin I, Ayerbe L (1999) Barley yield in water-stress conditions. The influence of precocity, osmotic adjustment and stomatal conductance. Field Crops Res 62:23–34

Gonzalez EM, Galvez L, Royuela M et al (2001) Insights into the regulation of nitrogen fixation in pea nodules: lessons from drought abscisic acid and increased photoassimilate availability. Agronomie 21:607–613

Hall AE, Richards RA, Condon AG et al (1994) Carbon isotope discrimination and plant breeding. Plant Breed Rev 12:81–113

Hammer G, Cooper M, Tardieu F et al (2006) Models for navigating biological complexity in breeding improved crop plants. Trends Plant Sci 11:587–593

Hammer G, Dong Z, McLean G et al (2009) Can changes in canopy and/or root system architecture explain historical maize yield trends in the U.S. corn belt? Crop Sci 49:299–312

Haque MM, Mackill DJ, Ingram KT (1992) Inheritance of leaf epicuticular wax content in rice. Crop Sci 32:865–868

Harris K, Subudhi PK, Borrell A et al (2007) Sorghum stay-green QTL individually reduce post-flowering drought-induced leaf senescence. J Exp Bot 58:327–338

Harvey HP, van den Driessche R (1997) Nutrition, xylem cavitation and drought resistance in hybrid poplar. Tree Physiol 17:647–654

Hauck B, Gay AP, Macduff J et al (1997) Leaf senescence in a non-yellowing mutant of festuca pratensis – implications of the stay-green mutation for photosynthesis, growth and nitrogen nutrition Plant. Cell Environ 20:1007–1018

Haussmann BIG, Obilana AB, Ayiecho PO et al (1999) Quantitative-genetic parameters of sorghum [*Sorghum bicolor* (L) Moench] grown in semi-arid areas of Kenya. Euphytica 105:109–118

Haussmann BIG, Mahalakshmi V, Reddy BVS et al (2003) QTL mapping of stay-green in two sorghum recombinant inbred populations. Theor Appl Genet 106:133–142

Holbrook FS, Welsh JR (1980) Soil water-use by semi-dwarf and tall wheat cultivars under dryland conditions. Crop Sci 20:244–247

Hooker TS, Millar AA, Kunst L (2002) Significance of the expression of the CER6 condensing enzyme for cuticular wax production in *Arabidopsis*. Plant Physiol 129:1568–1580

Horie T, Matsuura S, Takai T et al (2006) Genotypic difference in canopy diffusive conductance measured by a new remote-sensing method and its association with the difference in rice yield potential. Plant Cell Environ 29:653–660

Horner TW, Frey KJ (1957) Methods for determining natural areas for oat varietal recommendations. Agron J 49:313–315

Hsiao TC (1973) Plant responses to water stress. Ann Rev Plant Physiol 24:519–532

Huang BR, Fry J, Wang B (1998) Water relations and canopy characteristics of tall fescue cultivars during and after drought stress. HortScience 33:837–840

Huang Y, Xiao B, Xiong L (2007) Characterization of a stress responsive proteinase inhibitor gene with positive effect in improving drought resistance in rice. Planta 226:73–85

Hurd, EA (1974) Phenotype and drought tolerance in wheat. Agric Meteor 14:39–55

Hyoun Chin J, Lu X, Haefele SM et al (2010) Development and application of gene-based markers for the major rice QTL Phosphorus uptake. Theor Appl Genet 120:1073–1086

Innes P, Blackwell RD (1983) Some effects of leaf posture on yield and water economy of winter wheat. J Agric Sci Camb 101:367–376
Irvine RB, Harvey BL, Rossnagel BG (1980) Rooting capabilities as it relates to soil moisture extraction and osmotic potential of semi-dwarf and normal statured genotypes of six-rowed barley. Can J Plant Sci 60:241–248
Islam MA, Du H, Ning J et al (2009) Characterization of Glossy1-homologous genes in rice involved in leaf wax accumulation and drought resistance. Plant Mol Biol 70:443–456
Ito K, Tanakamaru K, Morita S et al (2006) Lateral root development, including responses to soil drying, of maize (*Zea mays*) and wheat (*Triticum aestivum*) seminal roots. Physiol Plant 127:260–267
Izanloo A, Condon AG, Langridge P et al (2008) Different mechanisms of adaptation to cyclic water stress in two South Australian bread wheat cultivars. J Exp Bot 59:3327–3346
James AT, Lawn RJ, Cooper M (2008) Genotypic variation for drought stress response traits in soybean. II. Inter-relations between epidermal conductance, osmotic potential, relative water content, and plant survival. Aust J Agric Res 59:670–678
Jefferson PG (1994) Genetic variation for epicuticular wax production in Altai wild rye populations that differ in glaucousness. Crop Sci 34:367–371
Jenkins MT (1932) Differential resistance of inbred and crossbred strains of corn to drought and heat injury. Agron J 24:504–506
Jenks MA, Hasegawa PM, Mohan Jain S (eds) (2007) Advances in molecular breeding towards drought and salt tolerant crops. Springer, Dordrecht
Jiang YW, Huang BR (2001) Physiological responses to heat stress alone or in combination with drought: a comparison between tall fescue and perennial ryegrass HortScience 36:682–686
Johnson DA, Asay KH (1993) Viewpoint – selection for improved drought response in cool-season grasses. J Range Manag 46:194–202
Johnson GR, Frey KJ (1967) Heritabilities of quantitative attributes of oat (*Avena* sp) at varying levels of environmental stress. Crop Sci 7:43–46
Johnson DA, Richards RA, Turner NC (1983) Yield water relations gas exchange and surface reflectance of near isogenic wheat lines differing in glaucousness. Crop Sci 23:318–321
Jones HG (1998) Stomatal control of photosynthesis and transpiration. J Exp Bot 49:387–398
Jordan WR, Monk RL, Miller FR et al (1983) Environmental physiology of sorghum. I. Environmental and genetic control of epicuticular wax load. Crop Sci 23:552–555
Jung C, Seo JS, Han SW et al (2008) Overexpression of AtMYB44 enhances stomatal closure to confer abiotic stress tolerance in transgenic *Arabidopsis*. Plant Physiol 146:623–635
Kebede H, Subudhi PK, Rosenow DT et al (2001) Quantitative trait loci influencing drought tolerance in grain sorghum (*Sorghum bicolor* L Moench). Theor Appl Genet 103:266–276
Kholov J, Hash CT, Kakkera A et al (2010) Constitutive water-conserving mechanisms are correlated with the terminal drought tolerance of pearl millet [*Pennisetum glaucum* (L)]. J Exp Bot 61:369–377
Kim KS, Park SH, Jenks MA (2007) Changes in leaf cuticular waxes of sesame (*Sesamum indicum* L) plants exposed to water deficit. J Plant Physiol 164:1134–1143
Kiniry JR (1993) Nonstructural carbohydrate utilization by wheat shaded during grain growth. Agron J 85:844–849
Koonjul PK, Minhas JS, Nunes C et al (2005) Selective transcriptional down-regulation of anther invertases precedes the failure of pollen development in water-stressed wheat. J Exp Bot 56:179–190
Kubo K, Jitsuyama Y, Iwama et al (2004) Genotypic difference in root penetration ability by durum wheat (*Triticum turgidum* L *var.* durum) evaluated by a pot with paraffin-Vaseline discs. Plant and Soil 262:169–177
Kubo K, Jitsuyama Y, Iwama K et al (2005) The reduced height genes do not affect the root penetration ability in wheat. Euphytica 141:105–111
Kuchel H, Williams K, Langridge P et al (2007) Genetic dissection of grain yield in bread wheat. II. QTL-by-environment interaction. Theor Appl Genet 115:1015–1027
Kuhbauch W, Thome U (1989) Nonstructural carbohydrates of wheat stems as influenced by sink-source manipulations. J Plant Physiol 134:243–250
Kumar R, Sarawgi AK, Ramos C et al (2006) Partitioning of dry matter during drought stress in rainfed lowland rice. Field Crops Res 96:455–465

Kumar R, Venuprasad R, Atlin GN (2007) Genetic analysis of rainfed lowland rice drought tolerance under naturally-occurring stress in eastern India: heritability and QTL effects. Field Crops Res 103:42–52

Lafitte HR, Courtois B (2002) Interpreting cultivar × environment interactions for yield in upland rice assigning value to drought-adaptive traits. Crop Sci 42:1409–1420

Lafitte HR, Edmeades GO, Johnson EC (1997) Temperature responses of tropical maize cultivars selected for broad adaptation. Field Crops Res 49:215–229

Lal S, Gulyani V, Khurana P (2008) Overexpression of HVA1 gene from barley generates tolerance to salinity and water stress in transgenic mulberry (*Morus indica*) Transgen Res 17:651–663

Lambert L, Beach RM, Kilen TC et al (1992) Soybean pubescence and its influence on larval development and oviposition preference of lepidopterous insects. Crop sci 32:463–466

Landi P, Sanguineti MC, Conti S et al (2001) Direct and correlated responses to divergent selection for leaf abscisic acid concentration in two maize populations. Crop Sci 41:335–344

Laporte MM, Shen B, Tarczynski MC (2002) Engineering for drought avoidance: expression of maize NADP-malic enzyme in tobacco results in altered stomatal function. J Exp Bot 53:699–705

Lascano HR, Antonicelli GE, Luna CM et al (2001) Antioxidant system response of different wheat cultivars under drought: field and in vitro studies. Aust J Plant Physiol 28:1095–1102

Leport L, Turner NC, French RJ et al (1999) Physiological responses of chickpea genotypes to terminal drought in a Mediterranean-type environment. Eur J Agron 11:279–291

Leport L, Turner NC, Davies SL et al (2006) Variation in pod production and abortion among chickpea cultivars under terminal drought. Eur J Agron 24:236–246

Levitt J (1980) Response of plants to environmental stresses water, radiation salt and other stresses. Academic, New York

Liang GHL, Heyne HG, Walter TL (1966) Estimates of variety × environment interactions in yield tests of three small grains and their significance on the breeding program. Crop Sci 6:135–139

Lilley JM, Fukai S (1994) Effect of timing and severity of water deficit on four diverse rice cultivars 1. Rooting pattern and soil water extraction. Field Crops Res 37:205–213

Lilley JM, Ludlow MM, Mccouch SR et al (1996) Locating qtl for osmotic adjustment and dehydration tolerance in rice. J Exp Bot 47:1427–1436

Liu F, Andersen MN, Jensen CR (2004a) Root signal controls pod growth in drought-stressed soybean during the critical, abortion-sensitive phase of pod development. Field Crops Res 85: 159–166

Liu F, Jensen CR, Andersen MN (2004b) Drought stress effect on carbohydrate concentration in soybean leaves and pods during early reproductive development: its implication in altering pod set. Field Crops Res 86:1–13

Liu JX, Liao DQ, Oane R et al (2006) Genetic variation in the sensitivity of anther dehiscence to drought stress in rice. Field Crops Res 96:87–100

Lopatecki LE, Longair EI, Kasting R (1962) Quantitative changes of soluble carbohydrates in stems of solid- and hollow- stemmed wheats during growth. Can J Bot 40:1223–1228

Lopezcastaneda C, Richards RA, Farquhar GD (1995) Variation in early vigor between wheat and barley. Crop Sci 35:472–479

Lu ZM, Radin JW, Turcotte EL et al (1994) High yields in advanced lines of pima cotton are associated with higher stomatal conductance, reduced leaf area and lower leaf temperature. Physiol Plant 92:266–272

Ludlow MM, and Bjorkman O (1984) Paraheliotropic leaf movement in *Sirato* as a protective mechanism against drought-induced damage to primary photosynthetic reactions: damage by excessive light and heat. Planta 161:505–510

Ludlow MM, Santamaria JM, Fukai S (1990) Contribution of osmotic adjustment to grain yield in *Sorghum-Bicolor* (L) Moench under water-limited conditions 2. Water stress after anthesis. Aust J Agric Res 41:67–78

Lynch J (1995) Root architecture and plant productivity. Plant Physiol 109:7–13

Lynch PJ, Frey KJ (1993) Genetic improvement in agronomic and physiological traits of oat since 1914. Crop Sci 33:984–988

Ma BL, Dwyer LM (1998) Nitrogen uptake and use of two contrasting maize hybrids differing in leaf senescence. Plant Soil 199:283–291

Maes B, Trethowan M, Reynolds MP et al (2001) The influence of glume pubescence on spikelet temperature of wheat under freezing conditions. Aust J Plant Physiol 28:141–148

Malinowski DP, Kigel J, Pinchak WE (2009) Water deficit, heat tolerance, and persistence of summer-dormant grasses in the US Southern Plains. Crop Sci 49:2363–2370

Manschadi AM, Christopher J, deVoil P et al (2006) The role of root architectural traits in adaptation of wheat to water-limited environments. Funct Plant Biol 33:823–837

Manschadi AM, Hammer GL, Christopher J et al (2008) Genotypic variation in seedling root architectural traits and implications for drought adaptation in wheat (*Triticum aestivum* L). Plant Soil 303:115–129

Maroco JP, Rodrigues ML, Lopes C et al (2002) Limitations to leaf photosynthesis in field-grown grapevine under drought – metabolic and modelling approaches. Funct Plant Biol 29:451–459

Martineau JR, Williams JH, Specht JE (1979) Temperature tolerance in soybeans II. Evaluation of segregating populations for membrane thermostability. Crop Sci 19:79–1979

Marulanda A, Barea J-M, Azcón R (2009) Stimulation of plant growth and drought tolerance by native microorganisms (AM fungi and bacteria) from dry environments: mechanisms related to bacterial effectiveness. J Plant Growth Regul 28:115–124

Masle J, Farquhar GD, Wong SC (1993) Transpiration ratio and plant mineral content are related among genotypes of a range of species. Aust J Plant Physiol 19:709–721

Mathews KL, Malosetti M, Chapman S et al (2008) Multi-environment QTL mixed models for drought stress adaptation in wheat. Theor Appl Genet 117:1077–1091

May OL, Kasperbauer MJ (1999) Genotypic variation for root penetration of a soil pan. J Sustain Agric 13:87–94

Mccaig TN, Morgan JA (1993) Root and shoot dry matter partitioning in near-isogenic wheat lines differing in height. Can J Plant Sci 73:679–689

Mckersie BD, Bowley SR, Harjanto E et al (1996) Water-deficit tolerance and field performance of transgenic alfalfa overexpressing superoxide dismutase. Plant Physiol 111:1177–1181

Mclaughlin JE, Boyer JS (2004) Sugar-responsive gene expression, invertase activity, and senescence in aborting maize ovaries at low water potentials. Ann Bot 94:675–689

Messmer R, Fracheboud Y, Bänziger M et al (2009) Drought stress and tropical maize: QTL-by-environment interactions and stability of QTLs across environments for yield components and secondary traits. Theor Appl Genet 119:913–930

Mian M, Bailey MA, Ashley DA et al (1996) Molecular markers associated with water use efficiency and leaf ash in soybean. Crop Sci 36:1252–1257

Miller G, Suzuki N, Ciftci-Yilmaz S et al (2010) Reactive oxygen species homeostasis and signaling during drought and salinity stresses. Plant Cell Environ 32:453–467

Miralles DJ, Slafer GA, Lynch V (1997) Rooting patterns in near-isogenic lines of spring wheat for dwarfism. Plant Soil 197:79–86

Mitchell JH, Fukai S, Cooper M (1996) Influence of phenology on grain yield variation among barley cultivars grown under terminal drought. Aust J Agric Res 47:757–774

Moinuddin KCR, Khanna-Chopra R (2004) Osmotic adjustment in chickpea in relation to seed yield and yield parameters. Crop Sci 44:449–455

Moinuddin KCR, Fischer RA, Sayre KD et al (2005) Osmotic adjustment in wheat in relation to grain yield under water deficit environments. Agron J 97:1062–1071

Morgan JM (1991) A gene controlling differences in osmoregulation in wheat. Aust J Plant Physiol 18:249–257

Morgan JM, Tan MK (1996) Chromosomal location of a wheat osmoregulation gene using RFLP analysis. Aust J Plant Physiol 23:803–806

Morgan JM, Hare RA, Fletcher RJ (1986) genetic variation in osmoregulation in bread and durum wheats and its relationship to grain yield in a range of field environments. Aust J Agric Res 37:449–457

Morgan JM, Rodriguezmaribona B, Knights EJ (1991) Adaptation to water-deficit in chickpea breeding lines by osmoregulation – relationship to grain yields in the field. Field Crops Res 27:61–70

Mungur R, Wood AJ, Lightfooot DA (2006) Water potential is maintained during water deficit in Nicotiana tabacum expressing the *Escherichia coli* glutamate dehydrogenase gene. Plant Growth Regul 50:231–238

Munns R (1988) Why measure osmotic adjustment? Aust J Plant Physiol 15:717–726

Munns R, Richards RA (2007) Recent advances in breeding wheat for drought and salt stresses. In: Jenks MA, Hasegawa PM, Jain SM (eds) Advances in molecular breeding toward drought and salt tolerant crops. Springer, Dordrecht

Muñoz-Perea CG, Terán H, Allen RG et al (2006) Selection for drought resistance in dry bean landraces and cultivars. Crop Sci 46:2111–2120

Nagel OW, Konings H, Lambers H (1994) Growth rate, plant development and water relations of the ABA-deficient tomato mutant *sitiens*. Physiol Plant 92:102–108

Nakashima K, Ito Y, Yamaguchi-Shinozaki K (2009) Transcriptional regulatory networks in response to abiotic stresses in *Arabidopsis* and grasses. Plant Physiol 149:88–95

Ney B, Duthion C, Turc O (1994) Phenological response of pea to water stress during reproductive development. Crop Sci 34:141–146

Nguyen HT, Babu RC, Blum A (1997) Breeding for drought resistance in rice – physiology and molecular genetics considerations. Crop Sci 37:1426–1434

Nielsen OC, Bind BL, Verma SB et al (1984) Influence of soybean pubescence type on radiation balance. Agron J 76:924–930

Nizam-Uddin M, Marshall DR (1988) Variation in epicuticular wax content in wheat. Euphytica 38:3–9

Norton MR, Volaire F, Lelievre F et al (2009) Identification and measurement of summer dormancy in temperate perennial grasses. Crop Sci 49:2347–2352

Okosun LA, Akenova ME, Singh BB (1998) Screening for drought tolerance at seedling stage in cowpea (*Vigna unguiculata* [L] Walp) II. Selecting for root length and recovery ability traits. J Arid Agric 8:11–20

Oliver SN, Dennis ES, Dolferus R (2007) ABA regulates apoplastic sugar transport and is a potential signal for cold-induced pollen sterility in rice. Plant Cell Physiol 48:1319–1330

O'Toole JC, Bland WL (1987) Genotypic variation in crop plant root systems. Adv Agron 41:91–145

O'Toole JC, Namuco OS (1983) Role of panicle exsertion in water stress induced sterility. Crop Sci 23:1093–1097

O'Toole JC, Hsiao TC, Namuco OS (1984) Panicle water relations during water stress. Plant Sci Lett 33:137–143

Ouk M, Basnayake J, Tsubo M et al (2007) Genotype-by-environment interactions for grain yield associated with water availability at flowering in rainfed lowland rice. Field Crops Res 101:145–154

Paleg LG, Stewart GR, Starr R (1985) The effect of compatible solutes on proteins. Plant Soil 89:83–94

Palta JA, Fillery IRP, Rebetzke GJ (2007) Restricted-tillering wheat does not lead to greater investment in roots and early nitrogen uptake. Field Crops Res 104:52–59

Pandy S, Bhandari H (2008) Drought: economic costs and research implications. In: Serraj R, Bennett J, Hardy B (eds) Drought fronteirs in rice crop improvement for increased rainfed production. World Scientific and IRRI, Singapore/Los Banos

Pantuwan G, Fukai S, Cooper M et al (2002) Yield response of rice (*Oryza sativa* L) genotypes to different types of drought under rainfed lowlands – Part 3. Plant factors contributing to drought resistance. Field Crops Res 73:181–200

Passioura JB, Spielmeyer W, Bonnett DG (2007) Requirements for success in marker-assisted breeding for drought-prone environments. In: Jenks MA, Hasegawa PM, Jain S (eds) Advances in molecular breeding towards drought and salt tolerant crops. Springer, Dordrecht

Pastore D, Trono D, Laus MN et al (2007) Possible plant mitochondria involvement in cell adaptation to drought stress; A case study: durum wheat mitochondria. J Exp Bot 58:195–210

Patterson RP, Hudak CM (1996) Drought-avoidant soybean germplasm maintains nitrogen-fixation capacity under water stress. Plant Soil 186:39–43

Pepe JF, Welsh JR (1979) Soil water depletion patterns under dryland field conditions of closely related height lines of winter wheat. Crop Sci 19:677–680

Perry MW, D'Antuono MF (1989) Yield improvement and associated characteristics of some Australian spring wheat cultivars introduced between 1860 and 1982. J Agric Sci Camb 112:295–301

Peters S, Mundree SG, Thomson JA (2007) Protection mechanisms in the resurrection plant *Xerophyta viscosa* (Baker): both sucrose and raffinose family oligosaccharides (RFOs) accumulate in leaves in response to water deficit. J Exp Bot 58:1947–1956

Peters PJ, Jenks MA, Rich PJ et al (2009) Mutagenesis, selection, and allelic analysis of epicuticular wax mutants in sorghum. Crop Sci 49:1250–1258

Pimratch S, Jogloy S, Vorasoot N et al (2009) Heritability of N_2 fixation traits and phenotypic and genotypic correlations between N_2 fixation traits with drought resistance traits and yield in peanut. Crop Sci 49:791–800

Pinheiro HA, Damatta FM, Chaves ARM et al (2005) Drought tolerance is associated with rooting depth and stomatal control of water use in clones of *Coffea canephora*. Ann Bot 96:101–108

Poormohammad KS, Talia P, Maury P et al (2007) Genetic analysis of plant water status and osmotic adjustment in recombinant inbred lines of sunflower under two water treatments. Plant Sci 172:773–787

Porter JR, Klepper B, Belford RK (1986) A model (WHTROOT) which synchronizes root growth and development with shoot development for winter wheat. Plant Soil 92:133–145

Prester T, Weltzien E (2003) Exploiting heterosis in pearl millet for population breeding in arid environments. Crop Sci 43:767–776

Price AH, Steele KA, Moore BJ et al (2000) A combined RFLP and AFLP linkage map of upland rice (*Oryza sativa* L) used to identify QTLs for root-penetration ability. TAG 100:49–56

Price AH, Cairns JE, Horton P et al (2002) Linking drought-resistance mechanisms to drought avoidance in upland rice using a QTL approach: progress and new opportunities to integrate stomatal and mesophyll responses. J Exp Bot 53:989–1004

Proctor MCF, Ligrone R, Duckett JG (2007) Desiccation tolerance in the moss *Polytrichum formosum*: physiological and fine-structural changes during desiccation and recovery. Ann Bot 99:75–93

Purcell LC, Serraj R, Sinclair TR et al (2004) Soybean N_2 fixation estimates ureide concentration and yield responses to drought. Crop Sci 44:484–492

Rahman H, Malik SA, Saleem M (2004) Heat tolerance of upland cotton during the fruiting stage evaluated using cellular membrane thermostability. Field Crops Res 85:149–158

Rajabi A, Griffiths H, Ober ES et al (2008) Genetic characteristics of water-use related traits in sugar beet. Euphytica 160:175–187

Ramos ML, Gordon AJ, Minchin FR et al (1999) Effect of water stress on nodule physiology and biochemistry of a drought tolerant cultivar of common bean (*Phaseolus vulgaris* L.). Ann Bot 83:57–63

Ray JD, Yu L, McCouch SR et al (1996) Mapping quantitative trait loci associated with root penetration ability in rice (*Oryza sativa* L.). TAG 92:627–636

Rebetzke GJ, Richards RA (1999) Genetic improvement of early vigour in wheat. Aust J Agric Res 50:291–301

Rebetzke GJ, Condon AG, Farquhar GD et al (2008a) Quantitative trait loci for carbon isotope discrimination are repeatable across environments and wheat mapping populations. TAG 118:123–137

Rebetzke GJ, van Herwaarden AF, Jenkins C et al (2008b) Quantitative trait loci for water-soluble carbohydrates and associations with agronomic traits in wheat. Aust J Agric Res 59:891–905

Rebetzke GJ, Condon AG, Farquhar GD et al (2009) Water-use efficiency in wheat – the use of surrogate traits for breeding improved biomass and yield under drought. In: Interdrought-III, Shanghai, in press

Rehman A, Nautiyal CS (2002) Effect of drought on the growth and survival of the stress-tolerant bacterium rhizobium sp NBRI2505 sesbania and its drought-sensitive transposon Tn5 mutant. Curr Microbiol 45:368–377

Reitz LP (1974) Breeding for more efficient water-use – is it real or a mirage. Agric Meteorol 14:3–10

Reynolds MP, Balota M, Delgado MIB et al (1994) Physiological and morphological traits associated with spring wheat yield under hot, irrigated conditions. Aust J Plant Physiol 21:717–730
Reynolds MP, Ortiz-Monasterio JI, McNab A (eds) (2006) Application of physiology in wheat breeding. CIMMYT, El Batan
Ribaut JM, Hoisington DA, Deutsch JA et al (1996) Identification of quantitative trait loci under drought conditions in tropical maize I. Flowering parameters and the anthesis-silking interval. Theor Appl Genet 92:905–914
Ribaut JM, Jiang C, Gonzalezdeleon D et al (1997) Identification of quantitative trait loci under drought conditions in tropical maize 2. Yield components and marker-assisted selection strategies. Theor Appl Genet 94:887–896
Richards RA, Passioura JB (1989) A Breeding program to reduce the diameter of the major xylem vessel in the seminal roots of wheat and its effect on grain yield in rain-fed environments. Aust J Agric Res 40: 943–950
Richards RA, Rawson HM, Johnson DA (1986) Glaucousness in wheat: its development and effect on water-use efficiency gas exchange and photosynthetic tissue temperatures. Aust J Plant Physiol 13:465–473
Richards RA, Rebetzke GJ, Condon AG et al (2002) Breeding opportunities for increasing the efficiency of water use and crop yield in temperate cereals. Crop Sci 42:111–121
Riga P, Vartanian N (1999) Sequential expression of adaptive mechanisms is responsible for drought resistance in tobacco. Aust J Plant Physiol 26:211–220
Ripley B, Frole K, Gilbert M (2010) Differences in drought sensitivities and photosynthetic limitations between co-occurring C_3 and C_4 (NADP-ME) Panicoid grasses. Ann Bot 105:493–503
Rivero RM, Shulaev V, Blumwald E (2009) Cytokinin-dependent photorespiration and the protection of photosynthesis during water deficit. Plant Physiol 150:1530–1540
Rodrigues SM, Andrade MO, Gomes APS et al (2006) Arabidopsis and tobacco plants ectopically expressing the soybean antiquitin-like ALDH7 gene display enhanced tolerance to drought, salinity, and oxidative stress. J Exp Bot 57:1909–1918
Rodriguezmaribona B, Tenorio JL, Conde JR et al (1992) Correlation between yield and osmotic adjustment of peas (*Pisum sativum* l) under drought stress. Field Crops Res 29:15–22
Romagosa I, Han F, Ullrich SE et al (1999) Verification of yield QTL through realized molecular marker – assisted selection responses in a barley cross. J Mol Breed 5:143–152
Rosenow DT, Ejeta G, Clark LE et al (1996) Breeding for pre- and post-flowering drought stress resistance in sorghum. In: Rosenow DT Yohe JM (ed) Proceedings of the international conference genetic improvement of sorghum and pearl millet, INTSORMIL, Lubbock
Rubio MC, Gonzalez EM, Minchin FR et al (2002) Effects of water stress on antioxidant enzymes of leaves and nodules of transgenic alfalfa overexpressing superoxide dismutases. Physiol Plant 115:531–539
Ruuska SA, Rebetzke GJ, Van Herwaarden AF et al (2006) Genotypic variation in water-soluble carbohydrate accumulation in wheat. Funct Plant Biol 33:799–809
Saadalla MM, Quick JS, Shanahan JF (1990) Heat tolerance in winter wheat 2. Membrane thermostability and field performance. Crop Sci 30:1248–1251
Saccardy K, Cornic G, Brulfert J et al (1996) Effect of drought stress on net CO_2 uptake by *Zea* leaves. Planta 199:589–595
Sadras VO, Connor DJ, Whitfield DM (1993) Yield, yield components and source-sink relationships in water-stressed sunflower. Field Crops Res 31:27–39
Saini HS, Aspinall D (1981) Effect of water deficit on sporogenesis in wheat (*Triticum aestivum* L.). Ann Bot 48:623–633
Saint Pierre C, Trethowan R, Reynolds M (2010) Stem solidness and its relationship to water-soluble carbohydrates: association with wheat yield under water deficit. Funct Plant Biol 37:166–174
Sanchez FJ, Manzanares M, de Andres EF et al (2001) Residual transpiration rate, epicuticular wax load and leaf colour of pea plants in drought conditions Influence on harvest index and canopy temperature. Eur J Agron 15:57–70
Sanchez AC, Subudhi PK, Rosenow DT et al (2002) Mapping QTLs associated with drought resistance in sorghum (*Sorghum bicolor* L Moench). Plant Mol Biol 48:713–726

Sanguineti MC, Tuberosa R, Landi et al (1999) QTL analysis of drought related traits and grain yield in relation to genetic variation for leaf abscisic acid concentration in field-grown maize. J Exp Bot 50:1289–1297

Santamaria JM, Ludlow MM, Fukai S (1990) Contribution of osmotic adjustment to grain yield in *Sorghum-bicolor* (L.) moench under water-limited conditions 1. Water stress before anthesis. Aust J Agric Res 41:51–65

Saranga Y, Menz M, Jiang CX et al (2001) Genomic dissection of genotype x environment interactions conferring adaptation of cotton to arid conditions. Genome Res 11:1988–1995

Sayar R, Khemira H, Kharrat M (2007) Inheritance of deeper root length and grain yield in half-diallel durum wheat (*Triticum durum*) crosses. Ann Appl Biol 151:213–220

Schnyder, H (1993) The role of carbohydrate storage and redistribution in the Source-Sink relations of wheat and barley during grain filling – a review. New Phytol 123:233–245

Schoper JB, Lambert RJ, Vasilas BL et al (1987) Plant factors controlling seed set in maize 1. The influence of silk, pollen, and ear-leaf water status and tassel heat treatment at pollination. Plant Physiol 8:121–125

Selote DS, Khanna-Chopra R (2004) Drought-induced spikelet sterility is associated with an inefficient antioxidant defence in rice panicles. Physiol Plant 121:462–471

Serraj R, Sinclair TR (1998) Soybean cultivar variability for nodule formation and growth under drought. Plant Soil 202:159–166

Serraj R, Sinclair TR (2002) Osmolyte accumulation: can it really help increase crop yield under drought conditions? Plant Cell Environ 25:333–341

Serraj R, Vadez V, Sinclair TR (2001) Feedback regulation of symbiotic N_2 fixation under drought stress. Agronomie 21:621–626

Shearman VJ, Sylvester-Bradley R, Scott RK et al (2005) Physiological processes associated with wheat yield progress in the UK. Crop Sci 45:175–185

Shen B, Jensen RG, Bohnert HJ (1997) Mannitol protects against oxidation by hydroxyl radicals. Plant Physiol 115(2):527–532

Shen L, Courtois B, McNally KL et al (2001) Evaluation of near-isogenic lines of rice introgressed with QTLs for root depth through marker-aided selection. Theor Appl Genet 103:75–83

Sheoran IS, Saini HS (1996) Drought-induced male sterility in rice: changes in carbohydrate levels and enzyme activities associated with the inhibition of starch accumulation in pollen. Sexual Plant Reprod 9:161–169

Sinclair TR, Purcell LC, King C et al (2007) Drought tolerance and yield increase of soybean resulting from improved symbiotic N_2 fixation. Field Crops Res 101:68–71

Singh TN, Aspinall D, Paleg LG (1972) Proline accumulation and varietal adaptability to drought in barley: a potential metabolic measure of drought resistance. Nature 236:188–189

Specht JE, Williams JH, Pearson DR (1985) Near-isogenic analyses of soybean pubescence genes. Crop Sci 25:92–96

Sreedhar L, Wolkers WF, Hoekstra FA et al (2002) In vivo characterization of the effects of abscisic acid and drying protocols associated with the acquisition of desiccation tolerance in alfalfa (*Medicago sativa* L.) somatic embryos. Ann Bot 89:391–400

Srinivasan S, Gomez SM, Kumar S et al (2008) QTLs linked to leaf epicuticular wax, physio-morphological and plant production traits under drought stress in rice (*Oryza sativa* L.). Plant Growth Regul 56:245–256

Steele KA, Price AH, Shashidhar HE et al (2006) Marker-assisted selection to introgress rice QTLs controlling root traits into an Indian upland rice variety. Theor Appl Genet 112:208–221

Steele KA, Virk DS, Kumar R et al (2007) Field evaluation of upland rice lines selected for QTLs controlling root traits. Field Crops Res 101:180–186

Stiller WN, Reid PE, Constable GA (2004) Maturity and leaf shape as traits influencing cotton cultivar adaptation to dryland conditions. Agron J 96:656–664

Taketa S, Chang CL, Ishii M et al (2002) Chromosome arm location of the gene controlling leaf pubescence of a Chinese local wheat cultivar 'Hong-mang-mai'. Euphytica 125:141–147

Tambussi EA, Bort J, Guiamet J-J et al (2007) The photosynthetic role of ears in C_3 cereals: metabolism, water use efficiency and contribution to grain yield. Crit Rev Plant Sci 26:1–16

Tang R-S, Zheng J-C, Jin Z-Q et al (2007) Possible correlation between high temperature-induced floret sterility and endogenous levels of IAA, GAs and ABA in rice (*Oryza sativa* L.). Plant Growth Regul 54:37–43

Tangpremsri T, Fukai S, Fischer KS et al (1991) Genotypic variation in osmotic adjustment in grain sorghum. 2. Relation with some growth attributes. Aust J Agric Res 42:759–767

Tangpremsri T, Fukai S, Fischer KS (1995) Growth and yield of sorghum lines extracted from a population for differences in osmotic adjustment. Aust J Agr Res 46:61–74

Tenkouano A, Miller FR, Frederiksen RA et al (1993) Genetics of nonsenescence and charcoal rot resistance in sorghum Theor Appl Genet 85:644–648

Teulat B, Monneveux P, Wery J et al (1997) Relationships between relative water content and growth parameters under water stress in barley – a QTL study. New Phytol 137(1):99–107

Teulat B, This D, Khairallah M et al (1998) Several QTLs involved in osmotic adjustment trait variation in barley (*Hordeum vulgare* L.). Theor Appl Genet 96:688–698

Thiaw S, Hall AE (2004) Comparison of selection for either leaf-electrolyte-leakage or pod set in enhancing heat tolerance and grain yield of cowpea. Field Crops Res 86:239–253

Thomas H, Howarth CJ (2000) Five ways to stay green. J Exp Bot 51:329–337

Toldi O, Tuba Z, Scott P (2009) Vegetative desiccation tolerance: is it a goldmine for bioengineering crops? Plant Sci 176:187–199

Tollenaar M, Wu J (1999) Yield improvement in temperate maize is attributable to greater stress tolerance. Crop Sci 39:1597–1604

Tomar SMS, Kumar GT (2004) Seedling survivability as a selection criterion for drought tolerance in wheat. Plant Breed 123:392–394

Toyofuku K, Loreti E, Vernieri P et al (2000) Glucose modulates the abscisic acid-inducible Rab16A gene in cereal embryos. Plant Mol Biol 42:451–454

Tripathy JN, Zhang J, Robin S et al (2000) QTLs for cell membrane stability mapped in rice (*Oryza sativa* L.) under drought stress. Theor Appl Genet 100:1197–1202

Tuinstra MR, Grote EM, Goldsbrough PB et al (1997) Genetic analysis of post-flowering drought tolerance and components of grain development in *Sorghum bicolor* (L) Moench. Mol Breed 3:439–448

Turner NC, Abbo S, Berger JD et al (2007a) Osmotic adjustment in chickpea (*Cicer arietinum* L.) results in no yield benefit under terminal drought. J Exp Bot 58:187–194

Turner NC, Palta JA, Shrestha R et al (2007b) Carbon isotope discrimination is not correlated with transpiration efficiency in three cool-season grain legumes (pulses). J Integr Plant Biol 49:1478–1483

Vadez V, Sinclair TR (2001) Leaf ureide degradation and N_2 fixation tolerance to water deficit in soybean. J Exp Bot 52:153–159

van Eeuwijk FA, Malosetti M, Xinyou Y et al (2005) Statistical models for genotype by environment data: from conventional ANOVA models to eco-physiological QTL models: modelling complex traits for plant improvement. Aust J Agric Res 56:883–894

van Oosterom EJ, Weltzien E, Yadav OP et al (2006) Grain yield components of pearl millet under optimum conditions can be used to identify germplasm with adaptation to arid zones. Field Crops Res 96:407–421

Vartanian N (1981) Some aspects of structural and functional modifications induced by drought in root systems. Plant Soil 63:83–92

Vassileva V, Simova-Stoilova L, Demirevska K et al (2009) Variety-specific response of wheat (*Triticum aestivum* L.) leaf mitochondria to drought stress. J Plant Res 122:445–454

Venuprasad R, Shashidhar HE, Hittalmani S et al (2002) Tagging quantitative trait loci associated with grain yield and root morphological traits in rice (*Oryza sativa* L.) under contrasting moisture regimes. Euphytica 128:293–300

Venuprasad R, Lafitte HR, Atlin GN (2007) Response to direct selection for grain yield under drought stress in rice. Crop Sci 47:285–293

Venuprasad R, Sta Cruz MT, Amante M et al (2008) Response to two cycles of divergent selection for grain yield under drought stress in four rice breeding populations. Field Crops Res 107:232–244

Venuprasad R, Dalid CO, Del Valle M et al (2009) Identification and characterization of large-effect quantitative trait loci for grain yield under lowland drought stress in rice using bulk-segregant analysis. Theor Appl Genet 120:177–190

Verma V, Foulkes MJ, Worland AJ et al (2004) Mapping quantitative trait loci for flag leaf senescence as a yield determinant in winter wheat under optimal and drought-stressed environments. Euphytica 135:255–263

Villnlobos-Kodruigez H, Shibles R (1985) Response of determinate and indeterminate tropical soybean cultivars to water stress. Field Crops Res 10:269–275

Volaire F (2002) Drought survival, summer dormancy and dehydrin accumulation in contrasting cultivars of *Dactylis glomerata*. Physiol Plant 116:42–51

Volaire F (2008) Plant traits and functional types to characterise drought survival of pluri-specific perennial herbaceous swards in Mediterranean areas. Eur J Agron 29:116–124

Volaire F, Lelievre F (2001) Drought survival in *Dactylis glomerata* and *Festuca arundinacea* under similar rooting conditions in tubes. Plant Soil 229:225–234

Volaire F, Norton MR (2006) Summer dormancy in perennial temperate grasses. Ann Bot 98:927–933

Volaire F, Norton MR, Norton GM et al (2005) Seasonal patterns of growth, dehydrins and water-soluble carbohydrates in genotypes of *Dactylis glomerata* varying in summer dormancy. Ann Bot 95:981–990

Volaire F, Norton MR, Lelievre F (2009a) Summer drought survival strategies and sustainability of perennial temperate forage grasses in Mediterranean areas. Crop Sci 49:2386–2392

Volaire F, Seddaiu G, Ledda L et al (2009b) Water deficit and induction of summer dormancy in perennial Mediterranean grasses. Ann Bot 103:1337–1346

Wang Z, Huang B (2004) Physiological recovery of Kentucky bluegrass from simultaneous drought and heat stress. Crop Sci 44:1729–1736

Wang XJ, Loh CS, Yeoh HH et al (2002) Drying rate and dehydrin synthesis associated with abscisic acid-induced dehydration tolerance in *Spathoglottis plicata* Orchidaceae protocorms. J Exp Bot 53:551–558

Wang Z, Huang B, Bonos SA et al (2004) Abscisic acid accumulation in relation to drought tolerance in Kentucky bluegrass. Hortscience 39:1133–1137

Wang F-Z, Wang Q-B, Kwon S-Y et al (2005a) Enhanced drought tolerance of transgenic rice plants expressing a pea manganese superoxide dismutase. J Plant Physiol 162:465–472

Wang Y, Ying J, Kuzma M et al (2005b) Molecular tailoring of farnesylation for plant drought tolerance and yield protection. Plant J 43:413–424

Wang Y, Jiang J, Zhao X et al (2006) A novel LEA gene from *Tamarix androssowii* confers drought tolerance in transgenic tobacco. Plant Sci 171:655–662

Wardlaw IF, Willenbrink J (2000) Mobilization of fructan reserves and changes in enzyme activities in wheat stems correlate with water stress during kernel filling. New Phytol 148:413–422

Watanabe N, Naruse J, Austin RB et al (1995) Variation in thylakoid proteins and photosynthesis in Syrian landraces of barley. Euphytica 82:213–220

Welcker C, Boussuge B, Bencivenni C et al (2007) Are source and sink strengths genetically linked in maize plants subjected to water deficit? A QTL study of the responses of leaf growth and of anthesis-silking interval to water deficit. J Exp Bot 58:339–349

Winzeler M, Monteil P, Nosberger J (1989) Grain growth of tall and short spring wheat genotypes at different assimilate supplies. Crop Sci 29:1487–1491

Xiao B, Huang Y, Tang N et al (2007) Over-expression of a LEA gene in rice improves drought resistance under the field conditions. Theor Appl Genet 115:35–46

Xiong L, Wang R-G, Mao G et al (2006) Identification of drought tolerance determinants by genetic analysis of root response to drought stress and abscisic acid. Plant Physiol 142:1065–1074

Xiong Y-C, Li F-M, Zhang T et al (2007) Evolution mechanism of non-hydraulic root-to-shoot signal during the anti-drought genetic breeding of spring wheat. Environ Exp Bot 59: 193–205

Xu Y, Crouch JH (2008) Marker-assisted selection in plant breeding: from publications to practice. Crop Sci 48:391–407

Xu W, Subudhi PK, Crasta OR et al (2000) Molecular mapping of QTLs conferring stay-green in grain sorghum (*Sorghum bicolor* L. Moench). Genome 43:461–469

Yadav OP, Bhatnagar SK (2001) Evaluation of indices for identification of pearl millet cultivars adapted to stress and non-stress conditions. Field Crops Res 70:201–208

Yan J, Wang J, Tissue D et al (2003) Photosynthesis and seed production under water-deficit conditions in transgenic tobacco plants that overexpress an *Arabidopsis* ascorbate peroxidase gene. Crop Sci 43:1477–1483

Yan J, He C, Wang J et al (2004) Overexpression of the *Arabidopsis* 14-3-3 protein GF14 in cotton leads to a "stay-green" phenotype and improves stress tolerance under moderate drought conditions. Plant Cell Physiol 45:1007–1014

Yang J, Zhang J (2006) Grain filling of cereals under soil drying. New Phytol 169:223–236

Yang WJ, Nadolskaorczyk A, Wood KV et al (1995) Near-isogenic lines of maize differing for glycinebetaine. Plant Physiol 107:621–630

Yang JC, Zhang JH, Wang ZQ et al (2001) Activities of starch hydrolytic enzymes and sucrose-phosphate synthase in the stems of rice subjected to water stress during grain filling. J Exp Bot 52:2169–2179

Yang JC, Zhang JH, Wang ZQ et al (2003) Involvement of abscisic acid and cytokinins in the senescence and remobilization of carbon reserves in wheat subjected to water stress during grain filling. Plant Cell Environ 26:1621–1631

Yang JC, Zhang JH, Ye YX et al (2004) Involvement of abscisic acid and ethylene in the responses of rice grains to water stress during filling. Plant Cell Environ 27:1055–1064

Yang J, Zhang J, Liu K et al (2007) Abscisic Acid and Ethylene Interact in Rice Spikelets in Response to Water Stress During Meiosis J Plant Growth Regul 26:318–328

Yu LX, Ray JD, O'Toole JC et al (1995) Use of wax-petrolatum layers for screening rice root penetration. Crop Sci 35:684–687

Yue B, Xiong L, Xue W et al (2005) Genetic analysis for drought resistance of rice at reproductive stage in field with different types of soil. Theor Appl Genet 111:1127–1136

Zahran HH (1999) Rhizobium-legume symbiosis and nitrogen fixation under severe conditions and in an arid climate. Microbiol Mol Biol Rev 63:968–989

Zaidi PH, Srinivasan G, Cordova HS et al (2004) Gains from improvement for mid-season drought tolerance in tropical maize (*Zea mays* L.). Field Crops Res 89:135–152

Zhang J, Zheng HG, Aarti A et al (2001) Locating genomic regions associated with components of drought resistance in rice: comparative mapping within and across species. Theor Appl Genet 103:19–29

Zhang J-Y, Broeckling CD, Sumner LW et al (2007) Heterologous expression of two Medicago truncatula putative ERF transcription factor genes, WXP1 and WXP2, in Arabidopsis led to increased leaf wax accumulation and improved drought tolerance, but differential response in freezing tolerance. Plant Mol Biol 64:265–278

Zhang J, Dell Be, Conocono E et al (2009) Water deficits in wheat: fructan exohydrolase (1-FEH) mRNA expression and relationship to soluble carbohydrate concentrations in two varieties. New Phytol 181:843–850

Zheng H, Babu RC, Safiullah P et al (2000) Quantitative trait loci for root-penetration ability and root thickness in rice: comparison of genetic backgrounds. Genome 43:53–61

Zhu H, Briceño G, Dovel R et al (1999) Molecular breeding for grain yield in barley: an evaluation of QTL effects in a spring barley cross. Theor Appl Genet 98:772–779

Zilberstein M, Blum A, Eyal Z (1985) Chemical desiccation of wheat plants as a simulator of postanthesis speckled leaf blotch stress. Phytopathology 75:226–230

Zou GH, Mei HW, Liu HY et al (2006) Grain yield responses to moisture regimes in a rice population: association among traits and genetic markers. Theor Appl Genet 112:106–113

Chapter 4
Phenotyping and Selection

Summary To the same extent that the phenotyping of disease resistance requires exposure of plants to the disease, drought resistance phenotyping requires exposure to a designed stress in a managed drought environment. This environment must be managed for the required drought stress severity and timing with respect to plant growth stage. A drought environment can be managed in the field or in protected environments such as a greenhouse or a rainout shelter.

Under the designated drought stress plants are measured for their response to drought in terms of yield and its components as well as specific symptoms of reproductive failure. A drought resistance index in terms of yield can be developed by comparing yield between stress and non-stress conditions.

Drought resistance is measured by phenotyping the specific and relevant attributes of dehydration avoidance and dehydration tolerance. Phenotyping methods in a plant breeding program must be relatively fast and economical. Protocols for dehydration avoidance include direct measurements of plant water status in terms of visual symptoms of wilting, leaf water potential, relative water content and osmotic adjustment. Dehydration avoidance can also be measured by indirect methods such as fast plant and crop remote sensing techniques.

Several constitutive traits affecting dehydration avoidance can be measured without exposure to drought stress. An example is maximum (potential) root depth in various root installations.

Dehydration tolerance can be phenotyped only on the basis of similar plant water status in all genotypes. The most prominent feature of whole plant dehydration tolerance in grain crops is the capacity for stem reserve utilization for grain filling which can be assessed by the chemical desiccation method. Other specific tests for dehydration tolerance include plant growth, cell membrane stability and plant survival and plant recovery from extreme desiccation.

Inappropriate phenotyping for drought resistance is considered as the main cause of past difficulties of molecular biology and genomics to contribute towards plant breeding for water limited environments.

Phenotyping is a protocol for expressing the genotype in its resultant phenotype. This is crucial in phenotypic selection where an attempt is made to capture the target genotype by observing its phenotype. The major problem in phenotyping and

A. Blum, *Plant Breeding for Water-Limited Environments*,
DOI 10.1007/978-1-4419-7491-4_4,

in phenotypic selection is that the genotype is not always perfectly expressed in the phenotype, for a variety of well known reasons. This discrepancy is defined as the "genotype-phenotype gap." One of the major reasons for the gap is that the phenotype results from the expression of the effects of both the genotype and the environment it is exposed to, resulting also in a genotype by environment interaction (GxE).

Drought resistance is often considered as a complex trait at least partly because of difficulties in making the conceptual shift from environmental drought to plant drought stress. Drought stress *in planta* is a three-dimensional entity with respect to its condition and expression (Fig. 4.1). The consequence and impact of any specific drought stress depends on the plant growth stage when it occurs, on the rate of the development of water deficit in the plant and on peak level of water deficit when stress is relieved or measured. These axes and their significance in the plant were discussed in previous chapters.

Genes that help the plant to cope with the stress environment by specifying a certain phenotype may therefore present GxE according to each of the axes. For example, genes that condition OA will be expressed better when the rate of plant stress development is relatively slow and when it reaches a certain water deficit, say at least 70% of RWC. Thus, phenotyping drought resistance requires the measurement of the specific plant trait in question under the *specific relevant stress environment*. A specific stress environment is defined by the three axes (Fig. 4.1) and the target ideotype (Sect. 3.7). Some of these principles are remindful of the disease analogy discussed earlier.

Xu and Crouch (2008) wrote "The quantity and quality of phenotyping is becoming the most significant factor affecting the accuracy of genetic mapping and thus the power of the resultant MAS, particularly for complex traits." It might be added that part of the assumed "complexity" of drought resistance is the misunderstanding of the makeup of plant drought stress and consequently its distorted phenotyping in the research. Confusing results out of a misguided research can also lead to the conclusion that the subject matter is "complex."

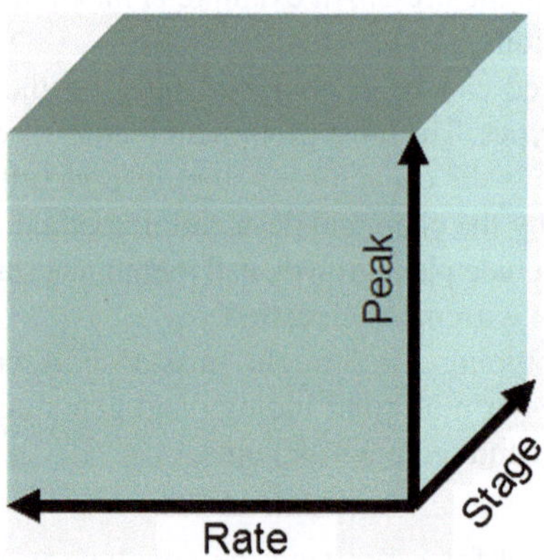

Fig. 4.1 Plant drought stress as a 3-D entity which is determined by the growth stage when drought occurs, the rate of water deficit development and the peak of plant water deficit reached (see text)

4.1 The Managed Stress Environment

Phenotyping drought resistance requires adequate control of the stress environment, when natural conditions do not provide a predictable stress level, rate and timing. This is achieved by managing and controlling the stress environment in which the breeding materials are tested. Two major factors determine success and failure in managing the field stress environment even before any actual phenotyping work begins: Site homogeneity and control over the water regime.

4.1.1 Site Homogeneity

Site homogeneity for any factor that affects plant growth is crucial. Poor homogeneity especially for soil moisture supply can destroy field screening, as can be seen for the dryland sorghum nursery in Fig. 4.2.

Following are the most important generators of site variability the impact of which increases as drought stress develops.

(a) *Topography* will drive variability by creating various spatial gradients influencing water runoff, snow cover and soil water infiltration. Topography can propel soil heterogeneity. When water table is high and potentially accessible to roots, even slight variation in topography can become crucial in creating large site variability in water supply to roots. This is a common fault in upland rice phenotyping sites. Poor topography will most likely undermine research on any subject. In my experience it is not uncommon to find experiment stations situated on sites with very variable topography. Avoid such sites at all cost.
(b) *Soil* variation throughout the site is crucial in its influence on soil moisture infiltration, runoff, retention and percolation. Soil texture and the effective soil depth are very important in this respect. Variation in mineral deficiencies on one hand and mineral toxicities on the other can be a major factor in certain soils. Soil acidity and Aluminum toxicity can be very variable in soils within a very short distance. Similarly for high soil pH and alkalinity. Whenever soil-inhabiting biotic stresses develop they tend to be highly patchy, typical of nematode infestation or soil born root diseases. Furthermore, certain diseases are preconditioned by drought stress and to the superficial observer they might appear as symptoms of drought stress, such as head blast in rice or charcoal rot in sorghum or bean (e.g., Frahm et al. 2004).
(c) *Bordering* trees, villages and animal compounds will affect the edges of the site. Eucalyptus trees are almost legendary for their horizontal root extensions. Past tracks and allies in the field retain their effect for a long time. Previous experiments on the site can ruin your work unless 2 or 3 years have passed under a homogeneous commercial crop.
(d) *Crop management* can create permanent or transient site variability. Poor distribution of plant growth inputs such as water (see below) and fertilizers are of main concern. Depending on soil water content and equipment, heavy traffic can leave clear strips of water deficiency due to excessive soil compaction.

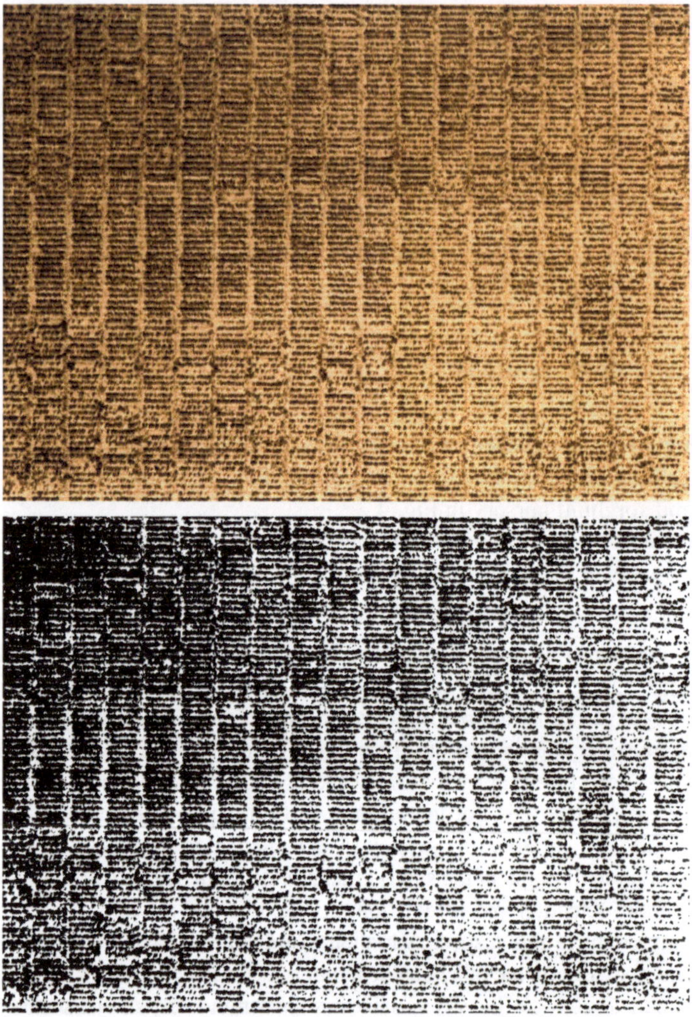

Fig. 4.2 *Top*: Color aerial photograph of poor homogeneity in a drought stressed sorghum nursery. *Bottom*: Same photograph with enhanced black and white rendition for improved contrast. A gradient of increasing stress can be seen across the nursery from the top left (favorable water status and growth) to the bottom right (poor water status and growth). Each small line represents a different genotype planted to one row 5 m long, arranged in blocks

4.1.1.1 Evaluating Field Heterogeneity

Any field intended for drought stress phenotyping work should be assessed for heterogeneity. The purpose of the evaluation is to accept or reject the field for phenotyping work or to map out sections which are suspected to have a soil problem.

One option is to create a soil heterogeneity map using equipment such as the Veris (www.veristech.com/) electrical conductivity (EC) mapping system. This system uses GPS and soil electrical conductivity (EC) scan to identify areas of contrasting soil properties. In non-saline soils, EC values largely represent soil texture. Soil texture is directly related to both water-holding capacity and cation-exchange capacity – key ingredients of productivity. As the Veris EC cart is pulled through the field by tractor, one pair of coulter-electrodes injects a known voltage into the soil, while the other coulter-electrodes measure the drop in that voltage. Smaller soil particles such as clay conduct more current than larger particles such as silt or sand. The Veris map has been shown to be reasonably reflected in the yield map (Fig. 4.3).

Another option is the noncontact, electromagnetic induction–based sensor such as that made by Geonics Ltd. While Veris and Geonics instruments were comparable in one performance test (Sudduth et al. 2003), it appears that the Veris direct measurement system of soil electro-conductivity has gained wider acceptance in agriculture.

Site heterogeneity can also be assessed by observing the heterogeneity of a preceding crop grown in the field, which offcourse involve "wasting" one season. The crop grown for this purpose should be managed under standard conditions as it would be in the phenotyping work. This concerns fertilization, weed, and complete disease and pest control. The crop should be drought stressed at a relevant growth stage since drought would amplify heterogeneity. If the crop is initially irrigated then irrigation should strictly follow practices for maintaining irrigation

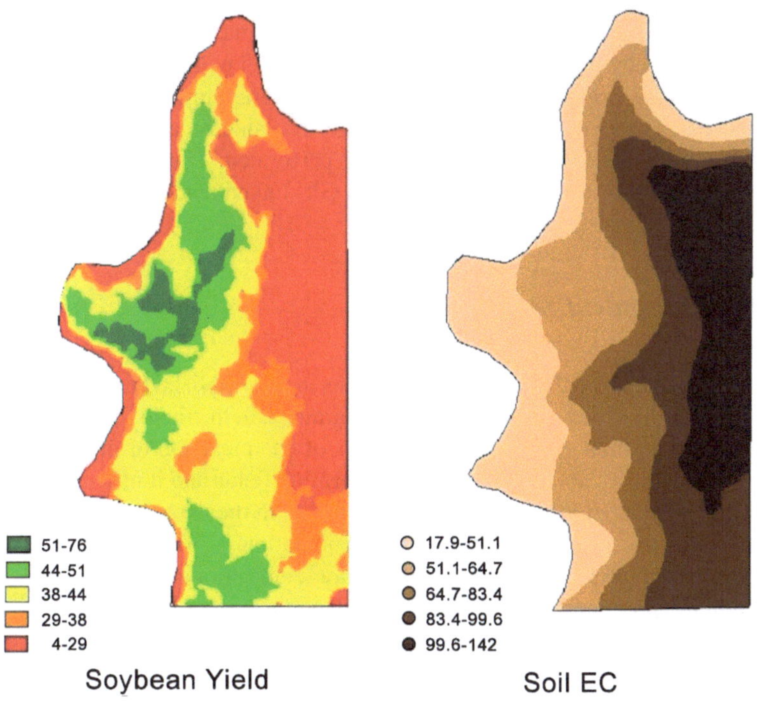

Fig. 4.3 Soil electroconductivity (EC) map (*right*) and the corresponding soybean yield in that field (*left*). Color scale represents EC values and corresponding yield levels (Graphic courtesy of Veris Inc.)

homogeneity (below). As drought stress develops the heterogeneity of this crop can be assessed by several methods.

Any experienced plant breeder can fairly well estimate general crop heterogeneity in a given field by walking through the field and inspecting it from a high point such as a hydraulic fruit picker. Repeated observations over time are desirable. It is then possible to stake out suspect patches of relatively poor crop performance or reject the field altogether. This is the simplest and cheapest but yet effective method, if you have a farmer level experience with the given crop.

Remote sensing techniques and tools such as satellite based global positioning systems (GPS) and differential global positioning system (DGPS) combined with various spectral imaging systems allow to obtain excellent data on crop spatial heterogeneity. These methods reached an advanced stage and are becoming commonplace in connection with precision agriculture applications (e.g., Srinivasan 2006). The availability and cost of the various methods varies with country and provider. In certain cases it was even possible to visually assess prominent field variability by inspecting historical crop images of the site in 'Google Earth' maps.

Aerial visible and infrared photography or spectrophotometry of the crop can provide data on heterogeneity for canopy color, leaf area, nitrogen status and biomass. General crop heterogeneity developing under drought stress as driven by soil factors can be very well represented by a single timely aerial color photograph (Fig. 4.2).

Finally, crop heterogeneity can be evaluated by developing a yield map for the given field using grid-based discrete sampling of biomass and or yield. While this can be achieved by manual labor and small harvesting equipment a more sophisticated method has been developed under precision agriculture technology. For most machine harvested crops special systems are available to acquire yield data for discrete grids in the field as harvest is being performed. The system comprises of sensors in the harvested product chute and a GIS linked readout.

4.1.2 Experiment Station Faults

It is widely recognized that growing conditions in an established experiment station are very often different from those in the farmer's field. Field stations are often characterized by high soil fertility after years of chemical and/or organic fertilization. High fertility is not typical of dryland farming fields and it might interact with genotypic responses to the imposed drought stress in the experiment.

Crop sequence practiced in the experiment station is often different from that used in the dryland target environment. A non-conventional crop grown before a phenotyping experiment with the target dryland crop might generate undesirable interactions with the genetic materials in the test. The drought phenotyping experiment might also follow a previous crop that has been irrigated. Such a crop might leave appreciable amount of deep soil moisture which can be atypical of the rainfed conditions in the target environment. This is a crucial fault that can create irrelevant phenotypic variations in the tested population.

Finally, because of extensive traffic in modern experiment stations, soil might be harder with higher bulk density or even a hardpan – which might be atypical of the farmer's field. In cases of crops where root growth constitutes a major component of drought resistance this can create irrelevant phenotypic variation for drought response in the population.

4.1.3 Controlling the Water Regime

It is perhaps a paradox that breeding for adaptation to a dryland environment can rarely use that environment for field drought phenotyping. The typical seasonal and annual variability in rainfall in the semi-arid region reduces the accuracy and effectiveness of phenotyping and phenotypic selection. Repeatability from 1 year to the other becomes a problem. Breeders encountering inconsistency in their phenotyping across years tend sometimes to blame the relevance of the measured trait for drought resistance rather than the variation between test years or sites in the drought stress profile and its consequences (e.g., Price et al. 2002).

Certain dryland environments with stable limited stored soil moisture or fairly repeatable terminal stress can be used effectively for phenotyping. However, just one year of unexpected shift in rainfall pattern can cause extensive waste of time and resources in the breeding program. Furthermore, in the semi-arid regions where dryland crops are grown under a mean annual rainfall of 300–400 mm, sever unpredictable drought can occur causing a total loss of breeding materials. More control over the field drought environment is essential for achieving effective results in field phenotyping and progress in selection. Control over the water regime is achieved by eliminating undesirable rainfall on one hand while controlling the desirable amount and timing of water input into the field on the other.

Since in most cases the managed field environment is also designated to include yield phenotyping, the general growing conditions of the site must suit the specific crop.

4.1.3.1 The "Desert" Environment

Perhaps the ideal option is to place the phenotyping site in a desert environment where no unexpected rainfall can interfere with the managed stress while the breeder is in total control of the water regime via irrigation. Under such conditions "Fetch" can be very important in affecting the test. Fetch is mainly the effect of a dry wind blowing across the crop (Sect. 1.3). Fetch creates a gradient of plant water deficit from the edge of the field towards its center. Therefore the phenotyping field should be surrounded by a sufficiently wide (at least 10 m) border crop. The border should be managed as the whole field. This requirement is relevant also for other types of managed stress environments according to the expected fetch effect.

4.1.3.2 The Offseason (Dry) Environment

Another solution to avoid unpredictable seasonal rainfall during phenotyping is to set up the managed stress environment out of the normal season – during the dry off-season. This is a common approach in the semi-arid tropics (Mahalakshmi and Blum 2006) and experience has been accumulated mainly with rice.

Besides the water regime, one has to consider the modified climatic and biotic conditions for the specific crop when planted off-season. The different conditions of temperature and photoperiod may or may not allow reasonably normal crop growth and development during the off-season. Crop cycle might change. Plants might become smaller. Phenotypic diversity in phenology might increase or decrease in the population as compared with the normal season. Certain diseases and pests can be highly prevalent during the off-season, making the test impossible unless chemical control is used. For grains that are favored by birds the availability of a grain bearing crop outside its normal season might attract birds and require protection measures. In certain environments the dry-season is also characterized by low air humidity. For crops with very permeable cuticle such as rice this might enhance leaf dehydration, which can be considered as a positive or a negative phenomenon towards drought phenotyping. In case of rice in the tropics, an irrigated dry-season crop can sometimes yield even better than normal season crop due to the higher radiation during reproductive stage (Yang et al. 2008). Careful and educated attention should be given to all possible off-season effects and the expected result in terms of effective drought phenotyping. One test run during offseason with typical germplasm might be required before entering actual phenotyping work for the first time.

Drought resistant rice materials and major effect QTLs for drought resistance at flowering were identified under a dry-season managed stress environment at IRRI in the Philippines (Bernier et al. 2007; Venuprasad et al. 2007). A positive phenotypic correlation has been established for diverse rice materials between their yield in the off-season nursery at IRRI and their yield in India under limited water supply (Atlin et al. 2009). This is a very encouraging and important observation.

Because of the various differences that might occur in crop growth between normal season and off-season it is not always recommended to perform actual selection by collecting seed from selected lines in the off-season nursery. Depending on the case, yield data from such a nursery may not always be a very reliable index of resistance. In such a case the off-season phenotyping nursery can be used mainly for collecting data on plant water status and plant responses to stress while actual selection of seed is performed from the preferred lines grown also during the normal season (Chap. 6). In other cases, after serious consideration (e.g., Venuprasad et al. 2007), selection for yield can be performed under the selection pressure of the off-season environment provides that the environmental effects on the phenotype are well understood.

4.1.3.3 Delayed Planting

This practice is becoming popular in cases where the final stages of plant development such as flowering and grain filling are being phenotyped. When the normal rainy season terminates towards the end of the crop growth cycle, delayed planting will place the final crop developmental stages into a dry environment. Due to temperature and photoperiod responses of the specific crop a given duration of delay in planting does not produce the same delay in flowering. The delay in flowering would typically be smaller than the delay in planting. Therefore the exact timing of delay in planting to achieve the desirable stress at plant reproduction depends on the materials grown and the local conditions.

While the delay in planting might achieve the desirable drought stress during the reproductive stage, this delay can result also in modified temperatures and air humidity during the reproductive growth stage as compared with the normal season. The extent of these changes should be understood and considered as discussed above for offseason planting.

4.1.3.4 Rainout Shelters

Rainout shelters are becoming more popular in recent years, especially with the recent boom in research grants targeted at "global warming," "water crisis," "water footprint" and "more crop per drop."

Rainout shelters are designed to protect a certain area of land from receiving precipitation so that controlled drought stress can be imposed on that area. Many types of rainout shelters were designed and used, with better or lesser results. Rainout shelters are not an economical solution for controlling drought stress on a large area and they are suitable for pre-breeding work such as evaluation of specific germplasm, parental lines or mapping populations.

There are two main designs: (1) moveable and (2) stationary. Within the moveable design there are automatic/motorized and manual versions. The rainout shelter is positioned on the "protected plot" when it rains and it is moved to the "parking area" when rain stops so that the protected plot will be exposed to ambient field conditions. The automatic version is signaled to move over the protected plot by a rain sensor and an electric circuitry which drive the motor. The manual version is moved either by manually switching on the drive ("manually driven") or by manually pushing it ("manually pushed") over the protected plot. The "manually pushed" version must by lightweight and hence it is cheaper but can cover only a limited land area. The automatic version is becoming less popular because of reliability problems and cost. Hence, many automatic types become "manually driven" after the first failure of the automatic drive system or the rain sensor. It takes only one failure to ruin an experiment.

The manual version is moved from its parking area onto the protected plot whenever a rain is expected rather than when the rain begins. It is moved into the parking space whenever rain cease completely and the forecast is clear. Good weather forecasting service is therefore important. If forecasting is unreliable, better have the shelter over the protected plot more time than expected. Therefore, the shelter construction must allow some light inside as well as some ventilation. The orientation of the protected plot and the parking place of the shelter should be designed so that the parked shelter will not shade the plot in the morning or the evening and that the direction of the wind would not allow rain to blow under the shelter.

The area around the shelter should be managed in consideration of surface runoff, drainage and the diversion of water flowing off the shelter especially under storm conditions. Hence ditches and other means of protection should be properly placed around the installation, with respect to topography. Consideration should be given also to the fact that horizontal flow of water in a saturated soil can be appreciable.

Often experiments under a rainout shelter involve control (non-stressed) plots. If the protected plot is not covered by the shelter for an extensive period of time, then the control plots may be placed outside the shelter and exposed to rainfall (or irrigation). However, if a perfectly controlled experiment is planned, then the control plots should be placed also under the shelter and irrigated. Drip irrigation (see below) is a good way to assure proper separation between control and stress plot treatments. The two treatments should constitute separate blocks under the shelter. Experiments were also performed by using containers (pots or large vessels such as PVC tubes for root studies) placed under a rainout shelter. While land under the shelter can be tilled and managed by hand labor, it is advisable to consider in the design the access of tractor operated machinery.

The Moveable Rainout Shelter

Many versions of motorized moveable shelters were constructed. The "classical" design used in many of these shelters consists of the same principle. A roof/walls structure is mounted on wheels on a track. The structure is driven by electric motors. The drive can be switched on and off manually or via an electronic signal from a rain sensor. The structure is relatively heavy and therefore costly.

Probably the largest and most elaborate moveable rainout shelter has been constructed at the National Key Laboratory of Crop Genetic Improvement Huazhong Agricultural University, Wuhan, China (Fig. 4.4). Besides covering a large area this installation also includes an elaborate soil drainage system and control which allows growing paddy rice and draining the paddy water at will in order to impose soil moisture stress. The installation is used in large scale screening and phenotyping of genetic and breeding material. The large structure is composed of two parallel units. Each unit is constructed of four relatively transparent roof sections which can be driven (two at a time one under the other) into the parking spaces on the two sides of the protected plot.

Large area rainout shelters can adopt a multi-cover system which acts like a bellows and expands over the protected area (Fig. 4.5). This system has the advantage of using small parking area for the moveable roof while avoiding very large and heavy roof construction.

Less elaborate and also less costly light, manually-pushed rainout shelters allow cost-effective control over smaller land area (Fig. 4.6). Being light and small, several units can be deployed instead of a large heavy one. The unit typically consists of a wheel-mounted light structure wrapped with polyethylene sheet. The unit travels on iron tracks. Front and back sides are open here, but this may not be allowed under conditions where the direction of rain and wind might cause rain to penetrate

Fig. 4.4 A large motorized moveable rainout shelter at the National Key Laboratory of Crop Genetic Improvement, Huazhong Agricultural University, Wuhan, China. Rice experimental materials grow in the protected area while the shelter is parked on the two edges of the protected area. *Inset*: The installation set at full cover

Fig. 4.5 Multi-cover installation combined of 4 roof sections set over the protected plot. Each section travels on its own tracks and sections are parked towards the far end one under the other when they are off the protected plot. Notice the water drains on each section (The Luoyang Station of the Luoyang Academy of Science, Henan China)

Fig. 4.6 Light-weight manually moveable rainout shelters at the International Crop Research Institute for the Semi-arid Tropics (ICRISAT), Patancheru, India

the shelter. Rolled up polyethylene front and back walls or any other arrangement might offer a solution. Always consider that rain can be driven by a strong wind.

Because of their light weight such rainout shelters can be very unstable under windy conditions. Strong winds can break the construction or lift the unit altogether off its tracks. More solid construction might be needed where strong winds are expected. Simple anchoring arrangements are essential in both the parking space and the protected plot to secure the structure in case of expected windy conditions. The simplest arrangement is the one used for anchoring light aircrafts in airports, using cables that can be fast-connect to a weight or a stake lodged into the ground. Another option is an arrangement to secure the structure onto its rails, as long as rails are well anchored to the ground. Never underestimate the power of wind.

Another simple and inexpensive version of a lightweight rainout shelter was seen at the experiment station in Western Australia (Fig. 4.7). Here the bare frame structure is static but the cover is provided by a moveable plastic "curtain" (not seen in photo). When it covers the protected area the structure becomes a large transparent tent. The portable frame is anchored to the ground with steel cables. Steel cables run along and over the protected plot. A strong curtain of transparent plastic (not shown) is connected to the cables so that it can be manually stretched over the plot or moved off horizontally. The "curtain" is attached to the horizontal cables by special runner clips that can smoothly move on the cable. This installation can be disassembled readily and transferred to another field. It is appropriate for a small plot.

The Stationary Rainout Shelter

This type of shelter is actually a well ventilated greenhouse that can be quickly rain-proofed or ventilated at will. When it is not closed up for rain exclusion it must be

Fig. 4.7 The portable structure and cables "tent" system seen in Merredin Western Australia (see text)

well ventilated to equilibrate with the external ambient conditions. Since it is permanently covered by a (transparent) roof, the water regime inside must be regulated by irrigation. Control (non-stress) plots are grown inside the shelter and are fully irrigated. Both the walls and part of the roof can be widely opened for ventilation when there is no rain. Irrigation can be provided by overhead low capacity sprinklers or by drip irrigation. See comments below on drip irrigation for this purpose.

Such shelters are usually constructed with strong frames and agricultural grade, light transmitting plastic walls and roof panels. Strong winds and the prevalence of hail should be considered in the construction of the roof. The plastic walls can be rolled up and down, manually or by electrical motor. The opening of the roof can be motorized or manual.

Both the roof and the ground surrounding the structure should provide full protection against vertical or horizontal leakage of water into the shelter. If possible, periodic measurements of temperature, humidity and light should be logged as well as soil moisture data – depending on the type of work performed under the shelter.

A simpler but less solid construction is the polyethylene covered "tunnel" that is used commercially for protected vegetable cropping in Europe and other places. The tunnel is based on a grounded semi-circular frames available commercially covered by agricultural grade transparent polyethylene sheets connected to the frame. The structure can be about 3 m high at the center. The lower sides can be opened for ventilation. A trench should be dug along both sides to collect and drain rainwater. Drip irrigation can be installed inside. The tunnel microclimate is less likely to equilibrate with the ambient conditions as compared with the larger ventilated greenhouse (above), especially in hot climates, but it is relatively cheaper and commercially available. This structure is not recommended in hot climates.

4.1.4 *Controlling the Severity and Timing of Stress in the Field*

4.1.4.1 The Principles of Achieving Control

Agronomists, micro climatologists and especially those who are involved with crop simulation models will advise that in order to be able to plan, execute and gage drought stress in a drought experiment one must measure all relevant atmospheric and soil variables in order to estimate daily crop water-use and thus to predict the timing and rate of the planned imposed drought. While this is a possible approach it is not absolutely necessary in phenotyping and selection work where one can access the plant and estimate directly or indirectly its water status with only a minimal reference to environmental variables. Therefore, in this case gauging and monitoring drought stress is usually done by directly observing the plants rather than indirectly by measuring their environment. This is not to say that environmental measurements in the experiment are useless. Since drought phenotyping is usually repeated in the same location during the course of breeding, experience gained can be an important lead for gauging stress treatments.

Drought stress is affected by stopping water supply, be it by terminating irrigation or by activating the rainout shelter. Stress will then develop gradually and it is crucial to be able to translate the number of days without watering into the desired level of plant stress. Several major factors will determine this relationship:

(a) The crop or plant species is of prime importance. Dehydration avoidant crops such as sorghum will take more time to dehydrate as compared with less avoidant species such as rice. Typically, when grown on deep soil of good water holding capacity, rice may take around 2 weeks to reach midday RWC of about 60–70% while roughly double this time or more might be required in sorghum, when both crops are around maximum LAI.
(b) Plant growth stage determines leaf area index and crop evapotranspiration. Hence, smaller plants will take more time to develop water deficit.
(c) A crop that has been pre-stressed will develop water deficit more slowly if subjected to a second drying cycle.
(d) High evaporative demand will accelerate stress development. The components of high evaporative demand are high irradiance, high temperature, low humidity (high VPD) and wind (Chap. 2).

It has been found by a survey performed among breeders (Mahalakshmi and Blum 2006) that they often use their own experience with the crop and the site in gauging drought stress by the number of days without watering while watching the crop for the appearance of drought stress symptoms, such as leaf wilting and desiccation. In some cases the total desiccation of the most susceptible genotypes in the nursery indicates peak stress. This is the reason behind the recommendation of Bänziger et al. (2000) to gain experience if needed by performing a simple experiment before phenotyping is attempted. In that experiment which should be performed in the same site, season and soil, seed of a particular crop cultivar is sown at different dates (say three weekly intervals) and irrigated at the same time during their growth. On a given day which is expected to result in stress at the desired growth stage of the middle planting date, irrigation is terminated. Watching and noting the rate of stress and time course of the development of plant stress symptoms (until total plant desiccation) will certainly inform on the date when irrigation should be terminated in the planned phenotyping experiment. Visual stress symptoms are quite indicative of the level of stress in most cases. However, occasional plant water status measurements (see below) can also be used for monitoring stress especially towards the peak. For example, the CIMMYT maize breeding program uses the following plant criteria for gauging the desirable level of plant stress: sever leaf roll at flowering; advancing leaf senescence; ASI of 3–8 days, all of which result in final yield of about 1–2 ton ha^{-1} (Atlin et al. 2009).

All of the above does not imply that environmental data are not important. The most important calculation is the balance between available soil moisture at root depth and evapotranspiration. Available soil moisture can be calculated from soil moisture measurement for the relevant crop root depth on the day following last irrigation or rain. Estimated crop water use at the site can be calculated from class-A pan evaporation times the crop coefficient. The crop coefficient adjusts pan evaporation according to crop development stage and leaf area. Coefficients for various crops are usually found in irrigation manuals and on the internet. As drought stress increases the crop is expected to reduce its water use below the estimate received from pan evaporation and the crop coefficient.

Planting at high density especially in row crops has been advocated as a method for promoting drought stress. However this is not required under most reasonably growing conditions.

As pointed out in Sect. 3.3.1, drought resistance become more important than yield potential in affecting yield under stress only when stress reduces yield below a certain level. For cereals such as wheat, barley and rice stress has to reduce yield to around a third of the potential yield. The question one might ask is how to formulate a stress level that would assure such yield reduction, when yield is measured only at the termination of the experiment. A simple answer is not at hand. Most breeders work by experience. It is however recognized that greatest effect of a given stress is achieved when applied at the most stress sensitive growth stage, namely the reproductive stage. Sensitivity to water deficit is generally increase in the following order: seedling > vegetative > pre-flowering > grain/fruit growth > Flowering/fertilization stages of development. The earlier developmental stages have lesser impact on yield for the same stress level also because options for recovery and repair of damage to future yield are still available when plants are younger. When stress is applied at pre-flowering vegetative growth stages, sometimes two cycles of stress are applied in order to achieve the desirable reduction in yield, without any stress at the reproductive stages.

When drought stress must be designed to peak exactly at a discrete growth stage, such as flowering, a major difficulty is in achieving this goal with a plant population that is inherently variable in phenology. It practice, phenotypic variation in flowering date beyond 3–4 days in the phenotyped population cannot allow accurate timing of stress at flowering in all tested accessions. In a routine breeding program where agronomic lines at F_4 and on are being phenotyped all materials are already at a homogenous plant phenology as dictated by the selected ideotype at earlier generations. However in the case of diverse germplasm, mapping populations etc., variability in flowering time is an inherent attribute of the population. The problem is reduced if phenotyping is performed at the vegetative growth stage, and if resistance at this stage is of interest. There may be specific cases where phenotyping at the vegetative stage might indicate also associated resistance at flowering. The problem of diverse phenology in drought phenotyping directed at flowering and grain/fruit development can be treated in several ways.

In certain cases, such as mapping populations which are designated for drought resistance phenotyping, the development of the population should initially consider minimal diversity in phenology without compromising general diversity. This can often be achieved by selecting parental lines of similar phenology while divergent for most other apparent traits.

Another solution sometimes deployed in practice is to subdivide the population into 2–3 subpopulations of similar phenology as previously expressed in a field trial under the same seasonal conditions. Each subpopulation is then phenotyped for drought stress at flowering independently in the same field with the others. A less popular approach is to plant each subpopulation at a different date ("serial planting") so that all subpopulations will flower on a similar date and will then be phenotyped for drought stress at flowering as one population. This is possible if one has gained sufficient experience at the site to be able to predict the shift in flowering date corresponding to a shift in planting date.

4.1.4.2 Measuring Soil Moisture and Water Table

Methods and sensors for soil moisture measurement vary according to accuracy, applicable soil depth, range of water status, ease of operation, cost, durability and reliability. A detailed discussion of all available methods and their pros and cons is beyond the scope of this book. A most comprehensive information resource for methods and equipment is http://www.sowacs.com web site. That resource provides also comparative studies of various sensors.

Not all methods listed in that web site are appropriate for drought phenotyping work in the field or in containers. For example, the tensiometer is a common instrument for soil moisture monitoring in irrigation scheduling. However it is useful only for measuring high soil moisture potential down to about 0.07 MPa. It is not suitable for a drying soil appropriate to plant stress.

The more appropriate and widely used systems for the field or in various container experiments are as follows:

(a) The classical benchmark gravimetric method
(b) The neutron probe
(c) Time domain reflectometry (TDR)

(d) Electrical capacitance sensors
(e) Electrical resistance sensors (e.g., gypsum blocks)

Methods generally require calibration for the specific soil to be measured.

The determination of water table depth and its seasonal fluctuation is crucial *before* selecting a site for phenotyping and then also during the experiment, if it might pose a problem. The simplest method is having several permanent bore-holes in the field and periodically measuring water level with a dip-stick. More sophisticated gadgets are available, even ones which will ring your telephone in case of a problem at midnight.

4.1.4.3 Irrigation Systems for Drought Control

Next to soil variability, faulty irrigation is a major factor in generating variability in drought experiments. Standard irrigation practice in commercial production rarely reveals slight malfunctions in water spreading. These are inflated mainly when water deficit develops. There are four major irrigation systems used in general plant breeding.

Flood irrigation is the least controlled system where water is released from a source point and spread into a large area, usually bounded and sometimes leveled. Even when the plot is leveled, water infiltration into the soil is always greater near the source and lesser at the farthest point. Only full saturation of the soil well below root depth will overcome the impact of this variability. Therefore, this method is unsuitable for managed stress phenotyping.

Furrow irrigation is more popular than flooding especially with row crops and where water is supplied by gravity. Here also water infiltration into the soil is greater close to the source and it decreases with distance. This is why furrows are always limited in length to reduce this variation. The method is not very commendable for drought phenotyping because of the variation it tends to create. Furrows can be used in drought phenotyping if their length is reduced drastically. This implies the development of a rather extensive network of water channels in the field. On the other hand it is conceivable that the most downstream part of the furrow will become water-logged if the control over water flow and discharge is poor. Placing replications downstream helps to account for the irrigation gradient. Furrow irrigation requires high level of proficiency on the part of the worker in the field. Where this method of irrigation has been traditionally used for generations, the level of accuracy and homogeneity is high, such as in China.

Sprinkler irrigation is more appropriate for drought field experiments because of its many known advantages. Still, the method is not perfect. The rate of water delivery should not over extend soil infiltration rate in order to avoid puddling and runoff. Sprinkler irrigation is extremely susceptible to wind. When portable aluminum irrigation pipes are used the system must be flushed in order to avoid clogging the sprinklers. During irrigation the performance of all sprinklers and lines should be checked frequently for clogs and leaks. Sprinklers should be mounted high enough above the canopy. The measurements of water output can be done by hooking capacity gages to the system or simply by placing rain gauges or even tin cans on selected spots in the field. The latter method will indicate irrigation homogeneity. Since there is always a certain gradient of water discharge perpendicular to the line it is advisable to place the sprinkler irrigation lines in parallel to the blocks (replicates) in the experiment.

In many studies related to selection and phenotyping for drought resistance, a fully irrigated control is used in order to compare genotypic performance between stress and non-stress conditions. An important application has developed by the deployment of the water spreading gradient inherent to sprinkler irrigation. The individual sprinkler does not discharge water uniformly whereas output is higher closer to the sprinkler. This is why sprinkler lines are placed close enough so that overlapping of sprinklers discharge reduces the effect of this gradient. The method, often termed the "line-source irrigation," is effectively being used in drought phenotyping. Individual accessions are planted in rows or as narrow and long plots perpendicular to a single irrigation line (i.e., the line-source). Thus, each plot/accession is subjected to a gradient of water deficit that increases with the distance from the line. Accessions may be replicated along one side of the line or at opposite sides of the line. Observations on each genotype are performed along the gradient at any desirable space intervals. Even a harvesting method was developed for this system (Gerards and Worrall 1986) whereby yields are automatically recorded as a function of the distance from the source in each genotype.

When two adjacent line sources are set and the test material is planted also between them, plots may be irrigated by both lines to meet their full water requirement at early growth stages. When a stress gradient is desirable at a later growth stage, only one of the lines is operated. Additional technical considerations in affecting the performance of the system were discussed by Willardson et al. (1987).

Drip irrigation has also been used in drought phenotyping with the evident advantage of being unaffected by wind and capable of supplying moisture to small area. Several major considerations are crucial with this system especially when used for drought studies. Water delivered into the system must be clean to avoid clogging the emitters and therefore water filtering should not be compromised. Recommended pressure should be observed with regard to length of lines in order to avoid a pressure and delivery rate reduction along the line. Distance between lines is important with respect to delivery rate and irrigation homogeneity. In row crops such as maize or sunflower each row is supplied by a line (Fig. 4.8).

Drip irrigation was designed for low rate of delivery. One of the important advantages of drip irrigation in commercial production is that watering is applied at high time frequency to a limited volume of soil (and root). This irrigation profile is unsuitable for managed drought environments. Concentration of roots in a small upper volume of soil due to irrigation method must be avoided. Here, more normal rates of delivery approaching standard rain or sprinkler irrigation are desirable to assure soil profile soaking to depth within a reasonable time period. This is technically possible if the system is initially designed for the purpose.

Fig. 4.8 Empresa Brasileira de Pesquisa Agropecuária (EMBRAPA) drought managed maize phenotyping site at Janauba, Brazil, using drip irrigation. Note the evident difference between the two genotypes in their drought resistance. Note also that the first plant in the susceptible genotype appears relatively well watered due to the border effect

4.1.5 Managed Drought in Protected Environments

4.1.5.1 The Pot Experiment

Experiments performed with pot-grown plants are probably the most common in plant research. The need to perform "controlled" experiments often replaces observations on plants grown in their native environment by bringing plants for closer observation in the laboratory. Plants are then grown in an artificial environment in various containers with various media and placed under growing conditions which can range from growth chambers and greenhouses to the laboratory window sill.

Since drought stress tends to amplify variations in growing conditions, pot experiments under stress conditions are grounds for amplified error variance and artifacts.

The following comments bring to the attention of the beginner some of the more common problems and their remedies.

Pot size has a huge effect on plant growth, especially where the root environment is exposed to water limitation. Water use from a pot depends on plant size and its leaf area. Larger plants use more water daily as compared with smaller ones. When pot water content is limited a large plant will show symptoms of wilting before the smaller plant does. There is no harm in this occurrence unless the researcher fails to recognize it to be the result of the interaction between plant size and pot size and not a result of the experimental treatment per se. Pot experiments are a very common method of researching drought resistance in plant transgenics work. In more than a few cases the transgenic plant is smaller in size than the wild type. This difference is sufficient to express less wilting in the transgenic than in the wild type on a given day after water has been withheld, irrespective of the expression of the studied gene. This artifact (e.g., Serrano et al. 1999; Hsieh et al. 2002; Dai et al. 2007) is not always recognized by the authors or the journal review process (Fig. 4.9).

The question is how to account for the genetic difference in plant size and leaf area in different genotypes when stressed in a pot experiment? There is no formal research-based qualified answer. One voiced suggestion was to grow the two plants of the transgenic and the wild type together in one pot with the assumption that whatever is the rate of transpiration in any genotype, both will be exposed to the same soil water status in the pot. However, even with the same soil water status in the common pot, the large plant will still express a lower leaf water status because of its greater demand for water. A better thought is to plant the two genotypes in different pots on staggered planting dates so that among the different planting dates one can find plants of comparable size in both genotypes and use these pots for the dry-down experiment. This would involve a difference in plant age, which might be a lesser problem than dealing with different plant size, as long as non-flowering plants are tested.

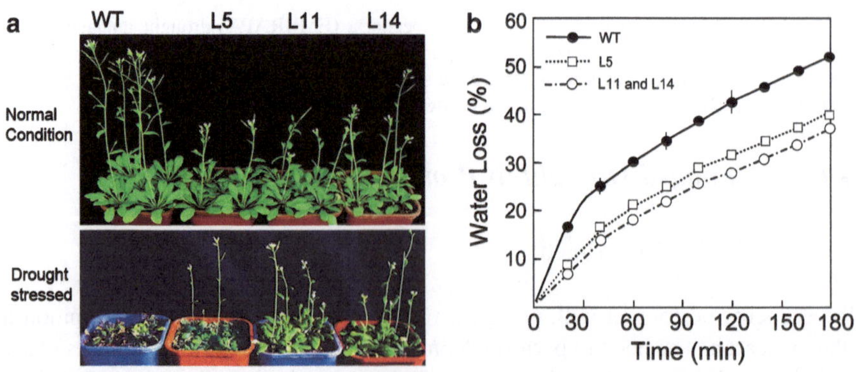

Fig. 4.9 Representing Fig. 6 in Dai et al. (2007) which was their basis for claiming better drought resistance in the transgenic *Arabidopsis* lines as compared with the wild type (WT). At the same time this figure represents an apparent effect of plant size on the differences between genotypes. (**a**) The WT plants were larger than the transgenic plants resulting in relatively earlier development of wilting; (**b**) figure clearly shows the relatively greater water use from the pot by the larger wild type plant. (With permission of The American Society of Plant Biologists)

Often in cases of limited facility space, such as a growth chamber, the smallest pots possible are chosen so as to accommodate more pots and plants inside the chamber. The conventional wisdom is that if nutrients and water are supplied almost daily, root space and size is of no consequence. However, the physical restriction of roots in a small container in itself can have a negative effect on plant growth and functions (e.g., Ismail and Davies 1998; Zaharah and Razi 2009). Even under well watered conditions such effects can involve reduced stomatal conductance and various shoot morphological and anatomical modifications, probably mediated by root ABA signaling.

Potting media such as soil, all kind of ad-hoc mixtures and commercial potting mixtures are an important determinant of plant growth. The pros and cons of the different media used are tightly linked to the purpose of the experiment. For example, if one wishes to recover clean roots by washing them out of the pot, a potting mixture containing a peat and other organic materials is a nuisance because these cling tightly to the roots. Commercial potting mixtures sometimes tend to be coarse so as to optimize aeration and drainage. This would reduce the likelihood of water-logging but on the other might induce fast dehydration when irrigation is stopped. Chances for water-logging in a soil-filled pot can be reduced if the pot is relatively tall, say >40 cm. On the other hand soil of high clay content should be avoided since it tends to shrink, crack and involve serious root damage upon drying. Shrinkage of the medium upon drying is common in some commercial mixture, which should be avoided. Do not use a commercial mixture before testing it for your purpose. Whatever potting medium is used it must be absolutely free of any soil inhabiting plant parasitic factors such as root diseases or nematodes. Sterilization is a standard precaution. Where sterilization equipment is unavailable, soil solarization is also an effective method (Katan and DeVay 1991).

One precaution is common to all container experiment and that is a standard routine for filling up the container. Usually a technician or student is sent over to the shed with the instruction to fill up a number of pots for an experiment, which he might do by scooping the mixture into the pots. Much more consideration should be given by caring to achieve standard soil (mix) bulk density in all pots. This is important with the heavier mixtures which contain more soil and with larger containers. Bulk density should be high enough to avoid air pockets especially around the walls of the pot. These capture roots which tend to grow along a path of least resistance. The simplest way to standardize soil bulk density is by taking care that the potting mix is very homogenous and by filling the pot in a standard manner and to a given weight. Eliminating air pockets in the pot is not done by pressing the mix into the pot with your fingers. The best way is simply to pound the pot against the ground in a standard way in order to settle the mix evenly in all pots.

Pot color is often not given any special consideration besides color-coding experimental treatments, sometimes. When the experiment is performed in a greenhouse exposed to sunlight, color makes a huge difference. Black pots will heat up several degrees above white pots when exposed to direct sun for several hours. This effect is magnified in smaller pots. However, when pots are crowded in a growth chamber, pot color will make no difference.

Stress treatment in most pot experiments is affected by stopping irrigation. This is a legitimate protocol as long as the pot is not too small and plants do not dehydrate and wilt too fast within a few days. The importance of time duration in the stress protocol has been repeatedly discussed above. Drought stress must proceed slowly, for at least a week to first symptoms (depending on the experiment). The main control of this time duration is pot size relative to plant size.

Another approach to control pot water deficit is sometimes found in the literature where an attempt is made to maintain a "fixed" level of drought stress in the pot. This is presumably done by calculating the water holding capacity of the pot and by maintaining less than that amount (say 50%) at all times. This is done by weighing the pot and replenishing by daily increments the amount of water loss up to the predetermined level. This results in a frequent supply of small amounts of water to the surface of the soil of a drought stressed plant. Only the top soil of the pot exchange water with the environment and the plant and only that layer receives small amount of water. These plants are not exposed to a normal stress as under natural conditions. This book-keeping protocol might be correct mathematically but it is incorrect physiologically.

Finally, always remember that the plant in the field might have other options for coping and different reasons for failing under drought stress, as compared with its performance in your pot experiment.

4.1.5.2 The Greenhouse as a Source of Experimental Error

The greenhouse is probably the most common installation for performing convenient in-house controlled experiments while it is a most variable environment. Here reference is made to the standard greenhouse which is not a fully regulated phytotron but rather designed to be cooled in the summer and heated in the winter while having some form of ventilation. The structure might consist of glass or any kind of plastic materials. It is usually located on the campus between other buildings.

The purpose of this little section is not a review of greenhouse technology and applications but rather to indicate some common problems typical of greenhouse experiments which the practitioner should watch for. These problems are usually accentuated in experiments involving drought stress with plants grown in the ground or in containers.

(a) When cooling or heating is applied there is usually a gradient of temperature from the source, downstream. If replications do not account for this gradient it can introduce serious variability into the experiment.
(b) Open sides or side fans used for forced ventilation can also develop a gradient of atmospheric conditions inside the greenhouse.
(c) It is very common to build the greenhouse close to a main building, as matter of convenience or economy. It is not uncommon for this building to shade the greenhouse differentially; namely shade only part of the greenhouse every day. This is a serious fault, especially with plant water relations studies.

The greenhouse should be inspected several times during the day for possible gradients in ambient conditions inside the structure and try to design the experiment to account for this variability, if possible. The perfectionists use various systems to rotate the pots inside the greenhouse in order to reduce the effect of spatial variability. Advanced systems use special conveyors which permanently rotate the potted plants inside the installation. Never take it for granted that a greenhouse is a "controlled environment."

4.1.5.3 Hydroponics in Drought Research

Growing plants in nutrient solutions does not appear to be a logical system for drought phenotyping. However, when the solution is mixed with an osmoticum to reduce solution water potential, then plant roots are subjected to a simulated state of a dry soil. This is not a true simulation of a drying soil where soil moisture gradients are developed and other factors typical of soil are involved. However, a nutrient solution of low water potential can be useful as a comparative screen for differences in shoot water status symptoms such as wilting, low RWC, arrested growth etc. The primary condition is that the addition of the osmoticum will not introduce any foreign root environmental factor or plant growth factor that can bias the results from representing true plant water relations effect.

Polyethylene glycol (PEG) is a polymer produced in a range of molecular weights. Lagerwerff et al. (1961) were the first to indicate that PEG can be used to modify the osmotic potential of nutrient solution culture and thus induce plant water deficit in a relatively controlled manner. It was assumed that PEG of large molecular weight did not penetrate the plant and thus was an ideal osmoticum for use in hydroponics culture.

During the 1970s and 1980s PEG of higher molecular weight (4,000–8,000) was quite commonly used in physiological experiments to induce drought stress in nutrient solution cultures. Several papers also reported theoretical or measured calibrations of PEG water potentials against molecular weight and concentration. Examples for PEG calibration can be found in Michel and Kaufmann (1973) and Money (1989) (Table 4.1). However, experience gained by users indicated that calibrations can diverge to some extent depending on the lot or source of the specific PEG used. It is therefore advisable to measure the actual osmotic potential of the nutrient solution used.

There are important problems to consider in using PEG solution as a root medium.

PEG uptake by plants. Even PEG of high molecular weight, such as 6,000–8,000 was found to be taken up by certain plants. The problem can be in the plant or in the chemical, which is not a pure and can be a mixture of different molecular weights. PEG was taken up by maize and bean plants at a relatively slowly rate of 1 mg g^{-1} fresh weight per week. However, when roots were damaged or broken, the rate was higher. Cotton absorbed less PEG than maize (Lawlor 1970). Pepper plants also took up PEG, where the higher molecular weight PEG was mostly concentrated in roots while the lower molecular weight fractions accumulated in leaves (Janes 1974).

Table 4.1 Osmotic pressures of PEG of different molecular weight (mw) as calibrated by Money (1989). OP = a.C + b.C2, where OP is the osmotic pressure in MPa and C is the molarity of the respective PEG

PEG (mw)	a	b
2,000	−5.2	128.1
3,000	−12.1	328.4
4,000	−8.5	435.2
6,000	−12.1	980.0

Yaniv and Werker (1983) presented striking photographs of PEG 1,500 to 6,000 deposited on leaves of various Solanaceous plants exposed to PEG in the root medium for 24 h or less. Again, greater deposition was seen in plants with physically damaged roots. PEG 6,000 was taken up by tomato plants and was found in older leaves and roots (Jacomini et al.). The critical finding was that leaves containing PEG behaved hydraulically differently from leaves without PEG when grown in PEG containing nutrient culture. It can also be suspected that plant water status might be affected by PEG in the root medium not via its osmotic effect but via uptake into and clogging of root xylem conduits. It can therefore be concluded that PEG, even of higher molecular weight, can sometimes be taken up by plants and the rate of uptake and concentration in roots and shoots depends on the species, on PEG, on time of exposure and on root damage. Therefore, before PEG culture is attempted, the specific plant species should be tested for PEG uptake into roots and leaves. Different methods of PEG analysis are available. The simplest is the turbidimetric method based on precipitation by trichloroacetic acid (Lawlor 1970; Janes 1974). High pressure liquid chromatography has also been used (Yaniv and Werker 1983).

Hypoxia in PEG solutions. Verslues et al. (1998) reported that plants grown in nutrient culture containing PEG suffered from hypoxia and if such a system is used the solution should be aerated. Standard aquarium pumps are often used.

Mineral contamination of PEG. PEG of 3,000–4,000 mw from two commercial sources was found to be contaminated by high concentrations of phosphorus, which could introduce a problem if used in experiments involving P interactions (Reid et al. 1978). Toxic metals such as aluminum were also found to contaminate PEG and pose a toxicity problem when PEG was used in culture (cited in Reid et al. 1978).

Tingey and Stockwel (1977) suggested to overcome most problems of PEG absorption by using a semi-permeable membrane to separate between the roots and the PEG solution culture. With this system the plants are grown in a "container" made of semi-permeable dialysis membrane (such as Spectrapor standard RC tubes) placed in a plastic test tube. These tube-like membranes come with different molecular weight cutoff. Membranes come in rolls of 5–15 m length, depending on their width and specs. A length of the tube (equal to double the length of the test tube) is cut and folded into a small bag and fitted into a plastic test tube which has a hole made in its base (Fig. 4.10). The membrane-tube diameter should fit the test tube diameter. Vermiculite is poured into the tubes and tubes are immersed in water in a tray. Seeds are germinated in the wet vermiculite. After emergence water is drawn out of the tray and the growing seedlings are allowed to dry the vermiculite to a moderate extent for several days. Then, nutrient solution containing PEG is poured into the tray. Water will penetrate the vermiculite but it will be under the

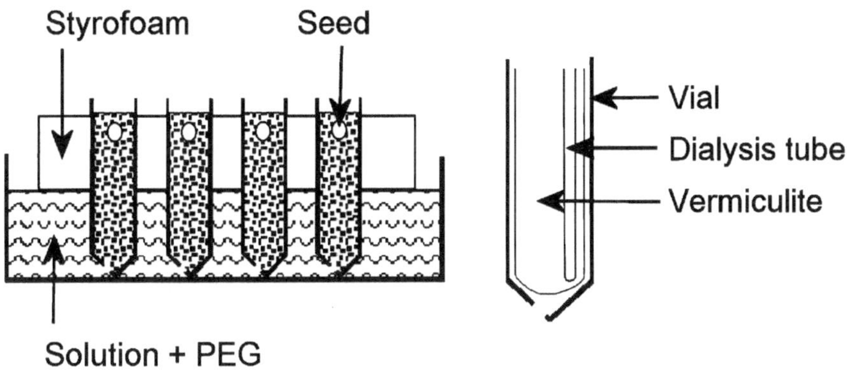

Fig. 4.10 A schematic representation of growing plants in polyethylene glycol (PEG6000) nutrient culture without exposing roots to direct contact with PEG as facilitated by use of a semi-permeable MCO3500 Spectrapor RC membrane (www.spectrapor.com) (see text)

negative potential ("suction") of the PEG outside the membrane. It may take up to a week or more to see initial symptoms of plant stress, depending on the initial water content of the vermiculite and ambient conditions. This slow development of stress is most desirable as it simulates natural conditions. The system requires no aeration because roots are aerated atmospherically by diffusion through the vermiculite. The PEG nutrient solution should be exchanged at least weekly. This system is suitable only with small plants.

Mannitol is sometimes used as an osmoticum in laboratory work to induce drought stress. Mannitol is a natural product that accumulates in certain lower and higher plant species. It should not be surprising that it is taken up by plants grown in mannitol solutions (e.g., Lipavska and Vreugdenhil 1996). Mannitol is therefore unsuitable for simulating drought stress in hydroponics.

4.2 Protocols for Drought Resistance

4.2.1 Plant Growth and Productivity

4.2.1.1 Yield

In most cases yield is the target of the breeding program and yield is also "the great integrator" of genetics and environment. As already discussed in detail above, the final expression of yield under drought stress is a complex integration of constitutive plant traits and stress-responsive processes which depend on stress intensity, duration and timing with respect to growth stage. Depending on the timing and intensity of stress, yield components may interact with and compensate for each other. Therefore absolute yield is of interest but measurement of yield components may provide additional insight into the expression of drought resistance in terms of productivity.

The rate of yield or biomass reduction by stress (e.g., yield under stress as percent of yield under non-stress) is a widely used measure of resistance in terms of plant production, in addition to absolute yield under stress. This ratio requires tests under both stress and fully irrigated conditions. Fischer and Maurer (1978) improved on this simple ratio by considering the ratio of reduction in a given genotype as compared to the mean reduction over all genotypes in a given test. They proposed a *drought susceptibility index*. Here a *Drought Resistance Index* (DRI) is given based on their calculation: DRI = (Ys/Yn)/(Ms/Mn), where Ys and Yn are the genotype yields (or biomass) under stress and non-stress respectively and Ms and Mn are the mean yields (or biomass) over all genotypes in the given test under stress and non-stress respectively.

Often claims were made that this index was not a true representation of drought resistance in terms of yield because it is biased by yield potential (yield under non-stress), namely genotypes of higher yield potential tend to express a lower DRI. This is true, but it is not necessarily a mathematical or a statistical bias. This bias has real biological grounds as discussed in Sect. 3.3. DRI is gaining popularity as a useful criterion in selection for drought resistance (e.g., Fukai et al. 1999).

Another drought resistance index has been proposed by Bidinger et al. (1987). This index is also derived from the comparison between yield under stress and yield under non-stress. However it also considers flowering time and attempts to normalize the yield results for variation in flowering date. Therefore, a multiple regression of stress yield on non-stress yield and days to flowering is performed over all genotypes in a test. Then the actual yield under stress is regressed across all genotypes on the predicted yield under stress. For every data point (genotype) it is possible to evaluate the deviation from the regression (by calculating the studentized residuals for example). Positive deviations of the actual yield indicate relative drought resistance independent of the effect of phenology and yield potential. Drought resistant genotypes of superior performance according to this analysis are found at the high end of the regression with a positive deviation from the regression.

4.2.1.2 Plant Growth Attributes

Since growth attributes may differ extensively among genotypes under potential conditions, growth response to stress in most cases should be measured and expressed in terms of deviation from the potential. This, again, requires phenotyping under stress and non-stress conditions.

Whole plant relative growth rate (RGR) is the classical parameter of plant growth analysis. Plant growth analysis dissects RGR into net assimilation rate (NAR, rate of dry matter gain per unit leaf area) and leaf area ratio (LAR, leaf area per unit total plant mass), where RGR = NAR × LAR.

NAR is determined primarily by the ratio of carbon gained through photosynthesis and carbon lost through respiration. LAR reflects the amount of leaf area a plant develops per unit total plant mass and it therefore depends on the proportion of biomass allocated to leaves relative to total plant mass (leaf mass ratio, LMR) and how much leaf area a plant develops per unit leaf biomass (specific leaf area, SLA), where LAR = LMR × SLA.

In its simplest application which is based on destructive plant sampling for whole shoot dry matter determination, RGR can be estimated as RGR = (ln W_2 – ln W_1)/(T_2 – T_1), where W_1 and W_2 are shoot dry weights at times T_1 and T_2 (Hunt 1990).This analysis should suffice for large scale phenotyping work, using a typical time interval of 5–7 days as slow stress develops.

When plant recovery from severe drought stress is phenotyped, visual estimates of plant re-growth and viability upon recovery may suffice in most cases where the extremes are screened for. However RGR will express well also the smaller effects of viable leaf area maintenance after stress in the different genotypes. Hence, estimating recovery growth by RGR is usually performed beginning at 2–3 days after recovery watering. Since large inter-plant variability exists in total shoot dry matter especially under drought stress, W_1 and W_2 should be estimated as a mean over several plant samples rather than by a single plant sample. Replications should also be employed. Therefore, RGR determination requires a sufficient number of plants for destructive sampling, over stress and non-stress treatments.

Extension and expansive growth is very sensitive to drought stress. When stress occurs during stem elongation, stem length is a sensitive expression of drought resistance in terms of growth. Thus, finite plant height reduction under drought stress as compared with non-stress conditions is a legitimate criterion of drought resistance in terms of growth during stem elongation.

In the cereals, the extension of the inflorescence peduncle or the exertion of the inflorescence is very sensitive to drought stress and it is frequently used as a manifestation of drought resistance during this growth stage, such as in rice (Subashri et al. 2009), wheat or sorghum (Fig. 3.9). In rice poor panicle exertion was linked to high rate of panicle sterility. Inflorescence exertion following a pre-flowering drought stress is therefore a very sensitive and clear stress symptom which can be quickly scored in the field. It does not require any comparison with non-stress conditions. However, variations in phenology within the population should be considered.

Wang and Bughrara (2008) concluded that leaf elongation under drought stress was a reliable selection criterion for drought resistance in Atlas fescue, perennial ryegrass, and their progeny. It seems that leaf elongation is a good integrator of whole plant capacity to cope with drought stress, as supported by detailed research in maize (Chenu et al. 2008). Leaf elongation measurement under drought stress requires repeated measurements of leaf length on the same expanding standard leaf in all genotypes. In non-cereal plants one has to decide on the most relevant linear measurement of the expanding leaf. This criterion may not be practical for screening large populations.

4.2.1.3 Flowering Delay

Depending on the crop, flowering might be delayed under drought stress and the extent of delay is a function of stress level and genotype. Flowering delay estimates can be developed if stress is applied before and into the flowering growth stage, in comparison with a non-stress treatment. In some respected flowering delay is an expression of inhibited inflorescence exertion which is a symptom of peduncle growth reduction under stress, as discussed above. Flowering delay has become a popular trait in rice breeding for drought-prone environments, probably because heading is exceptionally susceptible to drought stress. Two molecular marker loci from a drought resistant rice cultivar ('Apo') were found to ascribe reduced flowering delay under stress (Venuprasad et al. 2009).

A most significant development of flowering delay as an expression of drought susceptibility has been developed and used in maize in the form anthesis-silking interval (ASI) (Sect. 3.3.1). Drought stress during the reproductive stage delays silking and thus pollen shed does not correspond with silk receptivity. A short ASI involves minimal delay of silking under drought stress. ASI is estimated by measuring male flowering as the number of days from sowing to the first anther extrusion from the tassel glumes. Female flowering is measured as the number of days from sowing to the first visible silk. The difference in days between the two is ASI.

4.2.1.4 Reproductive Failure

Reproductive failure under drought stress can be driven by low flower water status, high ABA content or carbon starvation. The outcome is generally pollen sterility, poor ovule development and ovule abortion. The phenotype expresses poor seed or fruit set and genetic variation in this respect is eminent. Regretfully, despite the importance of reproductive failure as a component of drought resistance, we have no consensus on a fast and accurate method to rate reproductive failure on a single plant basis. Certainly yield will express this trait, but one has no indication on the part of fertility disruption in affecting yield.

One approach is to measure yield of fruit or grain per plant when drought stress has been targeted to the flowering growth stage, in comparison with non-stress conditions. Visual inspection of individual inflorescence in cereals or pods in legumes can lead to an estimate of fertility or sterility. Actual counts of failed florets or aborted grain are more accurate but take too much time. In some cases, such as rice, unfilled florets can be readily distinguished from filled ones, so that scoring for fertility is relatively fast. Ease and accuracy of fertility assessment therefore varies according to the species and more research is needed to develop phenotyping methods for reproductive failure.

4.2.2 Plant Water Status – The Expression of Dehydration Avoidance

4.2.2.1 Dehydration Symptoms in Leaves

The initial stage of plant dehydration may not always be distinguished, but as plant water status decreases and turgor reduces recognizable symptoms of dehydration appear. The most universal and popular symptom is wilting. Drooping leaves accompanied by some discoloration is common in broad-leaved species. A "slow-wilting" line has been recognized in soybean (Fletcher et al. 2007). In the cereals, leaf rolling is a well recognized symptom (Fig. 2.4). The rate of leaf rolling is being recommended and used as a reliable estimate of turgor loss in drought phenotyping of cereals (Fischer et al. 2003; Bänziger et al. 2000). Leaf rolling and wilting varies with leaf age and leaf insertion. The visual scoring of rolling or wilting of a given genotype must be based on an integrated impression of the symptom in the whole plant or even the whole plot. Therefore, in each case and type of materials the breeder himself decides on a scoring system based on his/her own experience and skills. Scores usually range from null to 5. Scoring is done several times during a drought cycle in order to obtain an average score or a score at peak stress. Scoring is done at peak daily stress during midday. Sometimes the lack of turgor recovery in the morning is an important observation to note. Low score for leaf rolling or wilting is an indication of relatively better dehydration avoidance.

As plant water deficit progresses, wilting is followed by leaf desiccation and death. Leaf death begins with lower leaves and proceeds upwards. To the same extent as with leaf rolling and wilting, leaf death is also being scored (usually 1–5) as an estimate of relative plant dehydration level. In this case variation in phenology within the phenotyped population must be considered and accounted for because natural leaf senesce is also expressed in desiccation of lower leaves after flowering.

Very small variations in leaf wilting or leaf death score, even if they are statistically significant, are of no real consequence in breeding for dehydration avoidance. Large and prominent differences are sought and these are easy to recognize visually (e.g., Fig. 4.8).

There is some debate about the physiological of role of early wilting and leaf death in cases when plant survival under extreme drought is considered. Minor views consider extensive leaf rolling and leaf death as a mechanism for reducing water use and for the extension of plant life in a prolonged drought. Such views would favor the selection of relatively severely wilted and desiccated genotypes. These views are generally unaccepted by breeders who prefer a non-wilted plant over a wilted one under drought stress.

4.2.2.2 Measuring Plant Water Status in Breeding Work

Direct Methods

The methods discussed here are only those applicable to phenotyping work, namely they are reasonably rapid and are suitable for assessing large numbers of genotypes. For example, the pressure probe method for measuring cellular turgor (Tomos and Leigh 1999) is not applicable here.

Sampling Leaves for Direct Measurement of Leaf Water Status

Plant water status changes with the daily march of the atmospheric conditions and solar radiation. The comparison of all genotypes must be done on the basis of equal atmospheric conditions in all. That is practically impossible. The accepted compromise is that measurements will be performed when the daily march of atmospheric condition is slow and there is almost a plateau in conditions for about a couple of hours. Depending on the locality and season this occurs at midday (solar noon) or immediately after midday, which is defined hereon as "midday". Measurements at midday are also desirable in terms of maximizing differences between genotypes in their water status, thus expressing well their capacity for dehydration avoidance. Variations in dehydration avoidance are minimized in the dark when plants re-hydrate. Therefore, plant water potential at sunrise nearly equates with soil water potential and genetic variations in the capacity for dehydration avoidance are not well expressed.

Since leaf water status is strongly affected by solar radiation, transient cloudy conditions during measurements in the field will inflate the experimental error. The effect of a passing cloud at midday on stomatal response and leaf water status can be noticed even within minutes.

Different leaves are at different water status, especially during peak stress around midday. Differences develop according to leaf insertion along the stem, leaf age, shade and sun leaves and leaf inclination with respect to the sun. Therefore, a standard leaf must always be sampled. Usually it is defined as the top-most fully expanded sun-lit leaf.

Water Potential by the Pressure Chamber

The pressure chamber affords relatively quick estimates of the water potential of whole leaves, small shoots or large leaf sections. This method was pioneered by Henry Dixon at Trinity College, Dublin, at the beginning of the twentieth century, but it did not come into widespread use until P. Scholander and coworkers at the Scripps Institute of Oceanography improved the instrument design and demonstrated its usefulness (Scholander et al. 1965) (Fig. 4.11).

The organ to be measured is excised from the plant and is sealed in a pressure chamber with the cut edge (a petiole in the case of a broad-leaf) protruding out. Before excision, the water column in the xylem is under tension. When the water column is broken by excision of the organ, water is pulled rapidly from the xylem into the surrounding living cells under the gradient of water potential existing between cells and xylem. The chamber is then pressurizes with compressed gas until the distribution of water between the living cells and the xylem conduits is returned to its initial, pre-excision, state. This can be detected visually by observing the water returns to the open ends of the xylem conduits as seen in the cut surface. The pressure needed to bring the water back to its initial distribution is called the balance pressure. It is equated with whole leaf water potential. This however is not absolutely correct since the balance pressure may also depend on tissue osmotic potential which is an unknown quantity at the time of measurement. Furthermore the measurement should be performed with a non-transpiring leaf so that wrapping the leaf in plastic bag before its excision and during measurement is recommended. However, for comparative and fast phenotyping work where large differences are sought these errors are of a lesser significance as with more accurate physiological studies.

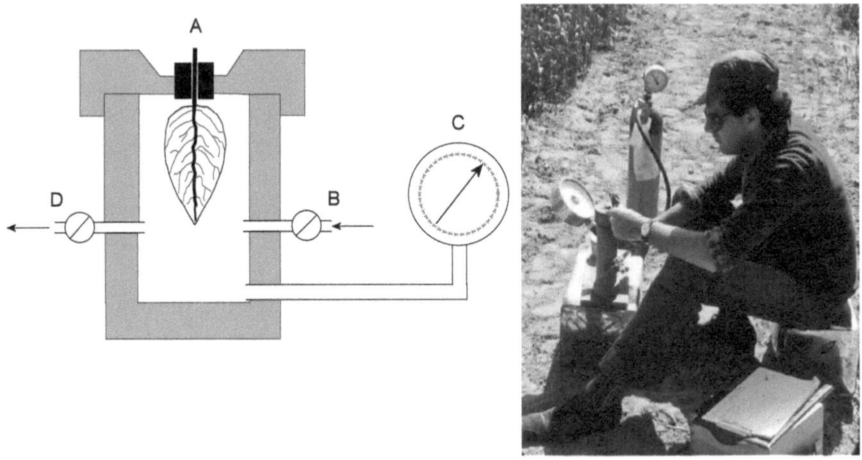

Fig. 4.11 *Left*: A schematic presentation of the pressure chamber, (**A**) leaf inserted into the sample holder; (**B**) chamber is pressurized; (**C**) Balance pressure is measured; (**D**) pressure is released (see text). *Right*: Pressure chamber operation in the field

The method is widely used in detecting genetic variation in dehydration avoidance in terms of leaf water potential. Care must be taken to minimize the time duration between leaf excision and its measurement to less than 30 s. Usually one person operates the chamber while another one fetches the leaf samples.

By comparing water potential and volumetric water content of leaves the pressure chamber has been used also for estimating osmotic potential, turgor loss point, symplastic fraction of water, and bulk modulus of elasticity (Schulte and Hinckley 1985). However such studies are slow and limited to a relatively small number of leaf samples and are beyond the scope of this text. Interested readers should search for "pressure-volume curve of leaves."

Water Potential by the Thermocouple Psychrometer

This method is probably the benchmark for leaf water potential. It is however very slow and meticulous. It is based on the vapor equilibrium between a leaf sample and a thermocouple within a small air-tight chamber.

A small leaf sample is enclosed in a small (e.g., 2–5 cm^3) air-tight chamber into which the tip of a thermocouple thermometer is exposed. After an equilibration period which may extend for 15 min to about an hour (depending on the instrument) a steady vapor pressure develops in the chamber, which is proportional to the sample water potential. A cooling current is briefly applied to the thermocouple. The cool thermocouple junction condenses water. The amount of water on the junction is proportional to the vapor pressure in the chamber which in itself is proportional to sample water potential. Once the cooling current is stopped the thermocouple will warm up. The rate of temperature change of the thermocouple depends on the rate of water evaporation from the junction which is proportional to the amount condensed. Thus, this rate of temperature change is calibrated against the water potential of standard salt solutions of known water potential placed in the psychrometer chamber.

The procedure is sensitive to ambient temperature and various methods were developed to account for temperature, ranging from placing the chambers in a thermostated water bath to various temperature compensation algorithms.

If the leaf sample is live then the resulting measurement is water potential. If the leaf is killed (by freezing the chamber with the enclosed sample, for example) all cellular solutes are released from the tissues, creating a free solution. A measurement performed after such treatment provides an estimate of tissue osmotic potential. Therefore the thermocouple psychrometer allows measuring leaf water and osmotic potentials of the same tissue in the same instrument, which permit the calculation a fairly accurate estimate of turgor potential.

It is quite apparent that the method is not suitable for large number of samples typical of breeding work but breeders should be aware of it and might use it with limited number of leaf samples or use it as an osmometer.

Leaf Relative Water Content (RWC)

This is a simple, standard and effective estimate of plant water status with respect to dehydration avoidance. It follows the initial principles set by Barrs and Weatherley (1962). It estimates the volumetric water content of the leaf relative to the water content at full turgor. The difference between leaf water potential and RWC is that while the former is a measure of the physical status of water, RWC measures volumetric water content and thus expresses also the effect of osmotic adjustment on leaf water content. Percent water in leaves has been used occasionally mainly in testing transgenic plants for drought resistance. This is not a legitimate measurement of water status since it is affected by leaf dry matter content variation across samples.

As discussed in Sect. 2.4.1 more water is held by cells when their osmotic pressure increases. Thus, for two different genotypes which may have the same midday leaf water potential, the one with a greater capacity for osmotic adjustment will express relatively higher RWC. RWC is a crucial measurement for identifying genotypes of a possible capacity for osmotic adjustment. However, difference in RWC can be developed between two genotypes in the field due to factors other than osmotic adjustment, which allow higher maintenance of leaf volumetric water content.

RWC is measured according to the same leaf sampling protocol as described above. If the leaves are small a whole leaf or a leaflet is sampled and placed in a pre-weighed, air-tight and dry vial with the cut petiole towards the bottom. In the cereals a short (5–7 cm) leaf section can be cut with scissors and placed with the basal cut surface towards the bottom of the vial. In large leaves one or more leaf discs are cut into the vial. Vials are collected in a cool picnic box (not frozen) and transferred to the lab after no more than 1 h. The number of samples collected is limited only by the number of workers. In the laboratory the vials and their content are weighed (FW). Small amount of water is added to all vials to allow the samples to re-hydrate at room temperature or in the refrigerator at about 10°. After a period of 4 h the leaf samples are taken out of the vial, surface-dried quickly with filter paper and weighed (TW). All samples are then oven-dried for their dry matter weight (DW). Sample RWC is calculated as: RWC = [(FW − DW)/(TW/DW)] * 100.

Error variance in RWC estimates can be large if the protocol and practice are not followed carefully. However, under well-executed conditions the least significant difference ($p = 0.05$) in RWC between genotypes can be as low as 2% of RWC.

Boyer et al. (2008) cautioned against excessive rehydration of samples (TW) which can result in excessive absorption of water by the leaf sample, beyond its normal full turgor capacity. This would bias estimated RWC downward. It seems that this problem might occur with leaf sections floated on water as with the original protocol of Barrs and Weatherley (1962). In the case of leaflets or cut leaf sections fed with water through a cut edge or similar cases where samples are not floated on water, the problem might be much less significant. It is recommended to establish the minimum duration of rehydration time required before any given experiment.

Osmotic Adjustment

A detailed comparative study of three main methods for measuring osmotic adjustment (OA) is described in Babu et al. (1999).

A correct estimate of OA begins with the pretreatment of the plants towards measurement. The rate of OA depends on the rate and extent of dehydration. Fast dehydration may not allow sufficient time for the expression of OA. Similarly, insufficient degree of dehydration may not express the full capacity for OA. Therefore, both a slow rate of stress and a standard endpoint of stress when measurement is performed in all tested genotypes are required. This cannot be achieved in the field. OA measurements are best performed in potted plants grown in a reasonably controlled environment. Growth chambers are usually unsuitable unless they can provide for high irradiance and large pots. Therefore, a uniformly sunlit greenhouse is an acceptable option.

Based on the experience of those who performed such measurements, genotypic differences in OA of crop plants are well expressed when young fully expanded leaves reach a RWC of 50–70%, depending on the species. Very generally, this level of RWC corresponds with pronounced symptoms of leaf wilting or rolling. This level of stress must be reached slowly, in at least 2 weeks after watering has been terminated. This period of time is regulated by fitting plant size to pot size. A fairly narrow range of time and RWC must be adopted in a single experiment, say measurement of OA at 65–70% of RWC as long as this RWC will be reached in no less than 2 weeks. It follows that for the given environment and plant, a preparatory test should be performed in order to establish the specific protocol.

There are three basic methods for measuring OA in crop plants subjected to controlled drought stress (Babu et al. 1999). A brief description without the detailed thermodynamic background of each method is given here.

Morgan's method (Morgan 1992) develops a relationship between RWC and osmotic potential (OP). This relationship is compared between two situations: when the relationship (a) is affected just by the concentration effect due to water loss from cells and (b) when the relationship is affected also by solute accumulation in cells. Therefore the difference between the two is OA. The method involves measuring leaves for their OP and RWC during a drying cycle to obtain a sufficient number of data points to develop regression of RWC on OP (b). The second regression which represents only the concentration effect (a) is developed be regressing RWC on the predicted value of OP (OP_0) from any measured value of RWC according to the following expression: OP_0 = OPi [(RWCi/100)/(RWC/100)], where OPi is the initial OP in well-watered plants and RWCi is the initial RWC in well-watered plants. An example of the two regressions is given in Fig. 4.12. OA is the difference in OP between the two regressions at a given chosen RWC where stress symptoms are generally observed (70% in this example).

Ludlow's method is comparatively simple and requires significantly fewer measurements of RWC and OP. By this method (Ludlow et al. 1983) OA is calculated as the difference in OP between non-stressed (a point measurement on the morning after last

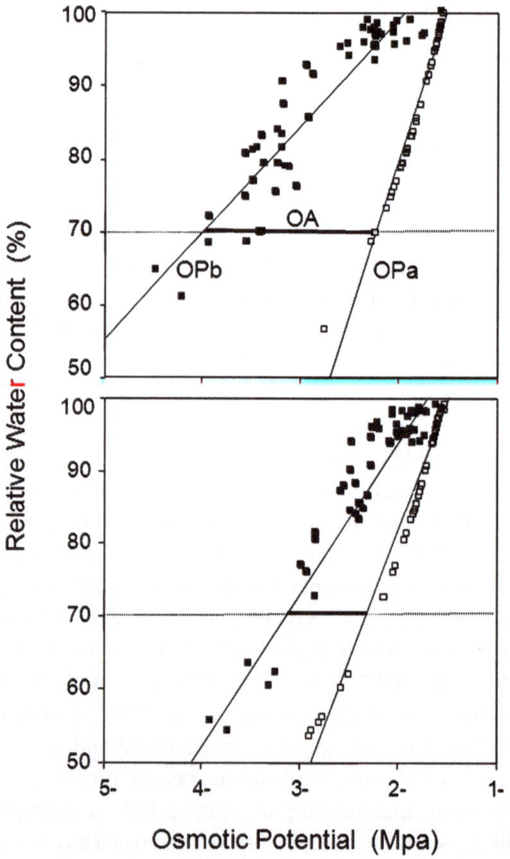

Fig. 4.12 The linear regressions of RWC on actual osmotic potential under drought stress (OP_b) and on the calculated osmotic potential under the "concentration effect" (OP_a). In two exemplary rice cultivars (*top* and *bottom*) differing in osmotic adjustment (OA). OA in this case is taken at RWC of 70% (see text)

irrigation) and stressed leaves both *calculated* to well-watered state (OP100). Data for the stressed plants consist of those taken at a RWC of around wilting in all genotypes (say at a RWC of 70%). OP100 for any value of measured RWC and OP (at non-stress point and stress point) is calculated as follows: OP100 = OP [(RWC – B)/(100 – B)], where B value is bound water. B of 18% was used in rice (Babu et al. 1999). However, since B does not vary with cultivars or levels of dehydration (e.g., Turner et al. 1986), it probably can be ignored when genotypes are compared in one experiment.

The rehydration method is a modification of Ludlow's method in that OP100 of stressed plants is not calculated but measured in re-hydrated plants. By this method, OA is calculated as the difference in measured OP between non-stressed and stressed leaves that are fully rehydrated (Babu et al. 1999). Stressed plants at RWC of about 60% or 70% (depending on the case) are irrigated in the evening and leaves are sampled next morning for measurement of OP. This value is compared with leaves sampled at full turgor in plants never experiencing drought stress. In most cases these leaves come from the same plants just after the last irrigation before stress is initiated.

The comparison of all three methods in rice and the correlations with values obtained with Morgan's method (Babu et al. 1999) indicated that the rehydration method was superior to Ludlow's method in accuracy. The rehydration method should be considered as appropriate in breeding related work.

Measurement of leaf osmotic potential (OP) for the calculation of OA can be done by two main methods.

1. Samples (such as leaf disks) are taken from standard leaves, placed in individually marked vials and frozen without delay. Freezing kills the tissue. Upon thawing all solutes are released from cells into the apoplast. The whole sample is then placed in a thermocouple psychrometer (see above) for the measurement of sample water potential which equals to osmotic potential.
2. A sufficient amount of live leaf is cut into pieces and placed in a syringe. The syringe is used as a press to extract several drops of tissue solute into a micro-osmometer cup for the measurement of osmotic potential.

There were in the past reported studies where OP of stressed leaves was taken as an estimate of osmotic adjustment. It is clear from the above discussion that while OP is measured for estimating OA, OP in itself is not an estimate of OA (Fig. 2.3).

Indirect Methods (Remote Sensing)

Remote sensing techniques are based on the spectral reflectance and radiation emittance from plant surfaces. These physical signatures of leaves and leaf canopies can shift under the effect of plant biotic and abiotic stress (Fig. 2.1). Correct sensing and its interpretation serve to evaluate plant stress and even yield.

Remote sensing technology combined with satellite platform allowed NOAA (National Oceanic and Atmospheric Administration) and cooperating agencies interested in crop health and productivity to develop a crop surveillance system. Firstly, Normalized Difference Vegetation Index (NDVI) was developed by using satellite Advanced Very High Resolution Radiometer (AVHRR) readings. If one band is in the visible region (VIS) and one is in the near infrared (NIR), then the NDVI is (NIR − VIS)/(NIR + VIS). NDVI is related to vegetation health while the visible channel provides some degree of atmospheric correction. The value is then normalized to the range −1<=NDVI<=1 to partially account for differences in illumination and surface slope. A vegetation health index (VHI) later combined also AVHRR thermal waveband sensing and three derived radiance (visible, near infrared and thermal) which were used as a proxy for satellite platform based assessment of vegetation health, moisture and thermal conditions.

VHI was developed into an important drought assessment tool as it directly senses the plant, where vegetation is offcourse the best integrator of the moisture environment at any point in time. For example while environmental data collection and models are very useful they cannot fully account for local conditions which affect crop water status. Such is the case for high ground water level that provides crop access to moisture despite short periods of drought. On the other hand soil impediments such as aluminum toxicity which cause reduced root growth and accelerated root death can severely accentuate environmental water shortage and the effect of mild drought.

Application of remote sensing principles and instrumentation towards phenotyping dehydration avoidance on the ground is free of many of the problems associated with aerial or satellite based sensing, such as need to differentiate between crop and soil or the atmospheric masking of crop signatures. Methods that can be used in phenotyping and selection are basically based on several wavebands and the associated sensors, as will be described below.

Infra-red Thermometry

Plant leaves which act almost like a black body emit long-wave infrared radiation according to their temperature (the Stephen Boltzmann law). Leaf temperature is linked to transpirational cooling. When stress develops and water status is low transpiration is reduced and canopy temperature rises. The infra-red thermometer is designed to sense long-wave infrared radiation emitted from its target and convert it to an average temperature display which can be related to transpiration.

The USDA research group in Phoenix Arizona (Idso et al. 1977) were the first to develop a crop stress index derived from ground-based canopy temperature measurement with the infra-red thermometer. The first application of infrared canopy temperature sensing as a screening technique for dehydration avoidance in a breeding program was developed by Blum et al. (1982) for wheat. Twenty five years later Olivares-Villegas et al. (2007) summarized their study with wheat as follows: "Field trials under different water regimes were conducted over 3 years in Mexico and under rainfed conditions in Australia. Under drought, canopy temperature was the single-most drought-adaptive trait contributing to a higher performance, highly heritable and consistently associated with yield phenotypically and genetically. Canopy temperature epitomizes a mechanism of dehydration avoidance expressed throughout the cycle and across latitudes, which can be utilized … as an important predictor of yield performance under drought."

Infrared thermometry of leaf canopies has been found to be very effective for field drought resistance phenotyping in wheat (Blum et al. 1982, 1990; Pinter et al. 1990; Olivares-Villegas et al. 2007), Rice (Ingram et al. 1990; Garrity and O'Toole 1995; Takai et al. 2010), Sunflower (Alza and Fernandezmartinez 1997), Oil seed Brassicas (Singh et al. 1985), Soybean (Harris and Schapaugh 1984; McKinney et al. 1989) and ryegrass (*Lolium rigidum* Gaudin) (Franca et al. 1998). Lesser success has been achieved with large leaf species such as maize and sorghum. There might be a basic problem in recording a truly representative ground-level canopy temperature in these large leaf crops where dark (cooler) spaces between the large leaves might bias the thermometer reading (Jones et al. 2009). This specific problem in large-leaf crops is expected to be resolved with further research.

Canopy temperature has been related to various facets of plant water status, as would be expected. High canopy temperature was related across breeding materials to high stomatal resistance (Jones et al. 2009) and low leaf water potential (Blum et al. 1982). Sensitivity of the infrared signal was found to increase with the increase of stress (Blum et al. 1982). High canopy temperature was associated with increased leaf rolling (Garrity and O'Toole 1995), poor soil moisture extraction (Olivares-Villegas et al. 2007) and low EC deposition (Sanchez et al. 2001). Low carbon isotope discrimination (high WUE) was positively correlated with high canopy temperature across diverse wheat (Zong et al. 2008) and wheatgrass (*Agropyron* sp.) accessions (Frank et al. 1997) because high WUE was driven by low stomatal conductance, low transpiration and reduced transpirational cooling under drought.

Low canopy temperature was related to better yield or yield stability under drought stress (Garrity and O'Toole 1995; Ingram et al. 1990; Blum et al. 1989, 1990; Blum and Pnuel 1990; Olivares-Villegas et al. 2007; Saint Pierre et al. 2010b). This relationship between canopy temperature, leaf water status and yield is generally stronger as drought stress increases. However, a relationship between high yield and low canopy temperature under non-stress conditions was initially seen also in cotton (Lu et al. 1994) and later also in wheat breeding materials (Reynolds et al. 1998; Olivares-Villegas et al. 2007) especially when grown under hot conditions. Variation in dehydration avoidance as a mechanism of drought resistance can be revealed only when the population is subjected to sufficient drought stress. Canopy temperature of wheat under drought stress and irrigate conditions was found to be controlled by additive, dominance and epistatic effects (Saint Pierre et al. 2010a) and therefore measurements in segregating populations should be delayed into later generations when heterozygosity is reduced.

The protocol for infrared thermometry should be observed very carefully so as to avoid all too common pitfalls which result in large error variance and non-repeatable results.

Different thermometers vary in their angle of view which determines the size of the target area in relations to the distance of the thermometer from the target. Typical desirable target area diameter is around 10–30 cm at a distance of 3–5 m. Close-up measurements should be avoided so that a reasonable sample of leaves and crop canopy can be integrated into one value. The crop should not be viewed with the sun in front of the instrument.

The most serious bias in measuring canopy temperature is if the thermometer views also the soil surface which is very different in temperature from the canopy. Maximum view of the soil is achieved when the thermometer views the crop from nadir. Off-nadir view of the target is desirable but the angle of view should be consistent throughout the experiment. When a hand-held thermometer is used to view the canopy, the angle of the thermometer with respect to the rows of the measured crop can be manipulated to avoid viewing the soil. This angle of view should be maintained throughout the experiment.

Canopy temperature should be measured under sunny conditions at midday or soon thereafter when evapotranspirational demand is high and water deficit is maximized. Transient clouds can seriously deviate temperature reading and increase experimental error. Windy conditions which are expressed in very noticeable leaf movements should also be avoided.

The breeder might be interested in recording canopy temperature at a certain growth stage. However, growth stage can have a specific effect on the reading. Reading at early growth stage before full ground cover is achieved should be avoided since the ground surface is too exposed. Flowering and heading will affect the temperature reading since the inflorescence is often warmer than the canopy. This is especially serious when the population varies for phenology. No relationship was found between dryland durum wheat yield and canopy temperature when temperature was recorded during heading and grain filling (Royo et al. 2002). A high rate of natural leaf senescence will also bias canopy temperature upwards, irrespectively of plant water status. To the same extent, diseased leaves can also result in warmer canopy temperatures.

Error variance in measured canopy temperature can be large depending on how well the protocol is followed and the level of experimental site spatial homogeneity. Some of the variability can be accounted for by normalizing data using methods such as nearest neighbor adjustment, a comparison to a running standard check variety, or converting temperatures data to percent of the mean of the block. However, taking steps to avoid the development of error variance initially is the correct approach.

Often canopy temperature measurements are reported as "temperature depression," namely the reduction in canopy temperature relative to air temperature. Temperature depression requires measurement of air temperature together with canopy temperature, which is possible with some thermometers. Still experience shows that in work under drought stress with a population diverse in canopy temperature one can use just canopy temperature data to compare genotypes for dehydration avoidance. Canopy temperature depression is more important in work with irrigated crops in a hot environment. Finally, not all thermometers are suitable to work under the sun and remain stable when heated up by the environment. Check instrument specifications before purchase.

Infrared Thermography (Thermal Imaging)

This method basically uses the same physical and biological principles of infrared thermometry, only that the sensor output is processed into a black and white or color digital image display of the different temperatures in the target area. Two types of instruments exist: cryogenic and no-cryogenic. In the cryogenic instrument the sensor is cooled to improve resolution, which makes the instrument balky and less portable for field work. The non-cryogenic instrument produces images of lower quality but it is lightweight and cheaper. There is initial use of thermography in plant science research and plant water relations (e.g., Wang et al. 2004; Kaukoranta et al. 2005; Leinonen et al. 2006). However, at this time thermal imaging has not yet reached the level of utility in screening work and field drought resistance phenotyping to the extent that the infrared thermometer has. For usable imagery that can compare different genetic materials in the field an image of a whole experiment or a large part thereof is needed. If singular images of individual plots are taken there is no advantage over using the infrared thermometry which results in one number per plot.

Spectroradiometry

The discrete spectral reflectance of leaves, from the visible (VIS) through near infrared (NIR) can be sensed with the appropriate spectroradiometer. Several canopy spectral reflectance indices were derived as discussed above. Additional indices were later developed such as the Water Index (WI = R970 nm/R900 nm) that assesses the water status of the canopy (Penuelas et al. 1993). With the appropriate canopy viewing protocol these indices or their derivatives were used to predict grain yield and biomass of various cereal crop genotypes (Araus et al. 2001).

The utility of spectral reflectance indices in the selection for yield was tested in common wheat under well-watered conditions (Babar et al. 2006), durum wheat under drought stress (Aparicio et al. 2000; Royo et al. 2003), and soybean (Ma et al. 2001). Strong phenotypic and genetic correlations were found between several NIR-based indices and grain yield of wheat cultivars and lines under irrigated conditions. Babar et al. (2006) found that NIR based indices gave the highest level of association with grain yield under reduced irrigation environments. In another study with wheat the water indices predicted better the yield of diverse genotypes as compared with the vegetation indices (Gutierrez et al. 2010). On the other hand, vegetation indices such as the NDVI proved especially useful in assessing drought resistance of diverse turfgrass genotypes (Merewitz et al. 2010).

Use of the mean spectral reflectance indices values averaged over growth stages and their progressive integration from booting to grain filling increased the capacity to explain variation among genotypes for yield under reduced irrigation conditions. The instrument is more expensive than the infrared thermometer and may require some training in its use, but significant benefits might be derived from its application to breeding and agronomy research.

Photography

Infrared (analog) photography is another use of the NIR reflectance to identify genotypic variations in canopy water status. This method has been developed and used before accurate portable field spectroradiometry was available (Blum et al. 1978).

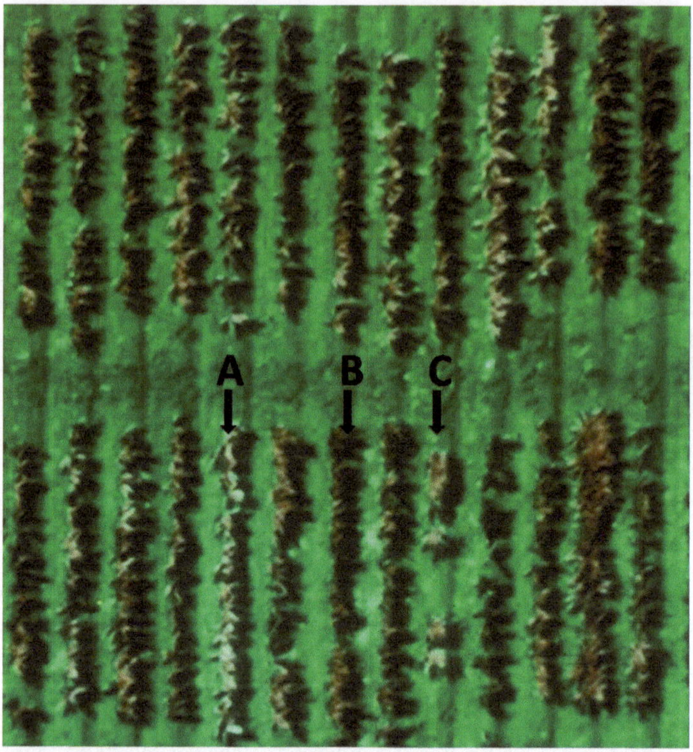

Fig. 4.13 Close-up section from a low altitude infrared color aerial photograph of a drought stressed sorghum breeding nursery of Texas A&M University at Chillicothe Texas. Line A with the whitish tinted color is drought susceptible as compared to the more resistant and color saturated line B. In line C, despite the low plant density, plants are still drought stressed. Noticeable differences between lines can be seen also on the right side of the upper block

Recent radiometric research of crops as described above support the reliability of this application in selection work.

Infrared film emulsion is sensitive to near UV, VIS and NIR radiation. Turgid leaves reflect more than wilted leaves at the NIR. The main reason for this difference is the well rounded shape of turgid mesophyll cells as compared with their collapsed shape at low turgor. When such leaves are photographed with filtered infrared film the turgid leaves appear color saturated as compared with non-turgid leaves. Figure 4.13 presents a close-up section from a low altitude aerial infrared photograph of a sorghum nursery planted to diverse germplasm each in one row 5 m in length. Light colored canopies in the photograph represent drought stressed lines (A and C) while the saturated colored lines are relatively more dehydration avoidant (B).

Infrared film (color or black and white) can be used in any camera (depending on film type and size) with several important precautions (see the specific film instruction sheet). UV and visible radiation should be filtered out by using the appropriate filter as recommended by the manufacturer. Both Kodak and Agfa supplied infrared film and film development chemistry but recently they both tend to discontinue certain types due to lack of demand. Check manufacturer's web site for current status. This option might become obsolete in the near future as infrared digital photography develops.

The CCD array in the standard commercial *digital camera* is basically sensitive to the near infrared and this sensitivity is reduced by the manufacturer by placing a filter in front of the CCD. Various photography web sites offer ways to modify the standard digital camera for near infrared imaging in a more or less reliable manner. Technically skilled photography enthusiasts are known to have removed the infrared blocker in front of the CCD and replace it with a filter that removes visible light. This filter is behind the mirror, so that the camera can be used normally. Light metering works but it is not accurate because of the difference between visible and infrared spectra.

Special infrared digital cameras are available commercially from Sony, Fuji and Nikon (at this time). Check company web site for further and current information. At this time there is no documented experience with infrared digital camera use in phenotyping dehydration avoidance, but studies towards this end are certainly forthcoming.

Regular *color photography* will record in an unbiased manner your visual impression of the crop. Obviously, lush green vegetation is indicative of better productivity than drying, yellowing and senescing vegetation. VIS color photography of dryland wheat was used to calculate vegetation indices with satisfactory relations to yield (Casadesús et al. 2007). These pictures were taken 2 weeks after anthesis which might have given more weight to the effect of senescence. Digital color photography images were used to calculate the number of days to 50% greenness of turf after last irrigation as a measure of drought resistance (Richardson et al. 2008). That paper refers to the details of the method which can suit various applications in field screening of drought stressed plots. Visible color and digital photography hold promise as relatively simple and cheap methods for drought resistance phenotyping and recording, pending more research.

4.2.2.3 Stomatal Activity and Transpiration

Leaf transpiration is partitioned into stomatal and non-stomatal (cuticular) transpiration. Transient stomatal transpiration rate as measured at any given time is the result of a delicate balance between leaf water status, substomatal cavity CO_2 concentration, and ABA produced in the leaf and/or imported from the root. The evidence provided above for low canopy temperature under drought stress as a marker for dehydration avoidance support the selection of phenotypes with high stomatal conductance. This ideotype is therefore very easily phenotyped by remote sensing methods.

Where detailed and direct stomatal information is still of interest (e.g., for parental materials etc.), two types of instruments are available for instantaneous measurements: The *diffusion porometer* and the *pressure drop porometer*. The former estimates leaf hydraulic conductivity by the response of a hygrometric sensor inside a small enclosure attached to the leaf. This instrument is available commercially in various configurations and costs. The pressure drop porometer (syn. "viscous flow porometer") estimates stomatal conductance by the measurement of air passage through the leaf (e.g., Rebetzke et al. 2003). It is relatively faster but less available commercially.

Measurement of whole plant transpiration in containers is generally performed by weighing container-grown plants over a given time interval. Measurement of whole plant transpiration in the field is sometimes done by the heat pulse/sap-flow method (e.g., Green et al. 2003) which is a slow procedure suitable mainly for trees and perhaps also large plants.

Oxygen Isotope Enrichment

As discussed in Sect. 3.5.1.1 oxygen isotope enrichment ($\Delta^{18}O_p$) is associated with plant transpiration and stomatal conductance especially under well watered conditions but also under certain water-limited conditions. According to theory, enrichment is proportional to the amount of water passing through the plant via transpiration. Oxygen isotope enrichment has been considered as a measure of total plant transpiration. The reliability of the method for drought resistance phenotyping depends on additional research in the specific crop and the environment in which it is to be phenotyped. There are various variables to consider when it is used to predict phenotypic variation in total crop transpiration under drought stress. Despite its potential it may not be taken yet as a routine phenotyping and screening test. The method is elaborate and requires expertise. An example of the procedure for plant materials is given in Cabrera-Bosquet et al. (2009).

Cuticular Transpiration

The importance of high cuticular resistance towards the effective use of water has been discussed in Chap. 3, section "Non-stomatal Control of Water Use and Canopy Energy Balance." Two major approaches to phenotyping cuticular resistance are recognized: Quantifying epicuticular wax load (EC) and assessing leaf water loss as a function of cuticular transpiration.

The determination of *EC load* on leaves can be performed by the relatively fast spectrophotometric method of Ebercon et al. (1977). The standard slower gravimetric method for EC is even simpler and it is described in that paper. EC deposition tends to increase under drought and heat stress. Therefore is advisable to precondition plants by drought stress before the analysis in order to express the full potential for EC. Field grown plants that were subjected to some stress are preferable to greenhouse or growth chamber plants in this respect. Standard leaf sampling should be adopted.

For measuring leaf *cuticular transpiration* the leaf (pre-conditioned as mentioned above) should be treated to induce full stomatal closure. While instrumental methods such as the diffusion porometer can be used to measure cuticular transpiration, the weighing of detached leaves over a time interval under standard conditions is a popular fast and cheap method. Stomatal closure can be induced by feeding

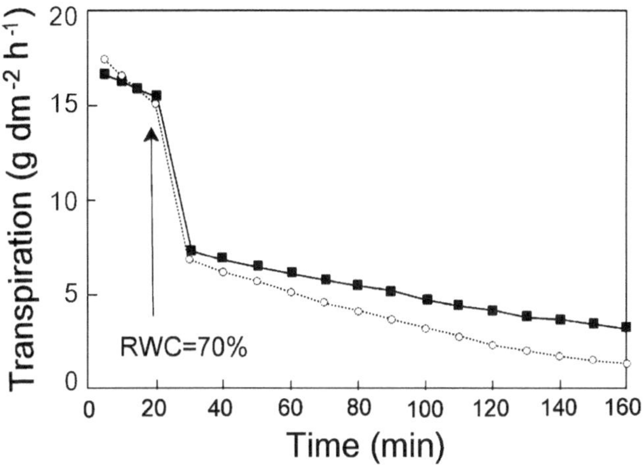

Fig. 4.14 Detached leaf transpiration under room conditions in two different rice cultivars (see text) (Data provided courtesy of Dr. R. Chandra Babu, Tamil Nadu Agricultural University, India)

ABA to the detached leaf or by placing it in the dark or both. Leaf area should be determined for the calculation of transpiration rate.

Figure 4.14 displays the march of leaf transpiration in untreated detached leaves of two rice cultivars. In both cultivars stomatal closure was initiated at a RWC of about 70% and full closure was apparently reached after about 20 min. After full stomatal closure, cuticular transpiration decreased steadily as a function of leaf drying. Measurement of cuticular transpiration in these leaves could therefore be estimated for example by leaf weight reduction between 40 and 60 min. While this method allows to verify stomatal closure it also has the advantage of measuring cuticular transpiration under realistic conditions of leaf water deficit rather than in a fully hydrated leaf under ABA-induced stomatal closure. Therefore, before planning extensive phenotyping by this method, the correct protocol should be established following the example in Fig. 4.14 as performed in several diverse genotypes.

A variation of this test was developed in wheat as the *leaf-water retention capacity* of detached leaves (Clarke and McCaig 1982). With that method water loss from leaves as percent of the leaf dry weight was determined over a time interval, including the stomatal transpiration phase. Later tests were performed after stomatal closure (Clarke et al. 1991). Additional studies of the association with yield of wheat (Winter et al. 1988; Gavuzzi et al. 1997) indicated that the method might have a potential as a screening method provided it is based on reasonable physiological grounds. For example, a major requirement for perfecting the method is a standardized protocol of hardening (Walker and Creighton Miller 1986) in order to account for genotypic variations in osmotic adjustment and EC accumulation. Osmotic adjustment would affect water retention as percent of leaf dry weight. The method raised a reasonable amount of interest at the time.

4.2.2.4 Stay-Green ('Non-senescence') (SG)

Distinct SG can be recognized visually. Post flowering drought stress enhance the expression of the trait as compared to normal senescence (Fig. 3.8). A visual senescence score of 1–5 on a whole plot basis has been used in general screening work as well as for molecular mapping of the trait in sorghum (e.g., Subudhi et al. 2000). The visual score as performed during grain maturation correlated very well with the Minolta Chlorophyll Meter SPAD-502 (Minolta Camera Co Japan) measurements in several top leaves as an indirect estimate of leaf chlorophyll content (Xu et al. 2000). The SPAD meter was also used in phenotyping a mapping population of rice for SG. It was used on standard leaves on a standard tiller on two dates during reproductive stage (Jiang et al. 2004).

Percent green leaf area is also an accurate estimate of SG in diverse genotypes. In a genetic study of the trait in sorghum (Haussmann et al. 2003) the following procedure was used: At the time of emergence of the flag leaf, three representative plants in each plot were tagged; the length and width of the upper six leaves were measured, and the area of each was estimated as: leaf length × leaf width × 0.70. This is a common equation for cereal leaves. Beginning at flag-leaf emergence, the percentage of each of the upper six leaves of each tagged plant remaining green was visually estimated at weekly intervals. The green-leaf area (GL) of each tagged plant was computed by multiplying the percent green-leaf area by the measured area of each leaf, and summing across the six measured leaves. The percentage of green-leaf area (% GL) for each plant, for each week, was calculated by dividing the estimated GL for that week by its measured leaf area at flowering. Plot values for % GL were derived by averaging the three individual plant values for each plot.

A major problem which can confound genetic studies and molecular mapping work is the phenotypic variation in phenology, as the case may be also for other traits involved with drought response at the reproductive growth stage. Measurements of SG should therefore be performed in each genotype at its own specific growth stage. Still, when SG is phenotyped under drought stress, large variation among genotypes in phenology might bias the results.

4.2.2.5 Observing and Measuring Roots in Breeding Work

Roots are a major heritable component of dehydration avoidance. However, roots are difficult to observe and measure. Root observations are generally work-intensive and expensive.

An important method for circumventing the problem and achieving selection for the desirable root ideotype in breeding work is to address root function in sustaining shoot dehydration avoidance rather than observing the root itself. This has been extensively discussed above. For example, wheat lines with greater root dry matter at depth expressed greater transpiration under drought stress and this was very well detected by infrared canopy temperature measurements (Lopes and Reynolds 2010). For root traits to be reflected in shoot function under drought stress the population should be tested under managed drought stress conditions with deep soil moisture.

The major and most common root attribute in dehydration avoidance is root depth, where deep soil moisture is available. A managed stress environment in the field or the greenhouse can be designed so as to ascribe a relative advantage in this respect to deep rooted genotypes. An almost classical field method used in rice has been the toposequence approach (Mambani and Lal 1983). Where groundwater exists, topographic differences create variable distance between the plant and the water level. Thus planting rice along a slope (a toposequence) allows comparing rice materials for root depth. However, a toposequence can create variability for other growth factors besides water supply, such as nutrients or even submergence. Still, the toposequence has been used effectively in selection for dehydration avoidance of upland rice. There were also attempts to simulate toposequence in the greenhouse in a form of a soil filled cement pool of variable soil depth with controlled water level at depth.

Root growth is affected by drought stress. Root architecture will be different in a drying or in a wet soil. It would seem that direct observation of genetic differences in roots with respect to their function in dehydration avoidance must be performed in a drying soil under drought stress. However for root studies it would be easier and simpler to phenotype roots under non-stress conditions. The crucial question is not whether root growth differs with the water regime but whether there is a widely observed genotype x drought stress interaction for root depth. Lilley and Fukai (1994) and Kato et al. (2006) did not find genotype by water regime interactions for root traits in rice. It is concluded that constitutive root growth capacity is the main driver of deep root development. Furthermore it was observed that thick adventitious root axes were associated with deep roots in rice, irrespective of the water regime. While the relationship is not well understood it seems that potential root thickness in rice can serve as a phenotypic marker for constitutively deep root growth. Thick roots are related to better soil penetration (Clark et al. 2008).

Reviews of various root observation methods are available in Taylor et al. (1991), Smit et al. (2000), Waisel et al. (1996) and Gregory (2006). Most of these methods are too slow or too costly for routine screening work in breeding. The most popular one among breeders is the tube method.

The Tube Method

Typically, plants are grown in PVC tubes filled with fertilized clean light soil or sand. Tube length fits the expected maximum length of the root of the specific crop. Tube width depends on the specific plant size, the planned stress treatment (if any) and the available soil moisture content in the tube. A typical tube used with small grains and rice for growing 1–2 plants in soil is 10–18 cm in diameter by 100–150 cm in length. The tube is generally sealed at the bottom with a hole for drainage. The seal is made so that it can be easily removed. Standard PVC white or grey drainage pipes and seals are often used.

The tube should be filled with soil in a standard fashion to a standard bulk density so as to avoid variability in soil conditions among tubes, as discussed for

Fig. 4.15 Split PVC tubes with growing wheat plants; the *inset* on right presents the exposed soil core with some of the roots on the surface

pot experiments (Sect. 4.1.5.1). A discussion of bulk density in tubes with respect to root growth and experiment homogeneity is presented by Cook et al. (1997).

Three types of tubes are used for root observations and root wash:

(a) Regular PVC tubes. When roots are to be washed the soil is saturated with water, the tube is laid out and then the plant is slowly pulled out of the tube with the aid of water jet.
(b) Before filling the PVC tubes are sawed lengthwise into two halves and then held together by metal clamps or a strong tape. To extract the roots the clamps or tape are removed and then the full soil core is revealed with the exposed roots which are easily washed out. Photographs of the exposed soil core can be taken as a record before washing (Fig. 4.15).
(c) Any of the above two methods can be used with transparent plexiglass tubes. This allows continuous observation and photographic record of roots seen through the wall. In this case a dark soil is desirable. Tubes must be protected

from light to avoid algae and other organisms growing on the exposed soil surface. The photographic records or scans of the roots themselves can than by subjected to image analysis for the development of numerical data. However, a correlation must first be established between actual root length density in the tube and the root image on its surface.

Various improvements and accessories can be added to any tube system, depending on the purpose of the study and the budget. For example, TDR probes for soil moisture measurement or other probes can be inserted through the tube wall into the soil. Tubes can be weighed by lifting them on a load cell and thus monitor water-use.

Very simple and inexpensive method has been used successfully for the selection of maximum root length. Single plants are grown in soil-filled disposable long polyethylene sleeves (typically 5–10 cm in diameter depending on the plant). After planting and emergence the soil filled sleeves are propped on a support constructed from wood or other materials. Roots can then be observed growing along the wall of the sleeve and their length can be recorded at any growth stage. At any point in time the sleeve can be slit open and roots quickly washed out for observations. The method is appropriate for resolving large genotypic diversity in root length.

Electrical Capacitance Method for Roots

Several attempts were made to estimate root size or root biomass on a relative basis, by measuring root electrical capacitance (Dalton 1995; van Beem et al. 1998). The method is based on a portable capacitance meter hooked to the base of the stem and a probe inserted into the soil near the plant. The basic requirements are that the soil should be similarly wet for all plants measured and the electrode clamped to the plant will be placed in a consistent manner in all plants. While the potential of the method for relative data in maize and tomato seemed to hold promise, others indicated that in maize the capacitance-root mass relationship was genotype-specific (McBride et al. 2008). Chloupek et al. (2006) performed a genetic study of barley roots using root capacitance measurement as the main root variable. It is not clear from that paper how well the method predicted root size. It is however seen from the report that variations in time of maturity and inter-plant contact can introduce a bias in the measurements.

At present the method requires more evaluation before it can be considered for root phenotyping work.

Root Penetration Ability

There are pronounced differences among and within species in root penetration ability. Despite a long period of research with several crops, the capacity of root to overcome resistance to penetration is not well understood. Work in rice suggested that thick roots and stem bending stiffness (Clark et al. 2008) were associated with better root penetration ability.

Fig. 4.16 Rice root penetration experiment using the wax layer. Plants were grown in pots in vermiculite in a nutrient solution where roots penetrated the wax layer at the bottom of the pot and grew into the nutrient solution. Bottom panel displays two different cultivars (Photograph courtesy of Dr. HT Nguyen)

Penetration ability can be assessed by the wax layer method. Hendrickson and Veihmeyer (1931) were probably the first to use the wax layer method in root penetration studies. The method was revived and improved as a screening method in rice by Yu et al. (1995) and later in cotton (Klueva et al. 2000) and wheat (Botwright-Acuna and Wade 2005).

Certainly there are genetic variations in root penetration of a wax layer (Fig. 4.16). The question is often raised how well results with the wax layer represent responses to soil impedance in the field. For example Clark et al. (2002) did not observe a very good association between field and laboratory data on root penetration. However it should be noted that results from penetration ability tests in the field are also subject to criticism.

The wax layer method as performed in rice is based on the number of roots penetrating the wax layer in proportion to the total number of roots per plant (Fig. 4.16). The wax layer is prepared by mixing wax and petrolatum white (sometimes defines as paraffin wax and soft paraffin, or paraffin wax and Vaseline, all depending on the conventions used). The produced layer should be tested with a soil penetrometer for its impedance. The common composition for a wax layer is 60% hard wax and 40% soft wax (Vaseline) with a typical impedance value of around −1.5 MPa at 25°C.

The pot layout for evaluating penetration may differ from one case to the other, depending on the crop and the desirable growing conditions. It might also differ if seminal or adventitious roots are concerned as the case is in the cereals. In principle pot construction should be designed so that roots will grow and penetrate the wax layer and that both total number of roots and penetrated number of roots can be counted. A fairly detailed description for rice is given by Yu et al. (1995). It is important to note the observation made by Clark et al. (2008) that roots penetrated better when the wax layer offered a gradual rather than an abrupt resistance. This would imply an acclimation effect towards root penetration capacity, which is probably the natural situation in soil. Such a layer can be constructed from several thin layers of gradual strength. Such a configuration has not yet been used in routine screening work.

4.2.3 *Dehydration Tolerance*

Since dehydration tolerance is defined as the ability to function at low plant water status, the primary axiom for phenotyping dehydration tolerance is that all tested materials for function must have comparable low water status. If materials differ in their water status then phenotypic variations among genotype in function cannot be ascribed to variation in dehydration tolerance. This has been discussed in detail in Sect. 3.4.2.2.

Therefore, in most cases where dehydration tolerance is phenotyped, plant water status should be also determined. In some cases, normalization for variations in plant water status may not be sufficient, especially when variations in leaf water status are large. In that case measurements of function should be done when the specific leaf water status has been reached in all genotypes under stress.

Several tests were used in defining dehydration tolerance in terms of various plant functions.

4.2.3.1 Photosynthesis

As depicted above, photosynthesis is often used as a criterion of dehydration tolerance or general "drought resistance." It became popular more recently in the evaluation of transgenic plants and as instrumentation became better and easier to use.

It should be emphasized again (Chap. 3, section "A Dehydration Tolerant Photosystem and Respiration?") that there is no hard evidence that real tolerance of the photosystem or photosystem biochemistry to low leaf water status is at hand to the extent that it should serve as a target in plant breeding. The few cases of implied genetic variation in photosystem function can be interpreted also as a result of variations in dehydration avoidance (e.g., Oukarroum et al. 2007; Hu et al. 2009). The phenotyping of photosynthesis is therefore briefly described here, under the above considerations.

Chlorophyll fluorescence parameters as measured in dehydrated leaves can be used to assess photosystem-II function. The OJIP fluorescence transients as developed by RJ Strasser (e.g., Oukarroum et al. 2007) are a solid and very detailed approach. However, for work with large population simpler and sometimes cheaper chlorophyll fluorescence systems are available on the market, and each should be considered according to the type of work required. Understanding of the phenomenon of chlorophyll fluorescence and its expressions under drought stress is essential before one attempt to use this measurement in phenotyping dehydration tolerance (Maxwell and Johnson 2000). Furthermore this area receives great attention in plant biology and new developments are frequently reported (e.g., Chaerle et al. 2007).

Photosynthesis in terms of leaf gas exchange in response to drought is evaluated by portable leaf gas exchange systems which have become more accurate and expensive in recent years. These systems are not suitable for phenotyping large populations but they can be considered for limited use such as in pre-breeding work. Leaf gas exchange measurements should be performed during midday, with standard leaves similarly exposed to sunlight. After measurement the leaf should be immediately sampled for the measurement of leaf water status, preferably in terms of RWC. Variation in dehydration tolerance can then be identified by genotypic deviations from the general regression of photosynthesis on RWC.

4.2.3.2 Plant Growth

Drought resistance in terms of growth and development is almost always a function of dehydration avoidance. However, if one is still bent on evaluating growth as an expression of dehydration tolerance then it must be normalized for plant water status. I am not tired of repeating this indication in view of the many repeated faults in this respect in the literature.

Whole plant growth as a measure of drought resistance is often done with juvenile plants when whole plant dry matter is used for estimating growth by calculating RGR (Sect. 4.2.1.2). In this case, a standard leaf (usually uppermost fully expanded) is to be sampled for the determination of RWC at midday of T1, after which plants are harvested for the determination of W1. Further work proceeds normally. The resulting RGR is related to leaf water status at T1. Genotypic deviations from the regression of RGR over RWC across all genotypes allow isolating cases of dehydration tolerance in RGR.

Irrespective of dehydration tolerance consideration, RGR of plants exposed to a standard managed drought stress is a valuable estimate of the integrated capacity of the plant to sustain growth, whether the cause might be dehydration avoidance or tolerance.

4.2.3.3 Cell Membrane Stability (CMS)

The CMS test is a rapid, simple and a well representative assay of cellular membranes dysfunction under stress. While the exact structural and functional modification caused by stress is not fully understood, cellular membrane dysfunction due to stress is well expressed in increased permeability and the leakage of ions, which can be readily estimated by the efflux of electrolytes. Hence the estimation of membrane dysfunction under stress by measuring cellular electrolyte leakage from affected leaf tissue into an aqueous medium is finding a growing use as a measure of CMS and as a screen for stress tolerance. The method was initially applied in the late 1960s by CY Sullivan (University of Nebraska) to the screening of sorghum and maize for heat tolerance. Variations of the method were developed also for cold and dehydration tolerance.

The general protocol involves the application of stress to the leaf after it has been subjected to hardening, followed by the measurement of electrolyte leakage by using the conductometric method. The most common application has been for heat tolerance and therefore the initial detail is given here for *heat stress*.

Plants must be exposed to moderate heat stress for at least 24 h before the test so as to allow for hardening (acclimation). The capacity for hardening is a major component of the capacity for tolerance. It is sometimes defined as "acquired heat tolerance." Hardening can be achieved in the natural field environment, if heat stress occurs, or in the greenhouse or a in the growth chamber. Exposure of intact plants for 24 or 48 h to 32°C (for cool season plants) or 36°C (for warm season plants) is sufficient, even at low light.

Leaf discs or leaf sections or even whole small leaves are placed in standard glass vials that can accommodate a conductivity electrode. The total area of leaf material per vial is about 15–25 cm^2. Sample size does not have to be the same for all vials. The sample is then washed two times with de-ionized water. The water is drained off but samples remain wet so that they would not desiccate. In the case of screening work, at least ten vials (samples) are prepared for each genotype. Five pairs would be taken each from five different plants (replicates). For each pair, one vial per plant is designated as treatment (T) and the other as control (C).

The treatment vials are subjected to the heat stress treatment in vitro. They are placed in racks and covered (not stoppered) with plastic wrap so as to avoid drying the samples. Racks are placed in thermostated water bath so that the leaf samples will be completely below the water surface level. Temperature is set to a predetermined stress temperature and the samples remain in the bath for 1 h. The covered control vials are placed at room temperature. The treatment temperature should be such that it will result in average population CMS values around 40–60% so as to obtain good separation of the accessions for CMS. This temperature will change with the species, the population and the hardening conditions. It is therefore required that some representative genotypes will be initially pre-tested for CMS at a range of heating temperatures (typically: 48–50–52–54°C) in order to determine the final protocol.

After treatment 20 cc of deionized water is added to each T and C vial making certain that all leaf materials are submerged. All vials are then placed for incubation at about 10°C (typically, on the lowest refrigerator shelf) for 24 h. After incubation samples are equilibrated for 1 h at room temperature and the conductivity of the medium is measured by inserting a conductivity electrode into each vial.

All vials covered with plastic wrap and placed in an autoclave for 15 min to kill all tissues. Conductivity of all samples is measured again after samples are equilibrated at room temperature. CMS is calculates as follows: CMS (%) = [1 – (T1/T2)/1 – (C1/C2)] × 100, where T1 and T2 are treatment conductivities before and after autoclaving and C1 and C2 are the respective control conductivities. Calculated results are often better when each T value is calculated against the average of all C values for the given genotype.

The method has been adopted for measuring *dehydration tolerance* in two ways.

(a) Treatment samples are placed in a vial containing a solution of PEG 6,000 or PEG 8,000 (T) at about 10°C as compared with non treated samples (in humid stoppered vials) (C) (Blum and Ebercon 1981; Premachandra and Shimada 1988). After 24 h the leaf samples are washed in deionized water and placed in deionized water for affecting leakage. The method then proceeds as described above for heat stress. Protocol must be first tested in order to establish PEG concentration for best separation of genotypes in CMS. It is not clear if hardening is required before leaf sampling as the case is for heat stress, but sampling field grown leaves is probably better than growth chamber leaves. Bajji et al. (2002) concluded that CMS obtained by this method represented the relative performance of durum wheat cultivars under stress. The method seems to offer a potential as a phenotyping tool but further work is needed to assess its relationship to drought resistance in term of yield or biomass in the field.
(b) With the second method the stress treatment constitutes actual drought stress applied to plants grown in pots or in the ground. The requirement for a standard level of stress is filled by sampling leaves at a comparable RWC in all genotypes. This protocol was used by Tripathy et al. (2000) in rice where leaves were sampled at a RWC of 60–65%. It means that all genotypes must be tested for leaf RWC to establish when sampling can be made. While this protocol dictates more work than with method A above, its advantage is in using real drought stress rather than PEG imposed stress.

A small pilot test should be performed before the actual screening work, in order to adjust the protocol for the specific crop and test.

4.2.3.4 Stem Reserves for Grain Filling

An important component of reproductive success under drought stress is the capacity for grain filling from stem reserve when transient photosynthesis is inhibited by stress. This is a dehydration tolerance mechanism since the transport of reserves from stem to grain takes place in dehydrated plants, in the case of severe drought in the field. This capacity is often conditioned, signaled or enhanced by plant dehydration (Chap. 3, section "Stem Reserve Utilization for Grain Filling"). It can be phenotyped in large populations by the chemical desiccation method.

Chemical Desiccation Method

The method was developed by Blum et al. (1983a, b) as a fast and relatively simple field assay for revealing the capacity for grain filling from stem reserves in the absence of transient photosynthesis. Its usefulness has been later verified in phenotyping and selection work in wheat (Hossain et al. 1990; Nicolas and Turner 1993; Haley and Quick 1993; Gavuzzi et al. 1997; Sawhney and Singh 2002), barley (Gavuzzi et al. 1997), triticale (Royo and Blanco 1998) and millet (Mahalakshmi et al. 1994).

The method is based on the application of a chemical desiccant to plant canopies after flowering as means for inhibiting plant photosynthesis and thus revealing the capacity for grain filling by stem reserves. The treatment does not simulate drought stress. It simulates the effect of stress by inhibiting current assimilation. With this method a chemical desiccant (magnesium chlorate or sodium chlorate; 0.4% w/v) is sprayed to complete wetting over the whole canopy, including the ears in the case of wheat. In the original study the treatment was applied to each wheat genotype at 14 days after anthesis, when kernel growth entered its linear phase. At maturity, kernel weight was compared between treated and non-treated (control) plants, calculating the rate of reduction in kernel weight due to the treatment. The rate of reduction was typically between 5% and 50% in different wheat materials. An important consideration in using this test is that plants must be free of any biotic or abiotic stress, simply because under any stress grain filling would also be reduced in the controls.

Nicolas and Turner (1993) confirmed the utility of the method and proposed the use of a leaf spray with potassium iodide (0.4% w/v) in wheat as a milder treatment which mainly destroy chlorophyll and simulates natural senescence. Potassium iodide was working well also for millet (Mahalakshmi et al. 1994) and triticale (Royo and Blanco 1998).

The positive correlation across diverse genetic materials between the rate of reduction in kernel weight by chemical desiccation and the rate of reduction by drought stress was found in several of the above cited experiments to be significant and reasonably high. Hossain et al. (1990) noted that winter wheat cultivars of stable kernel weight over years and locations sustained relatively less reduction under sodium chlorate desiccation of the canopy. Finally, the reduction in kernel weight by chemical desiccation was significantly correlated across different wheat cultivars ($r = 0.48^*$) with the reduction in kernel weight caused by late epiphytotics of Septoria leaf blotch disease (Zilberstein et al. 1985).

Chemical desiccation can be incorporated into breeding programs in two ways. Firstly, it can be used to assess responses of individual advanced lines or families, always compared with non-treated controls under non-stress conditions. This can be easily performed with standard nursery rows (Fig. 4.17a). Nursery rows must then be planted with sufficient space between rows so as to allow spray application

Fig. 4.17 (**a**) Wheat lines grown in a breeding nursery sprayed with 0.4% potassium iodide (left) as compared with a non-sprayed control (*right*). (**b**) Wheat F_2 mass selection under chemical desiccation with 0.4% magnesium chlorate (*left*) as compared with non-treated control (*right*)

to reach all plant parts. Spray is applied manually with a hand sprayer. Each genotype should be treated at its defined stage of initial grain growth.

Secondly, the method can be used in mass selection (Fig. 4.17b). Blum et al. (1991) performed mass selection, where six spring wheat F_2 bulks were chemically desiccated with magnesium chlorate after which grain were divergently selected for kernel size by mechanical sieving. After two or three cycles of selection, random lines were selected and tested for their response to chemical desiccation stress. Mass selection for large kernels under chemical desiccation significantly improved kernel weight and grain yield under chemical desiccation stress, as compared with controls where selection for kernel size was performed without chemical desiccation. There was no shift in phenology or plant height under chemical desiccation selection, probably because the variation in these traits within the populations used was small. Haley and Quick (1993) performed a similar selection program under chemical desiccation with sodium chlorate in winter wheat. Two cycles of mass selection produced F_4 bulks that were more resistant to chemical desiccation stress, hence more capable of grain filling from stem reserves.

Reserve Content

There were limited attempts to use stem reserve accumulation as an indicator of the capacity for grain filling from stem reserves. Large genotypic variation and high heritability was found for water soluble carbohydrates content (WSC) or total non-structural carbohydrates (TNC) in wheat stems at flowering (Ruuska et al. 2006; Rebetzke et al. 2008), but the relationship to grain yield under stress was not clear. A relationship to yield involves both accumulation and the capacity for *remobilization* when stress occurs, which is expressed under chemical desiccation treatment. Therefore, reserve content is only a partial condition for grain filling from stem reserves under stress. WSC and TNC are assayed by standard methods available in detail in the literature. Fractionation of the reserve into specific components such as starch, fructan or sugars is also straightforward.

4.2.3.5 Drought Survival

As discussed in Chap. 3, section "Plant Survival and Lethal Water Status" the rate of recovery as a measure of drought survival is to a large extent a function of plant water status at peak stress, with exceptions. There are several methods for measuring survival and recovery. However not all tests used in phenotyping discern between the role of avoidance and tolerance in survival. Survival can be driven by both components, depending on the case.

The Seedling Wilting Test

This is the simplest test being employed for decades by practitioners (Ashton 1948). It compares the relative survival of different genotypes in a drying soil in terms of visual assessment. Often, the visual rating for recovery after rehydration is used. Since plant size and variations in plant size among genotypes have a large effect on desiccation rate and the relative rate of survival, this test is appropriate for seedlings which usually do not differ much is leaf area. In order for plants to express their capacity for adaptation, stress must be slow, about a week or two after last irrigation to stress symptoms.

The most common and widely used assay is the "box" test where several genotypes are tested together in one box and different boxes serve as replicates. This test has long been used as part of the cowpea breeding program at IITA (Agbicodo et al. 2009). It is also used as one of the formal drought tolerance tests at the Chinese Academy of Agricultural Science, Beijing.

In the box, roots of different genotypes are intermixed resulting in more or less homogenous soil drying throughout the box. An example is seen it Fig. 4.18 where eight rice genotypes are at a point of peak stress. Clear differences can be seen between line A and those at B to D, where A appears as the most tolerant and D is the least tolerant. Irrigation and recovery within several days would allow to re-asses these differences. However, depending on the specific test, stress can be continued to the point where most genotypes will be killed.

The test is simple, cheap, low-tech and can be done anywhere. There is however very limited proof of relationship between the results of this test and field performance under stress, unless only the seedling growth stage is concerned. A positive relation to field performance under drought stress was seen for cowpea (Agbicodo et al. 2009), a crop subjected to sever drought in the Sahel Region and often requires good survival capacity from one rainfall event to the other.

Fig. 4.18 The "box" test for seedling survival in eight rice genotypes grown in a local soil subjected to sever drying. (A) to (D) are four different genotypes (photograph taken at the Luoyang Academy of Agriculture Science Station, Luoyang, China) (see text)

Lethal Leaf Water Status

Flower and Ludlow (1986) introduced the "lethal water status" (LWS) parameter as an accurate estimate of dehydration (or desiccation) tolerance and absolute survival. LWS is the leaf water potential or RWC at which a standard leaf dies. This is different from assessing if leaf is dead or alive after a period of *time* under stress. In diverse sorghum lines (Basnayake et al. 1993) lethal RWC and lethal leaf water potential ranged from 58% to 68% and −3.1 to −3.9 MPa, respectively.

The method involves repeated measurements of leaf water status either in terms of leaf water potential (with the pressure chamber) or RWC. Leaf death is determined by visual assessment such as becoming very discolored and brittle. As such the method is too elaborate for routine use in selection work. In another study (Lilley et al. 1996) a leaf part was sampled just as the whole plant began to appear dead. At that point turgor was null and leaf osmotic potential was equal to leaf water potential. Since osmotic potential is more stable at that point than leaf water potential, the former was measured (psychrometrically) and taken as the lethal leaf water potential.

It is not known if any study was done to correlate results between LWS and results from the simpler wilting test with the "box method," notwithstanding field performance under drought stress.

Lethal Soil Water Status

Compared with LWS, this test represents the lethal soil water content (LSWC) at which plants die in a pot experiment (Volaire 2002; Likoswe and Lawn 2008). This test overrides to a large extent the effect of plant size in a pot experiment, as compared with using time to plant death. With this method individual plants are grown in pots large enough to allow slow drying. When an individual plant is visually determined as dead, pot soil is sampled at several points to determine mean volumetric soil moisture content which is LSWC. LSWC can also be determined on the basis of water content in the whole pot after accounting for plant and tare pot weight. Gypsum blocks or other cheap soil moisture sensors can also be used to provide a continuous estimate of pot soil moisture status (Sect. 4.1.4.2). The problem is always the assessment of the true point of plant death. For example, in a grass plant all leaves appear dead but the internal apical meristem might still be alive (e.g., Volaire 2002). The best approach is probably to design a scoring system for plant appearance towards the final score of apparent plant death at which LSWC is measured. The verification of lethality is offcourse by irrigation and plant recovery – which can be done after soil is measured for moisture content. It should also be remembered that waiting for a few more days after plants appear to be dead will not change soil moisture status markedly if indeed the plant is dead. Dead plants do not use any soil moisture when the pot is evaporation-proof.

It remains that for large breeding populations visual scoring of plant death and recovery after extreme drought stress is for now the main practical large scale phenotyping alternative for drought survival. The open question remains whether this test represent any drought resistance capacity beyond the specific survival response at the given plant developmental stage tested. Based on all that has been discussed in this book to this point it can be speculated that good survival is most likely associated with favorable constitutive or adaptive genes that should support plant functions under drought stress in other vegetative stages and different stress profiles. This speculation warrants critical testing.

4.3 High Throughput Commercial Phenotyping Service

Commercial plant phenotyping facilities emerged in recent years, in parallel perhaps to phenotyping facilities serving medical research. These are also defined as "high throughput" systems because of their automated and fast plant sensing systems. These facilities are based on various remote and non-destructive techniques to assess traits of pot grown plants under highly controlled environments. Data can be collected (remotely) on plant growth, plant water status, plant architecture, root morphology and development, etc. These facilities hold great potential and some of the possibilities towards drought phenotyping are still work in progress. The phenotyping of plant drought response is based on the remote sensing techniques discussed in this chapter and should therefore encourage their use by the breeder in his own field. These facilities, being automated, can handle and monitor large number of plants especially during their earlier growth stages. Their value and cost-effectiveness towards plant breeding programs remains to be seen.

Some examples for these facilities at this time are:

Jülich Phenomics Centre, Germany
LemnaTec, Germany
The High Resolution Plant Phenomics Centre, Canberra, Australia
The Plant Accelerator, Adelaide, Australia.

Finally, the reader who is not well versed in general biology, genomics and the associated contemporary linguistics should realize that after genomics, metabolomics and proteomics we now also have "phenomics" which simply means dealing with phenotype and with phenotyping.

References

Agbicodo EM, Fatokun CA, Muranaka S et al (2009) Breeding drought tolerant cowpea: constraints, accomplishments, and future prospects. Euphytica 167:353–370

Alza JO, Fernandezmartinez JM (1997) Genetic analysis of yield and related traits in sunflower (*Helianthus annuus* l.) in dryland and irrigated environments. Euphytica 95:243–251

Aparicio N, Villegas D, Casadesus J et al (2000) Spectral vegetation indices as nondestructive tools for determining durum wheat yield. Agron J 92:83–91

Araus JL, Casadesus J, Bort J (2001) Recent tools for the screening of physiological traits determining yield. In: Reynolds MP, Ortiz-Monasterio JI, McNab A (eds) Application of physiology in wheat breeding. CIMMYT, Mexico

Ashton T (1948) Techniques of breeding for drought resistance in crops. Commonwealth Bureau of Plant Breeding and Genetics, Cambridge

Atlin GN, Kumar A, Ramaiah V et al (2009) Using information from managed-stress drought environments in practical cultivar development and drought tolerance gene detection. In: Interdrought-III international conference, Shanghai (in press)

Babar MA, Reynolds MP, van Ginkel M et al (2006) Spectral reflectance indices as a potential indirect selection criteria for wheat yield under irrigation. Crop Sci 46:578–588

Babu RC, Pathan MS, Blum A et al (1999) Comparison of measurement methods of osmotic adjustment in rice cultivars. Crop Sci 39:150–158

Bajji M, Kinet JM, Lutts S (2002) The use of the electrolyte leakage method for assessing cell membrane stability as a water stress tolerance test in durum wheat. Plant Growth Regul 36:61–70

Bänziger M, Edmeades GO, Beck D et al (2000) Breeding for drought and nitrogen stress tolerance in maize: from theory to practice. CIMMYT, Mexico

Barrs HD, Weatherley PE (1962) A re-examination of the relative turgidity technique for estimating water deficit in leaves. Aust J Biol Sci 15:413–428

Basnayake J, Ludlow M, Cooper M et al (1993) Genotypic variation of osmotic adjustment and desiccation tolerance in contrasting sorghum inbred lines. Field Crop Res 35:51–62

Bernier J, Kumar A, Ramaiah V et al (2007) A large-effect QTL for grain yield under reproductive-stage drought stress in upland rice. Crop Sci 47:507–516

Bidinger FR, Mahalakshmi V, Rao GD (1987) Assessment of drought resistance in pearl millet (*Pennisetum americanum* (L) Leeke). I. Factors affecting yields under stress. Aust J Agric Res 38:37–48

Blum A, Ebercon A (1981) Cell membrane stability as a measure of drought and heat tolerance in wheat. Crop Sci 21:43–47
Blum A, Pnuel Y (1990) Physiological attributes associated with drought resistance of wheat cultivars in a Mediterranean environment. Aust J Agric Res 41:799–810
Blum A, Schertz KF, Toler RW et al (1978) Selection for drought avoidance in sorghum using aerial infrared photography. Agron J 70:472–477
Blum A, Mayer J, Gozlan G (1982) Infrared thermal sensing of plant canopies as a screening technique for dehydration avoidance in wheat. Field Crops Res 5:137–146
Blum A, Poyarkova H, Golan G et al (1983a) Chemical desiccation of wheat plants as a simulator of post-anthesis stress. I. Effects on translocation and kernel growth Field Crops Res 6:51–58
Blum A, Mayer J, Golan G (1983b) Chemical desiccation of wheat plants as a simulator of post-anthesis stress. II. Relations to drought stress. Field Crops Res 6:149–155
Blum A, Mayer J, Golan G (1989) Agronomic and physiological assessments of genotypic variation for drought resistance in sorghum. Aust J Agric Res 40:49–61
Blum A, Shpiler L, Golan G et al (1990) Yield stability and canopy temperature of wheat genotypes under drought stress. Field Crops Res 22:289–296
Blum A, Shpiler L, Golan G et al (1991) Mass selection of wheat for grain filling without transient photosynthesis. Euphytica 54:111–116
Botwright-Acuna TL, Wade LJ (2005) Root penetration ability of wheat through thin wax-layers under drought and well-watered conditions. Aust J Agric Res 56:1235–1244
Boyer JS, James RA, Munns R et al (2008) Osmotic adjustment leads to anomalously low estimates of relative water content in wheat and barley. Funct Plant Biol 35:1172–1182
Cabrera-Bosquet L, Sánchez C, Araus JL (2009) Oxygen isotope enrichment reflects yield potential and drought resistance in maize. Plant Cell Environ 32:1487–1499
Casadesús J, Kaya Y, Bort J et al (2007) Using vegetation indices derived from conventional digital cameras as selection criteria for wheat breeding in water-limited environments. Ann Appl Biol 150:227–236
Chaerle L, Leinonen I, Jones HG et al (2007) Monitoring and screening plant populations with combined thermal and chlorophyll fluorescence imaging. J Exp Bot 58:773–784
Chenu K, Chapman SC, Hammer GL et al (2008) Short-term responses of leaf growth rate to water deficit scale up to whole-plant and crop levels: an integrated modelling approach in maize. Plant Cell Environ 31:378–391
Chloupek O, Forster BP, Thomas WT (2006) The effect of semi-dwarf genes on root system size in field-grown barley. TAG 112:779–786
Clark LJ, Cope RE, Whalley WR et al (2002) Root penetration of strong soil in rainfed lowland rice: comparison of laboratory screens with field performance. Field Crops Res 76:189–198
Clark LJ, Price AH, Steele KA et al (2008) Evidence from near-isogenic lines that root penetration increases with root diameter and bending stiffness in rice. Funct Plant Biol 35:1163–1171
Clarke JM, McCaig TN (1982) Excised-leaf water retention capacity as an indicator of drought resistance of *Triticum* genotypes. Can J Plant Sci 62:571–577
Clarke JM, Romagosa L, DePauw RM (1991) Screening durum wheat germplasm for dry growing conditions: morphological and physiological criteria. Crop Sci 31:770–775
Cook A, Marriott CA, Seel W et al (1997) Does the uniform packing of sand in a cylinder provide a uniform penetration resistance – a method for screening plants for responses to soil mechanical impedance? Plant Soil 190:279–287
Dai X, Xu Y, Ma Q et al (2007) Overexpression of an R1R2R3 MYB gene, OsMYB3R-2, increases tolerance to freezing, drought, and salt stress in transgenic *Arabidopsis*. Plant Physiol 143:1739–1751
Dalton FN (1995) In-situ root extent measurements by electrical capacitance methods. Plant Soil 173:157–165
Ebercon A, Blum A, Jordan WR (1977) A rapid colorimetric method for epicuticular wax content of sorghum leaves. Crop Sci 17:179–180
Fischer RA, Maurer R (1978) Drought resistance in spring wheat cultivars. I Grain yield responses Aust J Agric Res 29:897–905

Fischer KS, Lafitte R, Fukai S (eds) (2003) Breeding rice for drought-prone environments. International Rice Research Institute, Los Baños

Fletcher AL, Sinclair TR, Allen LH Jr (2007) Transpiration responses to vapor pressure deficit in well watered 'slow-wilting' and commercial soybean. Environ Exp Bot 61:145–151

Flower DJ, Ludlow MM (1986) contribution of osmotic adjustment to the dehydration tolerance of water-stressed pigeonpea (*Cajanus cajan* [L] millsp) leaves. Plant Cell Environ 9:33–40

Franca A, Loi A, Davies WJ (1998) Selection of annual ryegrass for adaptation to semi-arid conditions. Eur J Agron 9:71–78

Frahm MA, Rosas JC, Mayek-Perez N et al (2004) Breeding beans for resistance to terminal drought in the Lowland tropics. Euphytica 136:223–232

Frank AB, Ray IM, Berdahl JD et al (1997) Carbon isotope discrimination, ash, and canopy temperature in three wheatgrass species. Crop Sci 7:1573–1576

Fukai S, Pantuwan G, Jongdee B et al (1999) Screening for drought resistance in rainfed lowland rice. Field Crop Res 64:61–74

Garrity DP, O'Toole JC (1995) Selection for reproductive stage drought avoidance in rice, using infrared thermometry. Agron J 87:773–779

Gavuzzi P, Rizza F, Palumbo M et al (1997) Evaluation of field and laboratory predictors of drought and heat tolerance in winter cereals. Can J Plant Sci 77:523–531

Gerards CJ, Worrall WD (1986) An automated system for harvesting wheat cultivars grown under a line source sprinkler irrigation system. Agron J 78:348–349

Green S, Clothier B, Jardine B (2003) Theory and practical application of heat pulse to measure sap flow. Agron J 95:1371–1379

Gregory PJ (2006) Plant roots: growth, activity and interactions with the soil. Wiley-Blackwell, New York

Gutierrez M, Reynolds MP, Raun WR et al (2010) Spectral Water indices for assessing yield in elite bread wheat genotypes under well-irrigated, water-stressed, and high-temperature conditions. Crop Sci 50:197–214

Haley SD, Quick JS (1993) Early-generation selection for chemical desiccation tolerance in winter wheat. Crop Sci 33:1217–1223

Harris DS, Schapaugh WT, Kanemasu ET (1984) Genetic diversity in soybeans for leaf canopy temperature and the association of leaf canopy temperature and yield. Crop Sci 24:839–842

Haussmann BIG, Mahalakshmi V, Reddy BVS et al (2003) QTL mapping of stay-green in two sorghum recombinant inbred populations. TAG 106:133–142

Hendrickson AH, Veihmeyer FJ (1931) Influence of dry soil on root extension. Plant Physiol 6:567–576

Hossain ABS, Sears RG, Cox TS et al (1990) Desiccation tolerance and its relationship to assimilate partitioning in winter wheat. Crop Sci 30:622–627

Hsieh T-H, Lee J-T, Chang Y-Y et al (2002) Tomato plants ectopically expressing arabidopsis cbf1 show enhanced resistance to water deficit stress. Plant Physiol 130:618–626

Hu L, Wang Z, Huang B (2009) Photosynthetic responses of bermudagrass to drought stress associated with stomatal and metabolic limitations. Crop Sci 49:1902–1909

Hunt, R (1990) Basic growth analysis: plant growth analysis for beginners. Springer, Dordrecht

Idso SB, Jackson RD, Reginato RJ (1977) Remote-sensing of crop yields. Science 196:19–25

Ingram KT, Real JG, Maguling MA et al (1990) Comparison of selection indices to screen lowland rice for drought resistance. Euphytica 48:253–260

Ismail MR, Davies WJ (1998) Root restriction affects leaf growth and stomatal response: the role of xylem sap ABA. Sci Hort 74:257–268

Jacomini E, Bertani A, Mapelli S (1988) Accumulation of polyethylene glycol 6000 and its effects on water content and carbohydrate level in water-stressed tomato plant. Can J Bot 66:970–973

Janes BE (1974) The effect of molecular size concentration in nutrient solution and exposure time on the amount and distribution of polyethylene glycol. Plant Physiol 54:226–229

Jiang GH, He YQ, Xu CG et al (2004) The genetic basis of stay-green in rice analyzed in a population of doubled haploid lines derived from an indica by japonica cross. TAG 108:688–698

Jones HG, Serraj R, Loveys BR et al (2009) Thermal infrared imaging of crop canopies for the remote diagnosis and quantification of plant responses to water stress in the field. Funct Plant Biol 36:978–989

Katan, J, DeVay JE (1991) Soil solarization. CRC Press, Boca Raton

Kato Y, Abe J, Kamoshita A et al (2006) Genotypic variation in root growth angle in rice (*Oryza sativa* L) and its association with deep root development in upland fields with different water regimes. Plant Soil 287:117–129

Kaukoranta T, Murto J, Takala J (2005) Detection of water deficit in greenhouse cucumber by infrared thermography and reference surfaces. Sci Hort 106:447–463

Klueva NY, Joshi RC, Joshi CP et al (2000) Genetic variability and molecular responses of root penetration in cotton. Plant Sci 155:41–47

Lagerwerff JV, Ogata G, Eagle HE (1961) Control of osmotic pressure of culture solutions with polyethylene glycol. Science 133:1486–1490

Lawlor DW (1970) Absorption of polyethylene glycols by plants and their effects on plant growth. New Phytol 69:501–514

Leinonen I, Grant OM, Tagliavia et al (2006) Estimating stomatal conductance with thermal imagery. Plant Cell Environ 29:1508–1518

Likoswe AA, Lawn RJ (2008) Response to terminal water deficit stress of cowpea, pigeonpea, and soybean in pure stand and in competition. Aust J Agric Res 59:27–37

Lilley JM, Fukai S (1994) Effect of timing and severity of water deficit on four diverse rice cultivars 1. Rooting pattern and soil water extraction. Field Crop Res 37:205–213

Lilley JM, Ludlow MM, Mccouch SR et al (1996) Locating qtl for osmotic adjustment and dehydration tolerance in rice. J Exp Bot 47:1427–1436

Lipavska H, Vreugdenhil D (1996) Uptake of mannitol from the media by in vitro grown plants. Plant Cell Tissue Org Cult 45:103–107

Lopes MS, Reynolds MP (2010) Partitioning of assimilates to deeper roots is associated with cooler canopies and increased yield under drought in wheat. Funct Plant Biol 37:147–156

Lu ZM, Radin JW, Turcotte EL et al (1994) High yields in advanced lines of pima cotton are associated with higher stomatal conductance, reduced leaf area and lower leaf temperature. Physiol Plant 92:266–272

Ludlow MM, Chu ACP, Clements RJ et al (1983) Adaptation of species of *Centrosema* to water stress. Aust J Plant Physiol 10:119–130

Ma BL, Dwyer LM, Costa C et al (2001) Early prediction of soybean yield from canopy reflectance measurements. Agron J 93:1227–1234

Mahalakshmi V, Blum A (2006) Phenotyping in the field: global capacity accessible to the GCP – Inventory of phenotyping resources and capacity for the GCP. Final report, Generation Challenge Program, El Batan, Mexico

Mahalakshmi V, Bidinger FR, Rao KP et al (1994) Use of the senescing agent potassium iodide to simulate water deficit during flowering and grain filling in pearl millet. Field Crop Res 36:103–111

Mambani B, Lal R (1983) Response of upland rice varieties to drought stress II. Screening rice varieties by means of variable moisture regimes along a toposequence Plant Soil 73:73–94

Maxwell K, Johnson GN (2000) Chlorophyll fluorescence – a practical guide. J Exp Bot 51:659–668

McBride R, Candido M, Ferguson J (2008) Estimating root mass in maize genotypes using the electrical capacitance method. Arch Agron Soil Sci 54:215–226

McKinney NV, Schapaugh WT, Kanemasu ET (1989) Selection for canopy temperature differential in six populations of soybean. Crop Sci 29:255–259

Merewitz E, Meyer W, Bonos S et al (2010) Drought stress responses and recovery of Texas x Kentucky hybrids and Kentucky bluegrass genotypes in temperate climate conditions. Agron J 102:258–268

Michel BE, Kaufmann MR (1973) The osmotic potential of polyethylene glycol 6000. Plant Physiol 51:914–919

Money NP (1989) Osmotic pressure of aqueous polyethylene glycols relationship between molecular weight and vapor pressure deficit. Plant Physiol 91:766–769

Morgan JM (1992) Osmotic components and properties associated with genotypic differences in osmoregulation in wheat. Aust J Plant Physiol 19:67–76

Nicolas ME, Turner NC (1993) Use of chemical desiccants and senescing agents to select wheat lines maintaining stable grain size during post-anthesis drought. Field Crop Res 31:155–171

Olivares-Villegas JJ, Reynolds MP, McDonald GK (2007) Drought-adaptive attributes in the Seri/Babax hexaploid wheat population. Funct Plant Biol 34:189–203

Oukarroum A, El Madidi S, Schansker G et al (2007) Probing the responses of barley cultivars (*Hordeum vulgare* L) by chlorophyll a fluorescence OLKJIP under drought stress and re-watering. Environ Exp Bot 60:438–446

Penuelas J, Filella I, Biel C et al (1993) The reflectance at the 950–970 nm region as an indicator of plant water status. Int J Remote Sens 14:1887–1905

Pinter PJ, Zipoli G, Reginato RJ et al (1990) Canopy temperature as an indicator of differential water use and yield performance among wheat cultivars. Agric Water Manag 18:35–48

Premachandra GS, Shimada T (1988) Evaluation of polyethylene glycol test measuring cell membrane stability as a drought tolerance test in wheat. J Agric Sci 110:429–433

Price AH, Cairns JE, Horton P et al (2002) Linking drought-resistance mechanisms to drought avoidance in upland rice using a QTL approach: progress and new opportunities to integrate stomatal and mesophyll responses. J Exp Bot 53:989–1004

Rebetzke GJ, Read JJ, Barbour MM et al (2003) A hand-held porometer for rapid assessment of leaf conductance in wheat. Crop Sci 40:277–280

Rebetzke GJ, van Herwaarden AF, Jenkins C et al (2008) Quantitative trait loci for water-soluble carbohydrates and associations with agronomic traits in wheat. Aust J Agric Res 59:891–905

Reid CP, Bowen GD, McCleod S (1978) Phosphorus contamination in polyethylene glycol. Plant Physiol 61:708–709

Reynolds MP, Singh RP, Ibrahim A et al (1998) Evaluating physiological traits to complement empirical selection for wheat in warm environments. Euphytica 100:85–94

Richardson MD, Karcher DE, Hignight K et al (2008) Drought tolerance and rooting capacity of Kentucky bluegrass cultivars. Crop Sci 48:2429–2436

Royo C, Blanco R (1998) Use of potassium iodide to mimic drought stress in triticale. Field Crop Res 59:201–212

Royo C, Villegas D, del Moral LFG et al (2002) Comparative performance of carbon isotope discrimination and canopy temperature depression as predictors of genotype differences in durum wheat yield in Spain. Aust J Agric Res 53:561–569

Royo C, Aparicio N, Villegas D et al (2003) Usefulness of spectral reflectance indices as durum wheat yield predictors under contrasting Mediterranean conditions. Int J Remote Sens 24:4403–4419

Ruuska SA, Rebetzke GJ, Van Herwaarden AF et al (2006) Genotypic variation in water-soluble carbohydrate accumulation in wheat. Funct Plant Biol 33:799–809

Saint Pierre C, Crossa J, Manes Y et al (2010a) Gene action of canopy temperature in bread wheat under diverse environments. TAG 120:1107–1117

Saint Pierre C, Trethowan R, Reynolds M (2010b) Stem solidness and its relationship to water-soluble carbohydrates: association with wheat yield under water deficit. Funct Plant Biol 37:166–174

Sanchez FJ, Manzanares M, de Andres EF et al (2001) Residual transpiration rate, epicuticular wax load and leaf colour of pea plants in drought conditions. Influence on harvest index and canopy temperature. Eur J Agron 15:57–70

Sawhney V, Singh DP (2002) Effect of chemical desiccation at the post-anthesis stage on some physiological and biochemical changes in the flag leaf of contrasting wheat genotypes. Field Crops Res 77:1–6

Scholander PF, Bradstreet ED, Hemmingsen EA et al (1965) Sap pressure in vascular plants. Negative hydrostatic pressure can be measured in plants. Science 148:339–346

Schulte PJ, Hinckley TM (1985) A comparison of pressure-volume curve data analysis techniques. J Exp Bot 36:1590–1602

Serrano R, Culianz-Macia FA, Moreno V (1999) Genetic engineering of salt and drought tolerance with yeast regulatory genes. Sci Hort 78:261–269

Singh DP, Singh P, Kumar A et al (1985) Transpirational cooling as a screening technique for drought tolerance in oil seed brassicas. Ann Bot 56:815–820

Smit AL, Bengough AG, Engels C (eds) (2000) Root methods: a handbook. Springer, Dordrecht

Srinivasan A (ed) (2006) Handbook of precision agriculture: principles and applications. CRC Press, Boca Raton

Subashri M, Robin S, Vinod KK et al (2009) Trait identification and QTL validation for reproductive stage drought resistance in rice using selective genotyping of near flowering RILs. Euphytica 166:291–305

Subudhi PK, Rosenow DT, Nguyen HT (2000) Quantitative trait loci for the stay green trait in sorghum (*Sorghum bicolor* L. Moench): consistency across genetic backgrounds and environments. TAG 101:733–741

Sudduth KA, Kitchen NR, Bollero GA et al (2003) Comparison of electromagnetic induction and direct sensing of soil electrical conductivity. Agron J 95:472–482

Takai T, Yano M, Yamamoto T (2010) Canopy temperature on clear and cloudy days can be used to estimate varietal differences in stomatal conductance in rice. Field Crops Res 115:165–170

Taylor HM, Upchurch DR, Mcmichael BL (1991) Applications and limitations of rhizotrons and minirhizotrons for root studies. Plant Soil 129:29–35

Tingey DT, Stockwell C (1977) Semipermeable membrane system for subjecting plants to water-stress. Plant Physiol 60:58–63

Tomos AD, Leigh RA (1999) The pressure probe: a versatile tool in plant cell physiology. Ann Rev Plant Physiol Plant Mol Biol 50:447–459

Tripathy JN, Zhang J, Robin S et al (2000) QTLs for cell-membrane stability mapped in rice (*Oryza sativa* L.) under drought stress. TAG 100:1197–1202

Turner NC, O'Toole JC, Cruz RT et al (1986) Response of seven diverse rice cultivars to water deficits. II. Osmotic adjustment, leaf elasticity, leaf extension, leaf death, stomatal conductance and photosynthesis. Field Crops Res 13:273–286

van Beem J, Smith ME, Zobel RW (1998) Estimating root mass in maize using a portable capacitance meter. Agron J 90:566–570

Venuprasad R, Lafitte HR, Atlin GN (2007) Response to direct selection for grain yield under drought stress in rice. Crop Sci 47:285–293

Venuprasad R, Dalid CO, Del Valle M et al (2009) Identification and characterization of large-effect quantitative trait loci for grain yield under lowland drought stress in rice using bulk-segregant analysis. TAG 120:177–190

Verslues PE, Ober ES, Sharp RE (1998) Root growth and oxygen relations at low water potentials. Impact of oxygen availability in polyethylene glycol solutions. Plant Physiol 116:1403–1412

Volaire F (2002) Drought survival, summer dormancy and dehydrin accumulation in contrasting cultivars of *Dactylis glomerata*. Physiol Plant 116:42–51

Waisel Y, Eshel A, Kafkafi U (eds) (1996) Plant roots: the hidden half, 3rd edn. Marcel Dekker Inc, New-York

Walker DW, Creighton Miller J Jr (1986) Rate of water loss of drought resistant and susceptible genotypes of cowpea. Hortscience 21:131–132

Wang JP, Bughrara SS (2008) Evaluation of drought tolerance for Atlas fescue, perennial ryegrass, and their progeny. Euphytica 164:113–122

Wang Y, Holroyd G, Hetherington AM et al (2004) Seeing 'cool' and 'hot' – infrared thermography as a tool for non-invasive, high-throughput screening of *Arabidopsis* guard cell signalling mutants. J Exp Bot 399:1187–1193

Willardson LS, Oosterhuis DM, Johnson DA (1987) Sprinkler selection for line-source irrigation systems Irr Sci 8:65–76

Winter SR, Musick JT, Porter KB (1988) Evaluation of screening techniques for breeding drought resistant winter wheat. Crop Sci 28:512–516

Xu Y, Crouch JH (2008) Marker-assisted selection in plant breeding: from publications to practice. Crop Sci 48:391–407

Xu W, Rosenow DT, Nguyen HT (2000) Stay green trait in grain sorghum: relationship between visual rating and leaf chlorophyll concentration. Plant Breed 119:365–367

Yang W, Peng S, Laza RC et al (2008) Yield gap analysis between dry and wet season rice crop grown under high-yielding management conditions. Agron J 100:1390–1395

Yaniv Z, Werker E (1983) Absorption and secretion of polyethylene glycol by Solanaceous plants. J Exp Bot 34:1577–1582

Yu LX, Ray JD, O'Toole JC et al (1995) Use of wax-petrolatum layers for screening rice root penetration. Crop Sci 35:684–687

Zaharah SS, Razi IM (2009) Growth, stomata aperture, biochemical changes and branch anatomy in mango (*Mangifera indica*) cv Chokanan in response to root restriction and water stress. Sci Hort 123:58–67

Zilberstein M, Blum A, Eyal Z (1985) Chemical desiccation of wheat plants as a simulator of postanthesis speckled leaf blotch stress. Phytopathology 75:226–230

Zong LZ, Liang S, Xu X et al (2008) Relationships between carbon isotope discrimination and leaf morphophysiological traits in spring-planted spring wheat under drought and salinity stress in Northern China. Aust J Agric Res 59:941–949

Chapter 5
Genetic Resources for Drought Resistance

Summary There is no comprehensive listing of genetic resources and potential donors for drought resistance. Sporadic data may sometime be found in seed banks but the quality of the data is uncertain. This chapter attempts to present the available groups of actual and potential genetic resources as derived from past and current literature. Five groups are recognized. These are listed by the order of their relative genetic compatibility with cultivated breeding germplasm.

While it is to be expected that breeders tend to search for donors of drought resistance among distant germplasm, it is quite apparent today that normal *agronomic breeding germplasm* may often carry latent genetic variation for drought resistance and this should be the first resource of choice. *Landraces* from dry habitats have been used successfully in breeding for water limited environments, whether towards developing open pollinated varieties or hybrids. *Wild species* and progenitors of our cultivated crops were always on the agenda as possible donors for drought resistance. Attempts at using these resources have increased in recent years and this chapter evaluates their potential and real contribution. *Transgenic plants* are first developed as a tool in functional genomics. They can constitute a realistic step towards transferring useful genes into a target crop plant, and as such they are an important genetic resource. Their importance is increasing as their drought phenotyping improves. Lastly, *resurrection plants* which survive extreme desiccation under harsh environments have always excited the imagination of biologists. Research on the nature of their tolerance may open new avenues for their use as donors of important genes for drought resistance.

Both the professional and the popular literature warn against the narrowing of the genetic base of our crop plants and the danger of reaching a situation of depleted genetic variation for plant breeding. This is often argued on the basis of some historical difficulties (which were overcome) in isolating sources of resistance to certain disease epidemics. This opinion also appears reasonable to the layman as he drives through the countryside seeing the very homogenous appearance of our crops.

This trendy thinking has migrated into the area of abiotic stress resistance. In the preface to their book (Jenks et al. 2007) Jenks state that "plant breeding was not able to develop drought resistant cultivars" and the reason was that "domestication

A. Blum, *Plant Breeding for Water-Limited Environments*,
DOI 10.1007/978-1-4419-7491-4_5, © Springer Science+Business Media, LLC 2011

has narrowed the genetic diversity within crops for stress tolerance thus limited options in traditional crop breeding." This statement is incorrect on two counts. Firstly, plant breeders were able to develop stress (and drought) resistant cultivars (Sect. 3.1). Most likely the bread he has for breakfast comes from a US Midwestern wheat variety possessing drought resistance trait(s). Secondly, there is no factual basis for stating that domestication has narrowed the genetic diversity within crops for abiotic stress resistance. Contrary to his statement as well as the popular mood, it was recently found that there was no reduction in the genetic diversity of crop varieties released during the twentieth century (van de Wouw et al. 2010). The specific case for drought resistance is discussed below.

Rich genetic resources for drought resistance breeding are available in five categories: (a) cultivated germplasm, (b) landraces, (c) wild species (or crop progenitors), (d) transgenic plants and (e) resurrection plants. The choice of genetic resource to use as donor of drought resistance depends on the probability of discovering the required genes as well as the expected difficulties and projected success in introgression of these genes into the chosen recurrent cultivar. In this respect, cultivated germplasm would be the first choice.

5.1 Cultivated Germplasm

Contrary to popular notions of depleted genetic variation in "domesticated" germplasm it has been found that cultivated crop germplasm may contain ample genetic variation for drought resistance. This has been first demonstrated by the high levels of dehydration avoidance within a relatively limited sample of elite wheat breeding materials (Blum et al. 1981). It was later confirmed by Warburton et al. (2006) who concluded that genetic variation for abiotic stress resistance in regular CIMMYT wheat breeding materials was no less than that found in a collection of landraces. In a later review from CIMMYT it was concluded (Trethowan and Mujeeb-Kazib 2008) that there is significant unexploited variation among landraces as well as modern wheat cultivars to improve the stress adaptation of cultivated wheat. In a study of 325 BC_2F_2 rice bulk populations developed by backcrossing of drought resistant donors to one of three elite recurrent parents (Lafitte et al. 2006) it was found that even drought susceptible rice cultivars carried ample cryptic genetic variation for drought resistance which was expressed in their progenies. Liu et al. (2004) found that alleles for improved root growth and distribution under drought stress were unexpectedly found in some cultivated japonica rice cultivars rather than in a collection of wild rice species. Work with maize (Ribaut et al. 1996) indicated that genetic segregation for short ASI (anthesis-to-silking interval) was transgressive with the drought-susceptible parent contributing alleles for reduced ASI at two QTL positions.

While these findings for wheat, rice and maize support the exploration of cultivated germplasm for drought resistance traits, the situation for other crops has not been similarly evaluated. However, one should notice that dozens of research papers present significant genetic variation for drought resistance among few crop cultivars or lines selected from cultivated breeding germplasm. This is true for cereals, pulses, forage crops, oil crops, range grasses, turf grasses, certain vegetables and other horticultural crops. Accordingly, the normal routine adopted by breeders still holds, namely searching for the desirable trait within the normal and often agronomic pool of breeding materials, before attempting to explore more exotic resources. Despite the popular notions it should never be flatly presumed that cultivated germplasm has no diversity for drought resistance.

Induced mutations have been for long an important part in the breeder's toolbox where elite breeding materials and cultivars were subjected to applied mutagenesis. There is even a dedicated unit at the IAEA (International Atomic Energy Agency) towards using irradiation in mutation breeding. Mutation breeding for drought resistance has been very backwards, as compared for example with disease resistance or plant morphological features. There has been a small trickle of reports on some results in the laboratory. However these were almost anecdotal in terms of assessing their real value towards breeding. There were also occasional declarations of intent as well as statements on the potential of the method. It can be assumed that earlier attempts to induce valuable mutations towards breeding for drought resistance were hampered by ineffective phenotyping of drought resistance. More recently, using both chemical and nuclear agents Cairns et al. (2009) produced and tested some 3,500 deletion mutants of rice cv. IR64. Several mutants performed better than the wild type under managed drought stress conditions. The advantage was seen in yield, biomass, transpiration and canopy temperature. Successful mutants tended to have deeper roots and improved soil moisture use under drought conditions. This singular study indicates great potential for induced mutations if performed on a large scale and tested accurately under managed drought stress protocols for the relevant variables.

5.2 Landraces

For reasons of genetic compatibility and potentially important genetic variation, landraces of crop species are an attractive genetic resource for drought resistance. It would be very reasonable to assume that landraces from dry regions and semi arid habitats are more likely to be drought resistant. The success of Yadav (see below) with pearl millet landraces from the arid Rajasthan region in India as resources for improved drought resistant cultivars and hybrids attends to this fact. Blum and Sullivan (1986) found that sorghum and pearl millet landraces from dry regions in India and Africa were more drought resistant in terms of their physiology than landraces from humid regions. When various landraces of barley (Ceccarelli et al. 1998) and pearl millet (Yadav and Weltzien 2000) were compared, the relatively highest yield under drought stress was obtained by the landraces from the most arid regions. Work with Mexican (Reynolds et al. 2007; Trethowan and Mujeeb-Kazib 2008) and Israeli (Blum et al. 1989) (Fig. 5.1) wheat landraces identified deep soil moisture extraction, use of stem reserves, and high biomass under stress as important traits for introgression into standard wheat cultivars.

Fig. 5.1 Variation in potential root depth (under non-stress conditions) between two landraces of durum wheat collected in the Northern Negev desert in Israel (Blum et al. 1989), in support of results for Mexican landraces (Reynolds et al. 2007; Trethowan and Mujeeb-Kazib 2008)

However when a large collection of wheat landraces from the arid Northern Negev desert of Israel was studied (Blum et al. 1989), extensive genetic variation in drought resistance was found in the collection contrary to the expectation that most landraces would be resistant. Resistance was expressed mainly in low canopy temperature under drought stress, osmotic adjustment, utilization of stem reserves for grain filling, biomass and grain yield under drought stress, but many landraces from this region were still relatively drought susceptible. Following this study and its publication, the growers of these landraces were queried for their practices in order to understand the possible reason for the high frequency of drought susceptible materials in the collection. An interesting and reasonable explanation emerged. When once in several years a very severe drought occurred and especially if it was followed by a second dry year, many farmers lost their crop and did not have any more seed of their own to plant. In that case they went up North to the more rainy regions and purchased local landrace seed for planting. These imported entries were very possibly the source of drought susceptibility in the studied collection. Therefore, not all landraces originating in an arid region are necessarily drought resistant although most are likely to be. A range of landrace materials should be studied in order to capture the most resistant types as donors in the breeding program.

A well documented successful use of landraces in breeding for water limited environments is the barley breeding program at ICARDA in Syria, ran by Ceccarelli and Grando (1991) with the help of many associates for many years. Wide evaluation of local (e.g., Ceccarelli and Grando 1991) and exotic (Lakew et al. 1997) landraces of barley was the foundation on which they developed the program. Participation by farmers in the selection process became an important component. The selection was mainly applied to yield and yield related traits in the low yielding fields of the target region. Both selections within landrace populations and crosses with high yielding materials served as basis for the released cultivars. In some cases selection within landraces from dry regions was effective in improving yield. While heritability for yield under drought stress is generally low (Sect. 3.3.2), they were still able to achieve progress in selection for yield under stress because of the apparently high frequency of drought resistance traits in their populations.

The sorghum conversion program at Texas A&M University during the 1980s was targeted at modifying tropical landrace materials from Africa (mainly) for use in breeding in the temperate region of the USA. Tropical sorghum does not generally flower or it flowers too late in the temperate region because of photoperiod requirement. Thus the conversion program consisted of extensive backcrossing program to convert this germplasm into short flowering materials under temperate conditions while retaining most of the original phenotype (Rosenow and Dahlberg 2000). This program provided valuable genetic resources which allowed a significant boost to dryland sorghum breeding in the USA and worldwide. Besides yield traits, disease resistance, and grain quality traits this program revealed important drought resistance traits used by breeders since then, including the well documented stay-green trait.

In India, pearl millet populations containing various proportions of landrace materials from the dry Rajasthan region performed better than high yielding cultivars under limited water supply (Yadav and Weltzien 2000). Landrace materials expressed better tillering and stable flowering under drought stress as compared with high yielding cultivars. Landraces were more productive under severe drought than elite materials and expressed less delay in flowering (Yadav 2008). When these drought resistant landraces were top crossed onto elite materials of large panicles it was possible to achieve greater yield and drought resistance in the hybrids.

5.3 Wild Species and Crop Plant Progenitors

The last conventional resort for identifying genetic resources for drought resistance is the wild species and the wild progenitors of the crop. This is a difficult option for two well known reasons: (a) a wide cross between the wild plant and the cultivated genotype can be difficult or impossible technically; and (b) much work is required to avoid the introgression of negative traits from the wild resource. There is also a third problem specifically concerning drought resistance.

Wild species from dry habitats are expected to carry genes for adaptation to drought stress. This adaptation is a consequence of evolution and natural selection.

Natural selection does not operate on productivity under water-limited conditions but on survival of the species. In annual plants an evolutionary survival depends on the production of at least one seed per season as compared with premature plant death under stress. Factors which support such survival are not necessarily compatible with crop production under drought stress in an agricultural context. The crop environment as created and controlled by the dryland farmer is different from that experienced by the wild species under natural conditions, concerning germination date, plant nutrition, herbivory, competition with other species, etc. For example, plant survival under natural drought might be driven by a small plant size and very early flowering. When this plant is used as a donor and then plant size and flowering date of the progeny are trimmed towards the required agronomic ideotype, the apparent drought resistance of that wild species might disappear from the progeny.

The value of a wild species towards drought resistance can be established by the performance of a reasonably agronomic phenotype after introgression of the genes of interest from the wild species, as done for example in wheat (Gororo et al. 2002), clover (Marshall et al. 2001) and lettuce (Johnson et al. 2000). Work with wild lentil collections (Hamdi and Erskine 1996) indicated that drought escape was unimportant in the wild lentil while it was important in cultivated lentil, Furthermore, performance of wild lentil under drought, measured in terms of dryland seed yield or drought susceptibility index, was randomly distributed among collection locations with little relation to collection site aridity. The authors concluded that the value of wild lentils for drought resistance must be evaluated after crossing with cultivated materials. These studies repeatedly confirm that the value of a wild species as a genetic resource for drought resistance can be established only after its cross with the cultivated germplasm.

Wild emmer (*Triticum dicoccoides*) collections are often hailed as potential donors for abiotic stress resistance of wheat (Carver and Nevo 1990; Nevo et al. 1992; Peleg et al. 2005; Baalbaki et al. 2006). Their value in this respect has been established mainly by testing the wild emmer collections themselves. The measurement of biomass, yield related traits or WUE under drought stress in the wild species (Peleg et al. 2005) do not provide a firm prediction of value in terms of drought resistance in the cultivated species. For example, WUE in the wild species bears very little significance towards the cultivated crop since it is a ratio affected by physiological, morphological and developmental traits that vary extensively between wild and cultivated plants. For example, the small leaf area typical of some wild species would tend to express higher WUE than in the cultivated wheat materials (Sect. 3.5.1).

The solution might be in the dissection of the exact and specific drought resistance trait and gene which ascribes resistance to the wild phenotype (e.g., Suprunova et al. 2007) and then introgressing it by backcross into an agronomic plant type. The accurate drought phenotyping of backcrossed agronomic derivatives of the wild species, such as the case for wild emmer (Blum et al. 1983) or *Triticum tauschii* (Gororo et al. 2002), should provide reasonable assessment of value in breeding towards water-limited environments. This is certainly a long process which is expected to be shortened by future genomic methods.

There has been some progress towards utilizing wild species in breeding for water limited environments, beyond the collection and evaluation of germplasm. Chromosome addition lines of *Agropyron elongatum* in common wheat cv. 'Chinese Spring' tested under drought stress indicated that chromosome 5E of *A. elongatum* carried important genes for drought resistance that are useful for breeding (Sutka et al. 1995). Putatively important genes for drought resistance were identified in wild barley (*Hordeum spontaneum*) (Suprunova et al. 2004, 2007) and wild chick-peas (*Cicer pinnatifidum*) (Bhattarai and Fettig 2005). Wild lettuce was found to have deeper roots than cultivated lettuce (Gallardo et al. 1996) and this was then verified in F_2-F_3 lines out of crosses between wild and cultivated lettuce (Johnson et al. 2000). A study of diverse rice cultivars and wild rice species (Liu et al. 2004) indicated that *Oryza longistaminata* and *O. rufipogon* could serve as sources of novel alleles for maintenance of leaf elongation, stomatal conductance, and membrane stability under drought stress.

A well documented case of using wild species materials in breeding for water limited environments is the introgression of *Festuca arundinacea* var. *glaucescens* drought resistance alleles into *Lolium multiflorum* (Humphreys et al. 2005) while conserving the forage quality of the latter (Zare et al. 2002).

On the other hand the introgression of some drought resistance traits from wild emmer (Blum et al. 1983) into common wheat produced very agronomic lines (Fig. 5.2). These performed as well but not better than the best dryland wheat cultivar at the time and therefore none was released as a cultivar.

Nevo and Chen (2010) published a detailed review of the use of wild relatives towards breeding drought resistance in wheat and barley. They describe various

Fig. 5.2 Wild emmer (*T. dicoccoides*) collections in front and wild emmer-derived agronomic common wheat F_{10} lines in the back, all planted on the same day (unpublished work of the wheat breeding team at the Volcani Center, Bet Dagan, Israel)

attempts to introgress traits and QTLs for drought resistance into cultivated breeding materials. They even indicate some favorable introgressions regarding yield performance. However despite their optimistic stance towards using these genetic resources the reader will recognize two major problems:

(a) Despite the intensive activity for the last 30 years with these wild progenitors as genetic resources for drought resistance the review does not specify any actual success in breeding and the delivery of a drought resistant cultivar as an end product. Most research is terminated with indications of potential for breeding.
(b) It emerges from the review that a major pitfall is the insufficient drought phenotyping in the attempts to derive drought resistant cultivars from the wild progenitors. This is reflected, as they indicate, by the low level of drought stress used in screening and the insufficient phenotyping and measurements.

Finally, a widely publicized case is the exciting story of the Nerica family of upland rice cultivars which has an ominous impact on rice production in Africa. Nerica has been developed from a cross between African rice *O. glaberrima* Steud. and the Asian rice (*O. sativa* L.) at WARDA (West Africa Rice Development Association, now called Africa Rice Center). This family of varieties carries various levels of disease and pest resistances as well as certain soil mineral deficiency and toxicity resistances. Drought resistance is also being mentioned as an important trait of Nerica. Nerica passports (available at http://www.warda.org/) do not indicate specific drought resistance traits. However its earliness ascribes drought escape. A major reason for Nerica yield advantage is its higher harvest index than the African glaberrima rice, which might be an outcome of Nerica earliness (Dingkuhn et al. 1998). Doto Leontine, a Nerica rice farmer in Benin was quoted on the WARDA web site, saying: "Our traditional rice is a long-duration variety, which takes 4–5 months to mature and if we grow Nerica, in 3 months we have the harvest. This can help us fight hunger. Nerica is very valuable in the hungry season, because it is ready to be harvested, while the traditional rice is not yet mature." Nerica development seems to have followed in the tracks of the past green revolution in many other cereals, by reducing photoperiod sensitivity and increasing harvest index, plus the incorporation of additional important traits.

5.4 Drought Resistant Transgenic Plants

Transgenic plants are the first major step in assessing the functionality of discovered genes at the whole plant level. Because of known technical and genetic considerations most transgenic work is performed with model plants, typically *Arabidopsis thaliana* and tobacco. However, transgenic crop plants are also becoming common and they certainly constitute a primary genetic resource and a bridge for utilizing the discovered gene in breeding. After its genetic stabilization and the affirmation of the desirable phenotype, the transgenic crop plant can be integrated into the breeding program in various fashions as any other genetic resource, pending regulatory issues. In certain cases the goal of transformation was to insert the gene directly into an elite agronomical material which can then be used to develop the drought resistant cultivar. There are also cases where the transgenic plant was used as parent in hybrids.

Xiao et al. (2007) outlined the major requirements for developing a transgenic drought resistant cultivar: Drought resistance should be improved significantly. No phenotypic changes for other traits should take place. No yield penalty under non-stress conditions should take place. Assure the over-expression of a single copy transgene since multiple gene copies can lead to instability of expression and inheritance or even gene silencing. Growth retardation and stunting is a common byproduct of the transgenic event. Such plants can be recognized immediately. However, impaired yield potential due to the transgenic event which is a likely occurrence can be verified only after a sufficient number of field tests.

In reviewing the potential of transcription factors towards breeding grasses (and cereals) for stress resistance Nakashima et al. (2009) concluded that "(a) an effective expression system, including suitable promoters, will be required for each grass because constitutive promoters are not always functional or can have negative effects on plant growth and development. (b) There is an urgent need to establish reliable systems to evaluate abiotic stress tolerance in transgenic grasses, especially under field conditions. (c) The collective and cooperative efforts of plant molecular biologists, physiologists, and breeders are required to generate stress-tolerant grasses through genetic engineering."

The important phase in assessing gene function is the accurate phenotyping of the transgenic plant after the insertion of the gene. Erratic phenotyping at this stage might bias the conclusion to the positive or the negative and possibly lead to waste of time and resources in pursuing a mirage or to loosing an opportunity.

Regretfully, from the onset, transgenic plants targeted for drought resistance were not always phenotyped properly (e.g., Blum et al. 1996). The following problems could be seen in reported phenotyping of transgenic plants for drought resistance:

1. The transgenic plant might sometime be modified by the transgenic event itself or by unexpected interactions of the inserted gene with the specific genome. The modification is not a direct effect of the gene, but it can affect plant water status under the drought conditions in the test protocol. Such modifications involved stunted growth, modified plant size, leaf size, leaf area and flowering time. Unless the test accounts for these modifications in order to resolve the net effect of the gene, no conclusions can be reached from the test.
2. Test protocols do not always adhere to the established physiology of plant water relations. For example in one study plant water status has been assessed by percent water in the leaf.
3. Understandably most tests are performed with potted plants which are dehydrated to induce drought stress. Pot experiments in water-relations and drought stress research are subject to well known pitfalls which can lead to gross artifacts (Sect. 4.1.5.1). These artifacts were not always recognized in transgenic research report.

Past critique by plant physiologists and breeders of transgenic research towards drought resistance is recently leading to more acceptable results. The best approach is reflected in work done in the private sector where tight collaboration exists between molecular biologists, breeders and physiologists. This approach appears to produce results in the field.

An updated listing of genes and transgenic plants developed for drought resistance appears in www.plantstress.com under "biotech issues." A short list of the genes appeared in Cattivelli et al. (2008).

An important part of the published public domain research with transgenic plants is done to resolve questions about gene expression, cellular expression of stress signals, metabolic pathways, etc. In this book the interest is in genes that indicate an immediate potential towards application in breeding. Such an assessment is offcourse quite subjective and open to debate. It is not a final judgment. Here the assessment is based firstly on whether tests of the transgenic plant already involved some measure of productivity under drought stress. It is also based on how well the genetic modification and its claimed effect on drought resistance were explained in terms of the known physiology. Quantification of the effect of the gene is also important in this respect and serious penalty of the gene effect under non-stress conditions is considered unacceptable. Other genes under various experiments which are not listed here might become very important in due course. The following list is therefore very transitory and should serve as an example of assessment towards breeding.

Late embryogenesis abundant (LEA) proteins were initially found in barley seed aleuron and were considered to have a role in conserving embryo life under the extremely dry environment of the dormant seed. Hydrophilins are a wide group of proteins whose defining characteristics are high hydrophilicity index (>1.0) and high glycine content. LEA proteins were found to belong to this group and they may have a role in protecting against enzyme degradation under water deficit (Reyes et al. 2005).

The HVA1 gene controls the synthesis of this protein in barley. LEA proteins are produced in maturing seeds, plants, animals and microorganisms. They are mainly low molecular weight (10–30 kDa) proteins, which are generally classified into six groups according to their amino acid sequence. Their expression is promoted by dehydration and ABA and it generally correlates with desiccation tolerance. HVA1 expressing wheat plants were tested in pots in the greenhouse. Transgenic lines had significantly higher total biomass as compared with non-transgenic controls under moderate water deficit conditions. No negative effect of HVA1 was seen under non-stress conditions (Sivamani et al. 2000). These and other HVA1 expressing transgenic wheat lines developed from different transgenic events were tested in nine field experiments over six cropping seasons (Bahieldin et al. 2005). Drought stress was not controlled or well described in these tests. Results were variable depending on test and line, indicating a potential in certain lines for better yield under drought stress.

Transgenic Chinese cabbage (*Brassica campestris* ssp. *pekinensis*) expressing a *B. napus* LEA protein gene were more drought resistant than the wild type as reflected by delayed development of damage symptoms and improved recovery from stress in a pot experiment (Park et al. 2005).

Transgenic rice plants expressing HVA1 LEA proteins maintained higher growth rates than wild type under drought stress conditions and displayed delayed stress symptoms. The extent of expressed drought resistance was correlated with the level of the HVA1 protein accumulated in the transgenic rice plants (Xu et al. 1996). Transformed Basmati rice overexpressing HVA1 LEA3 protein accumulation was more drought resistance than the control plants in terms of recovery from drought stress (Rohila et al. 2002). Rice transgenic lines overexpressing HVA1 LEA3 gene tested in very large pots (Babu et al. 2004) maintained higher leaf RWC and showed lesser reduction in plant growth under drought stress as compared to the wild type (cv. Nipponbare) (Fig. 5.3). Transgenic lines had relatively better cell membrane stability than the wild type. Transgenic rice lines overexpressing LEA protein gene OsLEA3-1 were tested in the field and in upright PVC pipes used for root studies. Most T2 and T3 families had better relative yield (yield ratio of stress/non-stress) under stress as compared with that of the wild type drought susceptible rice cultivar (Xiao et al. 2007). However, various levels of yield penalty were found in the transgenic families when grown under non-stress conditions.

Transgenic HVA1 mulberry (*Morus indica*) had enhanced drought resistance in terms of cellular membrane stability (CMS), photosynthetic yield, less photo-oxidative damage and better water use efficiency (Lal et al. 2008).

Taking all these studies together it can be concluded that LEA protein enhancing genes have a high potential for developing drought resistant crop species.

Protein farnesyltransferase (ERA1) is involved in the regulation of *ABA* sensing and drought response. Transgenic Canola carrying an ERA1 antisense construct

Fig. 5.3 Two transgenic HVA1 rice as compared with the wild type (cv. Nipponbare, center pot) under drought stress. Photograph was taken in Dr. HT Nguyen laboratory at Texas Tech University, Lubbock Texas when these materials were under a preliminary test. The transgenic HVA1 rice was produced by Dr. R Wu, Cornell University, Ithaca, NY. The complete study is reported by Babu et al. (2004)

driven by a drought-inducible rd29A promoter was examined with the expectation that their increased ABA sensitivity will enhance drought resistance. Transgenic canola showed enhanced ABA sensitivity, as well as significant reduction in stomatal conductance and transpiration under drought stress conditions. The antisense down-regulation of canola farnesyltransferase was facultative, depending on soil moisture deficit. Transgenic plants were more resistant to water deficit-induced seed abortion during flowering. Results from three consecutive years of field trials suggested that with adequate water, transgenic canola plants produced the same amount yield as the control. However, under moderate drought stress conditions at flowering, the seed yield of transgenic canola was significantly higher than the control. Further testing of the most promising transgenic line in diverse field environments in the Canadian Prairie proved its relative advantage in yield under water limited conditions as compared with control cultivars (Wang et al. 2009; Wan et al. 2009). It is not clear how enhanced ABA sensitivity reduced seed abortion.

High concentration of leaf *cytokinin* delays senescence. Leaf senescence is delayed in transgenic plants expressing *isopentenyltransferase* (IPT), an enzyme that catalyzes the rate-limiting step in cytokinin synthesis. Transgenic tobacco plants expressing an isopentenyltransferase gene driven by a stress- and maturation-induced promoter were generated and tested in large pots in the greenhouse (Rivero et al. 2007). Suppression of drought-induced leaf senescence resulted in drought resistance as shown by, among other responses, growth recovery after a long drought period that killed the wild type plants. The transgenic plants maintained better water status and retained photosynthetic activity under drought stress. The transgenic plants displayed minimal seed yield loss per plant when watered with only 30% of the amount of water used under control conditions. It was subsequently found (Rivero et al. 2009) that cytokinin induced enhanced photorespiration and that this had an important role in the maintenance of photosynthetic processes under drought stress. In some respects this gene expresses a strong stay-green phenotype.

Bacterial cold shock protein chaperones (CSP) were extensively tested in transgenic maize at Monsanto Inc. Over-expressing transgenic maize in the form of hybrids were tested for CspA from *Escherichia coli* and CspB from *Bacillus subtilis* (Castiglioni et al. 2008). Twenty-two separate CspB events were evaluated in water-limited field trials using commercial grade hybrid maize under drought stress. The water-deficit treatment resulted in an average reduction in growth rates to 50% of the well-watered control. The best performing events demonstrated growth rate increases of 12 and 24% over the wild type. The CspB-expressing plants also demonstrated significant improvements in chlorophyll content and photosynthetic rates. Several years of field trials have been conducted with a single CspB- event in three different hybrid backgrounds and evaluated under various water regimes at five replicated locations. Mean yield reduction under stress was 50% as compared with irrigated control. Yield advantage of CspB- in hybrids in these experiments ranged from 11 to 21%. Data indicated that CspB- might also have a positive effect on yield under non-stress conditions.

The *WXP1* gene enhances leaf *epicuticular wax* deposition. Enhanced epicuticular wax load is considered a priori to improve dehydration avoidance in crops with naturally limited wax deposition. Overexpression of WXP1 under the control of the CaMV35S promoter (Zhang et al. 2007) led to a significant increase (up to 38%) of epicuticular wax load on leaves of transgenic alfalfa. Transgenic leaves showed reduced water loss and less chlorophyll leaching. Transgenic alfalfa plants with increased epicuticular wax showed enhanced drought resistance demonstrated by delayed wilting under drought stress and better recovery upon rewatering. Expression of WXP1 in white clover under control of the Arabidopsis CER6 promoter (Jiang et al. 2010) resulted in higher leaf relative water content and leaf water potential which was reflected in higher net photosynthetic rates and higher efficiency of PSII under drought stress, as compared with the wild type.

Glycinebetaine is an osmoticum found in many organisms, including bacteria and certain higher plants such as spinach and barley. The bacterium *Escherichia coli* produces glycinebetaine by a two-step pathway where choline dehydrogenase (CDH), encoded by betA, oxidizes choline to betaine aldehyde which is further oxidized to glycinebetaine by the same enzyme, The second step, conversion of betaine aldehyde into glycinebetaine, can also be performed by the second enzyme in the pathway, betaine aldehyde dehydrogenase (BADH), encoded by betB. Various transgenic plant species engineered for enhanced glycinebetaine accumulation were tested for drought resistance with variable results. Results with high glycinebetaine transgenic plant did not always produce consistent results (e.g., Huang et al. 2000). On the other hand a study of four cotton lines expressing high glycinebetaine accumulation indicated an advantage in seed-cotton yield after drought stress in the field accompanied by better plant water status, higher osmotic adjustment and improved cell membrane stability under stress (Lv et al. 2007). Similarly, a glycinebetaine accumulating transgenic line of maize yielded better under drought stress in the field as compared with the wild type. The effect was possibly due to enhanced cell membrane stability (Quan et al. 2004).

It has often been argued that regulatory genes such as *DREB1* are more important than "structural genes" (such as HVA1 for example) because the former switch a cascade of stress responsive genes. However, again, it should be remembered that not all stress responsive genes which might be up-regulated have a role in drought resistance. Some can be involved in programmed cell death, for example. However, the *DREB transcription factor* has been repeatedly shown that when it is expressed under a drought induced promoter in transgenic plants some form of drought resistance was in evidence. A large part of this evidence was reported for transgenic *Arabidopsis* or tobacco model plants but also for rice and maize. Much of the work was done at the RIKEN Plant Science Center, Japan (e.g., Ito et al. 2006; Qin et al. 2007).

In trying to apply some of the expected benefit, work at CIMMYT tested DREB1A-transformed wheat plants which were found to express delayed wilting under drought stress in a greenhouse trial (Pellegrineschi et al. 2004). However, no effect was seen in biomass and yield under drought stress (Ortiz et al. 2007).

On the other hand, a DREB1A transcription factor ortholog was isolated from a xeric, wild barley (*Hordeum spontaneum* L.) accession originating from the Negev desert (James et al. 2008). It was over-expressed in bahiagrass (*Paspalum notatum* Flugge). The transgenic bahiagrass plants survived severe salt stress and repeated cycles of severe dehydration stress under controlled environment conditions, in contrast to the wild type.

NF-Y is a conserved heterotrimeric complex consisting of NF-YA (HAP2), NF-YB (HAP3), and NF-YC (HAP5) subunits (Nelson et al. 2007). Transgenic maize with promoted ZmNFYB2 were tested at Monsanto Inc and found to express drought resistance in terms of chlorophyll content, high stomatal conductance, lower leaf temperature, reduced wilting, and maintenance of photosynthesis under stress. These stress responses were associated with significant grain yield advantage in several field tests.

Lastly, in a very extensive project Xiao et al. (2009) produced different rice transgenic plants involving over-expression of CBF3, SOS2, NCED2, NPK1, LOS5, ZAT10, and NHX1. Thirty T1 families with expression of each transgene were tested for drought resistance at the reproductive stage in field, and ten of them were tested in PVC pipes with a defined stress protocol at the same stage. Detailed information was provided for each transgene on plant fertility and productivity. In general, LOS5 and ZAT10 showed relatively better effect than the other five genes in improving drought resistance of transgenic rice under field conditions.

All of the above are just chosen examples which appear potentially useful at this time. Undoubtedly in due course further progress will be made in this very active discipline and pending regulation certain results will be delivered to the farmer.

5.5 Resurrection Plants

These plants seem to constitute the most extreme native and exotic genetic resource for dehydration/desiccation tolerance. Resurrection plants are remarkable in that they can tolerate almost complete desiccation in their vegetative tissues. The desiccated plant can remain alive in the dried state for one or more years. Upon watering the plants rehydrate and are fully functional within 48 h or less. These plants therefore attracted great attention for understanding desiccation tolerance with some reference to potential use in crop plant breeding. Researched species in this context are mainly *Craterostigma plantagineum*, *Craterostigma wilmsii*, *Myrothamnus flabellifolius*, *Ramonda serbica*, *Reaumuria soongorica*, *Sporobolus stapfianus* and *Xerophyta viscose*. The mechanism of desiccation tolerance of these plants as much as it is understood has been reviewed by Moore et al. (2009) and Toldi et al. (2009). It is most interesting that in many respects there is a high level of metabolic homology for desiccation tolerance between resurrection plants and seed embryo.

In several resurrection plants ABA was found to be involved in desiccation tolerance by inducing the expression of many dehydration regulated genes. ABA is also involved in regulating functional stress proteins such is LEA. Extensive production of antioxidants is considered important in cellular protection under dehydration and rehydration. The accumulation of various sugars (mainly sucrose and trehalose) found in several species is considered important for osmotic adjustment and the protection of membranes and cellular integrity. Sugars may accumulate constitutively or in response to dehydration. In certain species sugar was transported from roots to leaves under dehydration stress. Conservation of the dehydrated cytoplasm in a "glassy matrix" is ascribed to high concentration of dissolved sugar in cells.

Dehydration-induced accumulation of structure-stabilizing proteins, typically LEA proteins, is often found in resurrection plants subjected to drying. Several protective functions were ascribed to LEA proteins, including roles in protecting DNA, stabilizing cytoskeletal filaments, acting as molecular chaperones and acting synergistically with sugars to prevent protein aggregation during desiccation. The conservation of cell wall flexibility when cells shrink under desiccation is extremely important for successful cellular recovery upon rehydration. Pronounced cell wall folding and unfolding capacity is typical of these plants. Cell wall flexibility is facilitated by cell wall xyloglucan modification, cell wall expansins and constitutively formed pectin-associated arabinans and/or arabinogalactan proteins.

In conclusion the capacity for extreme desiccation tolerance in most resurrection plants is typically associated with constitutive and/or adaptive and stress responsive metabolic traits. Most but not all stress adaptive responses can be ascribed to ABA accumulation at the onset of dehydration. In most cases desiccation tolerance is achieved when dehydration is slow. Cases for high desiccation tolerance under rapid dehydration are very few and might be perhaps associated with constitutive plant traits.

Regarding plant breeding, it should be reminded that these plant species are incompatible in terms of sexual transmission of genes to our crop plants, even by the most sophisticated methods. These plants are therefore candidates for transgenic methods. At this time there is no record of successful utilization of genes from resurrection plants in breeding crop plants. Certain genes known to affect desiccation tolerance in resurrection plants, such as those regulating LEA protein accumulation, are evaluated in transgenic plants for function and are close to becoming serious candidates for breeding.

References

Baalbaki R, Hajj-Hassan N, Zurayk R (2006) *Aegilops* species from semiarid areas of Lebanon: variation in quantitative attributes under water stress. Crop Sci 46:799–806

Babu CR, Zhang J, Blum A et al (2004) HVA1, a LEA gene from barley confers dehydration tolerance in transgenic rice (*Oryza sativa* L) via cell membrane protection. Plant Sci 166:855–862

Bahieldin A, Mahfouz HT, Eissa HF et al (2005) Field evaluation of transgenic wheat plants stably expressing the HVA1 gene for drought tolerance. Physiol Plant 123:421–427

Bhattarai T, Fettig S (2005) Isolation and characterization of a dehydrin gene from *Cicer pinnatifidum*, a drought-resistant wild relative of chickpea. Physiol Plant 123:452–458

Blum A, Sullivan CY (1986) The comparative drought resistance of landraces of sorghum and millet from dry and humid regions. Ann Bot 57:835–846

Blum A, Gozlan G, Mayer J (1981) The manifestation of dehydration avoidance in wheat breeding germplasm. Crop Sci 21:495–499

Blum A, Ebercon A, Sinmena B et al (1983) Drought resistance reactions of wild emmer (*T. dicoccoides*) and wild emmer x wheat derivatives. In: Proceedings of the 6th international wheat genetics symposium, Kyoto, pp 433–438

Blum A, Golan G, Mayer J et al (1989) The drought response of landraces of wheat from the Northern Negev desert in Israel. Euphytica 43:87–96

Blum A, Munns R, Passioura JB et al (1996) Genetically engineered plants resistant to soil drying and salt stress: how to interpret osmotic relations? Plant Physiol 110:1051

Cairns JE, Botwright Acun TL, Simborio FA et al (2009) Identification of deletion mutants with improved performance under water-limited environments in rice (*Oryza sativa* L). Field Crops Res 114:159–168

Carver BF, Nevo E (1990) Genetic diversity of photosynthetic characters in native populations of *Triticum-dicoccoides*. Photosynth Res 25:119–128

Castiglioni P, Warner D, Bensen RJ et al (2008) Bacterial RNA chaperones confer abiotic stress tolerance in plants and improved grain yield in maize under water-limited conditions. Plant Physiol 147:446–455

Cattivelli L, Rizza F, Badeck FW et al (2008) Drought tolerance improvement in crop plants: an integrated view from breeding to genomics. Field Crops Res 105:1–14

Ceccarelli S, Grando S (1991) Environment of selection and type of germplasm in barley breeding for low-yielding conditions. Euphytica 57:207–219

Ceccarelli S, Grando S, Impiglia A (1998) Choice of selection strategy in breeding barley for stress environments. Euphytica 10:307–318

Dingkuhn M, Jones MP, Johnson DE et al (1998) Growth and yield potential of *Oryza sativa* and *O. glaberrima* upland rice cultivars and their interspecific progenies. Field Crops Res 57:57–69

Gallardo M, Jackson LE, Thompson RB (1996) Shoot and root physiological responses to localized zones of soil moisture in cultivated and wild lettuce (*Lactuca* spp). Plant Cell Environ 9:1169–1178

Gororo NN, Eagles HA, Eastwood RF et al (2002) Use of *Triticum tauschii* to improve yield of wheat in low-yielding environments. Euphytica 123:241–254

Hamdi A, Erskine W (1996) Reaction of wild species of the genus *lens* to drought. Euphytica 91:173–179

Huang J, Hirji R, Adam L et al (2000) Genetic engineering of glycinebetaine production toward enhancing stress tolerance in plants: metabolic limitations. Plant Physiol 122:747–756

Humphreys J, Harper JA, Armstead IP et al (2005) Introgression-mapping of genes for drought resistance transferred from Festuca arundinacea var glaucescens into Lolium multiflorum. Theor Appl Genet 110:579–587

Ito Y, Katsura K, Maruyama K et al (2006) Functional analysis of rice DREB1/CBF-type transcription factors involved in cold-responsive gene expression in transgenic rice. Plant Cell Physiol 47:141–153

James VA, Neibaur I, Altpeter F (2008) Stress inducible expression of the DREB1A transcription factor from xeric, *Hordeum spontaneum* L in turf and forage grass (*Paspalum notatum* Flugge) enhances abiotic stress tolerance. Transgenic Res 17:93–104

Jenks MA, Hasegawa PM, Mohan Jain S (eds) (2007) Advances in molecular breeding towards drought and salt tolerant crops. Springer, Dordrecht

Jiang Q, Zhang J-Y, Guo X et al (2010) Improvement of drought tolerance in white clover (*Trifolium repens*) by transgenic expression of a transcription factor gene WXP1. Funct Plant Biol 37:157–165

Johnson WC, Jackson LE, Ochoa O et al (2000) Lettuce, a shallow-rooted crop, and *Lactuca serriola*, its wild progenitor, differ at QTL determining root architecture and deep soil water exploitation. Theor Appl Genet 101:1066–1073

Lafitte HR, Li ZK, Vijayakumar CHM et al (2006) Improvement of rice drought tolerance through backcross breeding: evaluation of donors and selection in drought nurseries. Field Crops Res 96:77–86

Lakew B, Semeane Y, Alemayehu F et al (1997) Exploiting the diversity of barley landraces in Ethiopia. Genet Resour Crop Evol 44:109–116

Lal S, Gulyani V, Khurana P (2008) Overexpression of HVA1 gene from barley generates tolerance to salinity and water stress in transgenic mulberry (*Morus indica*). Trans Res 17:651–663

Liu L, Lafitte R, Guan D (2004) Wild Oryza species as potential sources of drought-adaptive traits. Euphytica 138:149–161

Lv S, Yang A, Zhang K et al (2007) Increase of glycinebetaine synthesis improves drought tolerance in cotton. Mol Breed 20:233–248

Marshall AH, Rascle C, Abberton MT et al (2001) Introgression as a route to improved drought tolerance in white clover (*Trifolium repens* L). J Agron Crop Sci 187:11–18

Moore JP, Le NT, Brandt WF et al (2009) Towards a systems-based understanding of plant desiccation tolerance. Trends Plant Sci 14:110–117

Nakashima K, Ito Y, Yamaguchi-Shinozaki K (2009) Transcriptional regulatory networks in response to abiotic stresses in *Arabidopsis* and grasses. Plant Physiol 149:88–95

Nelson DE, Repetti PP, Adams TR et al (2007) Plant nuclear factor Y (NF-Y) B subunits confer drought tolerance and lead to improved corn yields on water-limited acres. Proc Nat Acad Sci U S A 104:16450–16455

Nevo E, Chen G (2010) Drought and salt tolerances in wild relatives for wheat and barley improvement. Plant Cell Environ 32:670–685

Nevo E, Gorham J, Beiles A (1992) Variation for Na-22 uptake in wild emmer wheat, *Triticum dicoccoides* in Israel – salt tolerance resources for wheat improvement. J Exp Bot 43:511–518

Ortiz R, Iwanaga M, Reynolds MP et al (2007) Overview on crop genetic engineering for drought-prone environments. J SAT Agr Res 40:1–30

Park B-J, Liu Z, Kanno A et al (2005) Genetic improvement of Chinese cabbage for salt and drought tolerance by constitutive expression of a *B. napus* LEA gene. Plant Sci 169:553–558

Peleg Z, Fahima T, Abbo S et al (2005) Genetic diversity for drought resistance in wild emmer wheat and its ecogeographical associations. Plant Cell Environ 28:176–191

Pellegrineschi A, Reynolds M, Pacheco M et al (2004) Stress-induced expression in wheat of the *Arabidopsis thaliana* DREB1A gene delays water stress symptoms under greenhouse conditions. Genome 47:493–500

Qin F, Kakimoto M, Sakuma Y et al (2007) Regulation and functional analysis of ZmDREB2A in response to drought and heat stresses in *Zea mays* L. Plant J 50:54–69

Quan R, Shang M, Zhang H et al (2004) Engineering of enhanced glycine betaine synthesis improves drought tolerance in maize. Plant Biotechnol J 2:477–486

Reyes JL, Rodrigo M-J, Colmenero-Flores JM et al (2005) Hydrophilins from distant organisms can protect enzymatic activities from water limitation effects in vitro. Plant Cell Environ 28:709–718

Reynolds M, Dreccer F, Trethowan R (2007) Drought-adaptive traits derived from wheat wild relatives and landraces. J Exp Bot 58:177–186

Ribaut JM, Hoisington DA, Deutsch JA et al (1996) Identification of quantitative trait loci under drought conditions in tropical maize I. Flowering parameters and the anthesis-silking interval. Theor Appl Genet 92:905–914

Rivero RM, Kojima M, Gepstein A et al (2007) Delayed leaf senescence induces extreme drought tolerance in a flowering plant. Proc Nat Acad Sci U S A 104:19631–19636

Rivero RM, Shulaev V, Blumwald E (2009) Cytokinin-dependent photorespiration and the protection of photosynthesis during water deficit. Plant Physiol 150:1530–1540

Rohila JS, Rajinder K, Wu JR (2002) Genetic improvement of basmati rice for salt and drought tolerance by regulated expression of a barley Hva1 cDNA. Plant Sci 163:525–532

Rosenow DT, Dahlberg JA (2000) Collection, conversion and utilisation of sorghum. In: Smith CW, Frederiksen RA (eds) Sorghum, origin, history, technology and production. Wiley, New York

Sivamani E, Bahieldin A, Wraith JM et al (2000) Improved biomass productivity and water use efficiency under water deficit conditions in transgenic wheat constitutively expressing the barley HVA1 gene. Plant Sci 155:1–9

Suprunova T, Krugman T, Fahima T et al (2004) Differential expression of dehydrin genes in wild barley, *Hordeum spontaneum*, associated with resistance to water deficit. Plant Cell Environ 27:1297–1308

Suprunova T, Krugman T, Distelfeld A et al (2007) Identification of a novel gene (Hsdr4) involved in water-stress tolerance in wild barley. Plant Mol Biol 64:17–34

Sutka J, Farshadfar E, Koszegi B et al (1995) Drought tolerance of disomic chromosome additions of *Agropyron elongatum* to *Triticum aestivum*. Cereal Res Commun 23:351–357

Toldi O, Tuba Z, Scott P (2009) Vegetative desiccation tolerance: is it a goldmine for bioengineering crops? Plant Sci 176:187–199

Trethowan RM, Mujeeb-Kazib A (2008) Novel germplasm resources for improving environmental stress tolerance of hexaploid wheat. Crop Sci 48:1255–1265

van de Wouw M, van Hintum T, Kik C et al (2010) Genetic diversity trends in twentieth century crop cultivars: a meta analysis. Theor Appl Genet 120:1241–1252

Wan J, Griffiths R, Yin J et al (2009) Development of drought-tolerant canola (*Brassica napus* L.) through genetic modulation of ABA-mediated stomatal responses. Crop Sci 49:1539–1554

Wang Y, Beaith M, Chalifoux M et al (2009) Shoot-specific down-regulation of protein farnesyl-transferase (-subunit) for yield protection against drought in canola. Mol Plant 2:191–200

Warburton ML, Crossa J, Franco J et al (2006) Bringing wild relatives back into the family: recovering genetic diversity in CIMMYT improved wheat germplasm. Euphytica 149:289–301

Xiao B, Huang Y, Tang N et al (2007) Over-expression of a LEA gene in rice improves drought resistance under the field conditions. Theor Appl Genet 115:35–46

Xiao B, Chen X, Xiang C-B et al (2009) Evaluation of seven function-known candidate genes for their effects on improving drought resistance of transgenic rice under field conditions. Mol Plant 2:73–83

Xu DP, Duan XL, Wang BY et al (1996) Expression of a late embryogenesis abundant protein gene, HVA1, from barley confers tolerance to water deficit and salt stress in transgenic rice. Plant Physiol 110:249–257

Yadav OP (2008) Performance of landraces, exotic elite populations and their crosses in pearl millet (*Pennisetum glaucum*) in drought and non-drought conditions. Plant Breed 127:208–210

Yadav OP, Weltzien E (2000) Differential response of landrace-based populations and high yielding varieties of pearl millet in contrasting environments. Ann Arid Zone 39:39–45

Zare AG, Humphreys MW, Rogers JW et al (2002) Androgenesis in a *Lolium multiflorum* × *Festuca arundinacea* hybrid to generate genotypic variation for drought resistance. Euphytica 125:1–11

Zhang J-Y, Broeckling CD, Sumner LW et al (2007) Heterologous expression of two Medicago truncatula putative ERF transcription factor genes, WXP1 and WXP2, in Arabidopsis led to increased leaf wax accumulation and improved drought tolerance, but differential response in freezing tolerance. Plant Mol Biol 64:265–278

Chapter 6
Breeding Considerations and Strategies

Summary Plant breeding for water limited environments is rarely performed as a separate program. In most cases it is part of a mainstream breeding program which is designed to release a cultivar appropriate to a designated ideotype and customer demand. The ideotype includes various specifications regarding morphological, developmental, phenotypic, biotic and marketing requirements. Drought resistance and abiotic stress resistance is part of the ideotype.

There is no consensus approach and design to the inclusion of drought resistance in a breeding program and the procedure is very specific to each case. The content of this book is designed to help the breeder in developing the specific procedure fitting his requirements. It is however emphasized that any novel procedure, test, or selection method should be pretested and critically evaluated by the breeder in his own working environment *before* its adoption as part of the program.

Regarding the flow of the program there might be two main approaches, namely phenotyping and selection for drought resistance during early generations or during late generations. The two options are discussed, favoring in general the case for late generation selection for drought resistance, with exceptions. A schematic example is presented for a simple pedigree breeding program integrating some components of selection for drought resistance.

Finally, the utilization of heterosis and hybrids towards water-limited environments is briefly discussed.

This chapter is not intended for repeating known methods and strategies used in general breeding work. It is assumed that the reader is versed in this respect. Here reference is made in the narrow sense to more specific issues related to breeding for water-limited environments. Only phenotypic selection is treated here. Readers interested in deploying molecular methods are referred to the technical literature on the subject as well as to information presented in previous chapters. As discussed above, MAS can be considered for certain hard-to-phenotype traits such as roots, pending the appropriate pre-requisites. There is always the consideration of the utility, speed, cost and reliability when phenotypic selection is compared with MAS and the solution lied with the specific trait and the breeder's working environment. Successful case studies to be followed in this respect are rare.

A. Blum, *Plant Breeding for Water-Limited Environments*,
DOI 10.1007/978-1-4419-7491-4_6,

Breeders are constantly reminded that various simulation and decision support systems are being developed in order to reduce the reliance on documented knowledge, experience and even intuition in planning and executing their breeding program. Certainly such tools would constitute no less than a revolution in breeding in general and even more so in breeding for water-limited environments.

Development of *crop simulation models* and decision support systems for use in breeding work have long been attempted (e.g., Spitters and Schapendonk 1990). Undoubtedly certain crop growth simulation models can predict yield more or less accurately with the appropriate environmental inputs. The literature is rich with reports of successful simulations. However, if crop simulation models are to service plant breeding they should incorporate the genetic component and its interaction with the environment. This is a difficult requirement. Without these components it would be impossible to predict genotypic response to drought stress or traits which might be crucial in certain stress scenarios. One example is the attempted prediction of soybean traits important for drought resistance in the US as based on a soybean crop growth simulation and by using historical and GIS data (Sinclair et al. 2010). This simulation concluded that "in over 0.7 of the years the rapid rate of root extension had no or a negative influence on yield." The reader therefore learns that extensive root growth is not an important component of drought resistance in US soybean. This is in contrast to present consensus that extensive root growth leading to deeper soil moisture extraction is an important trait of dehydration avoidance (Chap. 3, section "Roots and Dehydration Avoidance"). However, this simulation a priori limited root depth of soybean to 1 m. Whether roots in the simulation grew fast or slow, they were not allowed to grow deeper than 1 m. This would certainly reduce simulated yield in more than few water-limited environments. In their discussion the authors concede that "possibility remains that if the rapid root extension allows penetration to a maximum depth where there is stored soil water, this might prove to be a more desirable trait." If the authors were aware of this fact why root depth was set to 1 m in the simulation? The simulation also indicated that sensitive stomatal closure at relatively high leaf water status was advantageous to yield. Again, this is in contrast to a volume of information indicating that sustained stomatal conductance is conducive to higher yield also under drought stress (Sect. 3.5.1). If the model would have allowed roots to grow and extract soil moisture deeper than 1 m most likely early sensitive stomatal closure would not have emerged in the simulation as advantageous.

Dealing with genotype by environment interaction (G × E) is a major and a very serious component of breeding for drought resistance. Statistical tools to process multi-test yield data for G × E are available (e.g., Annicchiarico 2002). However, predicting G × E for yield by using genomic information is a vision. The problem is not only with the algorithm but with the insufficient biological understanding of the interaction between genes and the environment. This has often been described as the "genotype-phenotype gap."

A "stepping-stone" symposium on "Crop Modeling and Genomics" was held in the USA in 2000. This symposium set the ground for pressing forward towards applying crop simulation models to plant breeding. In his opening introduction

Weiss (2000) noted that what was limiting progress was not the ample availability of data on climate, soil or crop management or the capacity to maneuver the model algorithm. The obstacle to application of crop models to plant breeding was what crop modeling experts defined as the missing "genetic coefficients." In that symposium Chapman et al. (2003) reported a groundbreaking attempt to simulate selection for 15 genes controlling four adaptive traits which could affect sorghum yield under drought stress. This was done by combining a sorghum cropping system model (APSIM-Sorg) with a breeding program simulation model (QU-GENE). Results of these simulation exercises are most interesting and some are quite agreeable with what physiology would predict. However, the high level of complexity and gene interaction led to conclude that "given limitations in our current understanding of trait interaction and genetic control, the results are not conclusive. They demonstrate how the per se complexity of gene × gene × environment interactions will challenge the application of genomics and marker-assisted selection in crop improvement for dryland adaptation." Hammer et al. (2006) who basically constituted the same research group concluded later that "Challenges remain in finding improved means to connect model coefficients to tangible genomic regions (or genes), as organized on genetic and physical sequence maps." In plain words: we are not there yet.

Since that 2000 symposium great efforts were made to integrate molecular biology and crop simulation models into a useful tool that breeders can use in planning a breeding program towards water limited environment.

A notable approach is still work in progress at this time. With this approach simulation of leaf growth response to water deficit and temperature in maize has been scaled up to a simulation of silk growth, plant fertility and grain yield using growth controlling QTLs (e.g., Chenu et al. 2009). This and other attempts to merge QTL data with physiological or crop growth simulation models are summarized in Tardieu and Tuberosa (2010). At the same time we still recall the constant plight of molecular biologists and breeders (Sect. 3.2.2) that we require better understanding of the dependency of QTLs on genetic background or their sensitivity to the environment, coupled with a general lack of understanding of the biophysical bases of these context dependencies (Campos et al. 2004). In this respect perhaps the work by Tardieu and associates at Montpellier constitute sticking a foot in the doorway.

Until that time when these challenges will be met, breeders remain with the tools at hand. These tools constitute the combined knowledge and science developed by breeding experience and the lessons learned from the success and failure. The value of experience in plant breeding is so great that experience-based intuition can often be a key to success in selection work. Not to mention luck. When this profession involves experience, intuition and luck it is no surprise that some define plant breeding also as an art. The corollary is perhaps that a successful breeder is no less a farmer than a scientist. The breeding knowledge base is partly documented and partly remains with each generation of breeders and educators. There is a constant fear of losing this knowledge especially when "conventional breeding" is subjected to a declining public support and a receding education.

I must repeat here the wise recommendation by Reitz (1974) who warned against following a mirage in a selection program. New ideas and techniques that a breeder might consider using must not be immediately applied to the main stream of the program. These should be first tested prudently and correctly in an ancillary fashion for more than one season before being adopted as a main stream protocol. Errors made in selection work are usually revealed only after several years and the cost of such mistakes can be high. Breeders are generally conservative in their practice, and they are correct in that approach.

The first question to be asked by breeders when confronted with the issue of breeding for drought resistance is whether it should be undertaken to begin with. A standard crop breeding program is already heavily loaded with various responsibilities, ranging from various field observations, diseases and pest resistance or screening for product quality. The decision whether to include selection for drought resistance traits in the breeding program and the additional work load and learning that is requires depends on the available resources, the expected economic benefit of such involvement, pressure from administration, the professional standing of the breeder, and the availability of advice and consultation as the case might be. However, before anything else is considered the breeder must carefully review the discussion of Sect. 3.3.1. For grain crops there is sufficient evidence showing that if drought stress in the target environment does not normally reduce yield to below about 50% of the potential, breeding for high yield potential and general adaptation (optimized phenology, thermal adaptation, etc.) may suffice to assure progress in yield. This yield reduction benchmark is not available for all crops, such as vegetables, fruit, oil crops and most pulses. Information is lacking where biomass production constitutes yield. In certain crops where product quality is a prime consideration a slight stress during product development can still be detrimental. In other cases, slight stress can improve product quality by enhancing sugar accumulation and various components of taste, aroma or color. Therefore, the nature of the target environment and the expected effect of drought on total *crop value* is a prime question before deciding to embark on breeding for that environment. Considerations towards subsistence agriculture can again be totally different. Still with our present understanding of the genetics and physiology of yield variation in a changing environment it can be concluded that "mild" reductions in yield in a given target environment due to occasional drought stress may not justify the incorporation of drought resistance as a component in the breeding program, pending abovementioned considerations.

The second issue to consider is a possible *yield tradeoff* (crossover interaction) in a cultivar developed for water-limited conditions. It has already been discussed above that probability is low for improving yield simultaneously in both the potential and the severe stress environments, where yield level is already high. Previous chapters deal with this issue. It should be added here that the approach to yield tradeoff and the crossover interaction in a released cultivar depends also on social and economic issues such as wealth, credit, land ownership, crop insurance, etc. In most cases of developing countries and farmers of poor resources, a potentially low yielding cultivar of sustained stable performance over years with a high product value and acceptance is the generally preferred ideotype. In modern farming which is supported by cheap credit or by crop insurance farmers may opt for maximized yields in good years at the expense of a large yield loss in dry years. There is a range of physical or economical conditions between the two cases which should be considered by the breeder.

"*Pre-breeding*" is often defined as preparatory work which might offer the breeder superior germplasm or even selection techniques. This activity is essential before actual breeding for drought resistance is planned for the first time. Pre-breeding should involve a breeder and a stress physiologist who together should be able to identify the ideotype for the target environment, the germplasm to use for that ideotype and the phenotyping methods to adopt in the program. Pre-breeding is often taken as a necessary step for assembling appropriate germplasm before embarking on a crossing program. In the case of drought resistance it also involves the testing and preparation of the appropriate drought phenotyping work, beginning with the design of the managed stress environment down to the details of the measurement protocols. It also promotes learning. Unless these are already at hand, pre-breeding for drought resistance should not be compromised.

A major question often raised is at which point the selection for drought resistance is integrated within the general flow of the program. Two main schemes of approach are possible in principle:

(a) Early generation selection
(b) Advanced generation selection

A backcross program for drought resistance is not given special consideration here since the principles of its execution for drought resistance can also be derived from the following discussion.

Early generation selection is performed in highly segregating populations typically the F_2 where the desirable genotype is expected to be captured by phenotyping single plants. Sometimes such a genotype might also be phenotyped by the performance of its F_3 progeny. The general advantage with this strategy is well known. The success of this approach towards drought resistance depends on several conditions.

Firstly, it depends on the specific targeted drought resistance trait and whether such a trait can be reliably phenotyped in a segregating population of single plants. Selection for low canopy temperature, for example, would be very difficult technically, if not impossible. Furthermore, the genetic control of canopy temperature, at least in wheat, requires that phenotyping will be performed in the more homozygous advanced generations (Saint Pierre et al. 2010). On the other hand visual estimate of leaf wilting in a single plant is possible. Selection for root depth in tubes is possible. There were cases where segregating populations were subjected to a "killing stress" and only the surviving plants were selected. Survival of individual plants and its verification in the next generation can be an effective approach wherever such survival is relevant. Mass selection for grain filling under chemical desiccation to improve stem reserve support for grain filling was successful when performed in F_2-F_3 generations (Sect. 4.2.3.4).

Secondly, plant water deficit of a single plant depends also on competition with its close neighbors in the field. Hence, visual or instrumental estimates of plant water status of single F_2 plants in a highly segregating population can be biased. This problem is especially serious when the population segregates also for plant height, flowering date and leaf area, traits which enhance competition and also inherently involve plant water status. For example, a "killing stress" might favor the survival of early flowering or very small plants, which may or may not be the desirable outcome for the planned ideotype. Again, the option for verifying a single plant response by re-evaluating its progeny is an import support for an early selection program.

Thirdly, results by this method also depend on the genetic makeup of the population. In many cases drought resistance is introduced into the population from an exotic donor that lacks in the desirable agronomic traits. The F_2 generation will typically constitute of many undesirable single plants which might express poor agronomic phenotype together with drought resistance. Furthermore, important desirable agronomic traits may not be recognized in a segregating population of single plants. Selection for drought resistance in this population may result in improved drought resistance on a background of a useless advanced line of poor agronomic or marketing qualities. For this reason, early generation selection should be considered only if the F_2 population as a whole is expected to be highly agronomical.

Advanced generation selection (Fig. 6.1) begins by selecting for the agronomic ideotype in the early generations (F_2-F_3). In subsequent generations (beginning with F_4 for example) the routine selection program continues for yield and other desirable agronomic traits under optimal conditions. However selection for yield retains also a proportion of medium yielding lines, expecting a possible crossover interaction for yield in this population. The selected F_4 agronomic lines based on reasonable yield performance under optimal conditions will be continuously tested for yield and other agronomic traits. At the same time, *duplicate samples* of F_4 and F_5 lines will be phenotyped for drought resistance in the managed stress environment which can constitute of field, greenhouse, or any other appropriate facility or test protocol. Data obtained from drought phenotyping will be merged with all other data for the given materials (e.g., yield under optimal conditions, yield under stress conditions, disease reactions, quality data etc.) and then used to select the appropriate lines from the main nursery grown under optimal conditions (Fig. 6.1).

This procedure has one main known disadvantage and that is that desirable drought resistance genes might be lost by the time drought resistance phenotyping is performed at later generations. The probability for such loss can be reduced by growing a large F_2 population at the onset of the program.

This route is preferred if the frequency of the drought resistance genes in the population and their heritability is high. Work at advanced generation also allows to phenotype for traits that are best expressed in a homogenous and more homozygous progeny. Since in the earlier generations the agronomic type has already been optimized any recovered drought resistant line at later generations can be a potential release. In case of failing to recover drought resistant lines by this program, it allows at least to obtain novel high yielding materials under optimal conditions.

This approach is more common in advanced breeding program for water-limited environments. The collaborative network of the Indian rice breeding programs, in partnership with International Rice Research Institute (IRRI) used advanced generation

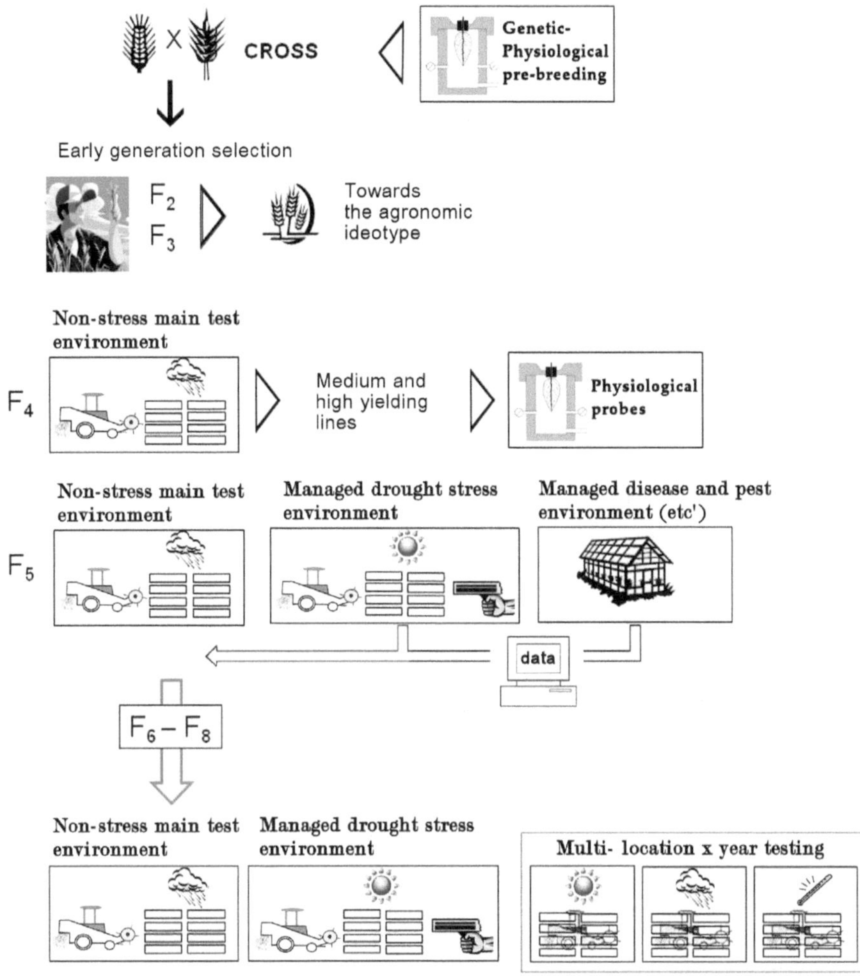

Fig. 6.1 A schematic example for a simple pedigree breeding program for yield and drought resistance in a typical cereal crop where phenotyping for drought resistance is performed in advanced generations (see text)

screening in multiple managed drought environments with excellent progress (Verulkar et al. 2010). Lines out of this program had good drought resistance without yield penalty under irrigated conditions. Another rice breeding program in China (He et al. 2010) used backcross breeding with several donors of drought resistance. A two-round-selection for yield under severe drought together with a two-round-selection for yield under irrigation resulted in 113 BC_2F_8 introgression lines which included superior yielding lines under both irrigated and drought stress conditions.

Breeding drought resistant *hybrids* is a constant and important issue on which active debate has been conducted for a long time. Hybrid performance depends on two key factors: heterosis (hybrid vigor) and specific parental effect. A discussion of hybrids as related to drought resistance has been presented in Sect. 3.3.1. Here only the final conclusions are drawn with respect to practice.

When hybrid breeding and heterosis are viewed in perspective it can be seen that the basic and primary heterotic effect per se for yield is finite and modern hybrid development relies heavily on continuous small improvements in parent performance and the identification of better specific or general parental combination. Fairly extensive research has been performed on heterosis in the past in public institutions in order to resolve the genetic and physiological basis of the phenomenon. While progress has been made on some important issues, most research was directed at the manifestations of the phenomenon rather dissecting its functional basis. Heterosis still remains an enigma. Today most of hybrid research and development is performed in the seed industry and for obvious reasons it is largely behind closed doors. Our present knowledge is based mainly on historical data and the current limited information. Maize, sorghum and pearl millet are by and large the main crops where hybrids are developed towards rainfed and water limited environments. Extensive and impressive work is being performed on hybrid rice in China but it is mostly directed towards irrigated conditions, at this time.

There has been a long debate on relative expression of heterosis under stress and non-stress conditions (Sect. 3.3.1). It seems that heterosis might have a specific benefit under non-optimal temperatures. There is strong evidence that heterosis in certain tropical maize materials is expressed in effective use of water and improved plant water status, probably as a result of large roots in the hybrids (Araus et al. 2010). At the same time there is strong evidence provided throughout this book that parental effect is crucial for hybrid performance under drought stress. This has already been observed by Jenkins (1932) when drought occurred in his maize breeding plots in 1930. He noted that "The 10 crosses of one of the lines were completely free from leaf burning, whereas those of another line ranged from some to many plants with burned leaves in the different crosses." This statement together with all the evidence produced since then allows concluding that that having one or two drought resistant parents is the main condition for achieving a drought resistant hybrid. In singular cases (e.g., Araus et al. 2010; Castiglioni et al. 2008; Yadav 2008, 2010) there was an indication that a recombination of parental drought resistance with high rate of heterosis for yield (namely high yield potential) might be possible. This allows concluding this chapter in an optimistic mood regarding a possible break from the crossover interaction for yield normally existing for drought resistant cultivars.

References

Annicchiarico P (2002) Genotype X environment interactions: challenges and opportunities for plant breeding and cultivar recommendations. Food and Agriculture Organization of the United Nations, Rome

Araus JL, Sánchez C, Cabrera-Bosquet L (2010) Is heterosis in maize mediated through better water use? New Phytol 187:392–406

Campos H, Cooper M, Habben JE et al (2004) Improving drought tolerance in maize: a view from industry. Field Crops Res 90:19–34

Castiglioni P, Warner D, Bensen RJ et al (2008) Bacterial RNA chaperones confer abiotic stress tolerance in plants and improved grain yield in maize under water-limited conditions. Plant Physiol 147:446–455

Chapman SC, Cooper M, Podlich D et al (2003) Evaluating plant breeding strategies by simulating gene action and dryland environment effects. Agron J 95:99–113

Chenu K, Chapman SC, Tardieu F et al (2009) Simulating the yield impacts of organ-level quantitative trait loci associated with drought response in maize: a "gene-to-phenotype" modeling approach. Genetics 183:1507–1523

Hammer G, Cooper M, Tardieu F et al (2006) Models for navigating biological complexity in breeding improved crop plants. Trends Plant Sci 11:587–593

He YX, Zheng TQ, Hao XB et al (2010) Yield performances of japonica introgression lines selected for drought tolerance in a BC breeding programme. Plant Breed 129:167–175

Jenkins MT (1932) Differential resistance of inbred and crossbred strains of corn to drought and heat injury. Agron J 24:504–506

Saint Pierre C, Crossa J, Manes Y et al (2010) Gene action of canopy temperature in bread wheat under diverse environments. Theor Appl Genet 120:1107–1117

Sinclair TR, Messina CD, Beatty A et al (2010) Assessment across the United States of the benefits of altered soybean drought traits. Agron J 102:475–482

Spitters CJT, Schapendonk AHCM (1990) Evaluation of breeding strategies for drought tolerance in potato by means of crop growth simulation. Plant Soil 123:193–203

Tardieu F, Tuberosa R (2010) Dissection and modeling of abiotic stress tolerance in plants. Curr Opin Plant Biol 13:206–212

Verulkar SB, Mandal NP, Dwivedi JL et al (2010) Breeding resilient and productive genotypes adapted to drought-prone rainfed ecosystem of India. Field Crops Res 117:197–208

Weiss A (2000) Introduction. Agron J 95:1–3

Yadav OP (2008) Performance of landraces, exotic elite populations and their crosses in pearl millet (*Pennisetum glaucum*) in drought and non-drought conditions. Plant Breed 127:208–210

Yadav OP (2010) Drought response of pearl millet landrace-based populations and their crosses with elite composites. Field Crops Res 118:51–56

Chapter 7
Epilogue

Historically, plant breeding has been effective in coping with water limited environments. This is well represented by the viable field crop production industry in the semi arid dryland regions and in the scientific achievements in this discipline. In the latter part of the twentieth century plant and crop physiology were very active in this respect. This could be seen in the adoption of physiological selection criteria and the evolvement of valid drought resistant physiological ideotypes which could be used by the breeder for planning his program. Crop physiology has often been criticized in the past as offering explanations only after the fact. This book presents numerous cases where crop physiology performed in the forefront of breeding and contributed significantly to breeding programs for water limited environments, before the fact. This can be seen in topics such as crop transpiration and carbon isotope discrimination, remote sensing and canopy temperature, stem reserve utilization, osmotic adjustment, root development and more.

Throughout the last several decades genetic resources for drought resistance were found to be more prevalent and usable than previously expected. While landraces were used successfully in breeding for water limited environments it was also realized that the run of the mill agronomic breeding germplasm contained important and sometimes latent genetic variation for drought resistance.

This book did not discuss extensively plant genomics in relations to drought resistance. The literature is rich with such texts ranging from edited books to frequent literature reviews and conference proceedings. Genomics was addressed in this book only by focusing on how it can overcome past mistakes in its approach to breeding and how it can contribute to future progress in plant breeding for water limited environments. This book addressed only those aspects of genomics and molecular biology which at the time when this text was written were ripe for application to breeding work. The potential of genomics towards plant breeding for water limited environments is amply discussed elsewhere.

The application of molecular techniques and bioinformatics towards improving drought resistance was initially attempted in isolation of plant breeding and plant physiology, which did not provide any progress in the field. At the same time breeding programs were cut back and the number of plant breeders in the field diminished. The inability of genomics to fulfill its past promise grew between these two

A. Blum, *Plant Breeding for Water-Limited Environments*,
DOI 10.1007/978-1-4419-7491-4_7, © Springer Science+Business Media, LLC 2011

failing trends. It is now realized also by the genomics community that results will be achieved only through close collaboration. Nakashima et al. (2009), a leading stress genomics research group in Japan concluded: "The collective and cooperative efforts of plant molecular biologists, physiologists, and breeders are required to generate stress-tolerant grasses through genetic engineering. It is hoped that, in the future, the collective efforts and results of collaborative studies will positively contribute to sustainable food production in the world and will help to prevent global-scale environmental damage that results from abiotic stress."

This book may have been overly critical of certain genomics research related to plant drought stress. This has been done with the positive intention to provide the necessary feedback and required education in order to enhance real contribution of genomics scientists and students to plant breeding for water limited environments. The initial achievements towards water limited environments are now emerging as a result of collaborations. It is no surprise that the private seed industry is making better stride in this respect, partly because management can enforce collaboration. This message should be acted upon by those sponsors of research who for long supported plant stress genomics research in isolation of any consideration for application, including institutions which under their mandate should have considered delivery to the farmer.

The general debate on agricultural adoption of transgenic plants in terms of science and regulation will continue. Transgenic plants developed into a hot social and political issue. It becomes clear that most countries with a full belly will continue to reject transgenic plants, especially those who anyway blame their farmers as evil doers. Countries which are still in pursuit of sufficient food supply for their people or are concerned with farmers' livelihood will adopt transgenic plants provided their politicians and media will be educated. This technology as part of a breeding program proved that it can bring about important progress in coping with water limited environments. Regulatory systems and checks are now well in place and are operational.

Plant environmental stress is the rule. Non-stress conditions are the exception. In the past this has rarely been grasped by decision makers and politicians. Only now, with the sudden explosion of the climate change dilemma it is publicly and politically realized that plants need to cope with abiotic stress. This is no news to the farmer and the agricultural scientist. Coping with plant abiotic stress has always been an important part of agricultural research, directly or indirectly. As the problem of water scarcity increases with urbanization and use in agriculture and as new land enter into crop production with limited resources the need for breeding programs towards limited water supply is becoming a clear reality, irrespective of climate change. More countries, CGIAR centers and NARS began expanding breeding solutions for dry conditions even before global warming became a hot agenda. The up side is that now agriculture research and plant stress research are being awarded greater support.

While funding and capacity building will be made available under this new mood of urgency, the required intellectual resources are not as abundant. After all has been said and done it remains that the key to progress in plant breeding for water limited environments is the well informed and intelligent plant breeder, linked to his associates in the ancillary disciplines. This human resource is not as abundant as we might expect. We are now experiencing the damage done by hurting education in these disciplines during the 1990s.

It has been my experience that in most cases of insufficient progress in breeding, education, training and accumulated knowledge was the major shortcoming and not necessarily always the lack of capacity building. I have come across individual plant breeders in certain developing countries who were knowledgeable, intelligent and hard-working enough to achieve progress with very limited resources.

Plant breeders and agronomists were always outstanding in their broad education in related disciplines such as plant physiology, soil science, climatology, plant pathology etc. With the knowledge explosion taking place in these modern times, the broad education required for the agricultural scientist is being compromised.

We certainly need dams, irrigation projects, soil conservation, mechanization, social programs, crop insurance programs and even satellites. However, at the end of the day, we always come back to the amazing solutions available in the DNA of the seed. Education in plant breeding for water limited environments must be placed with all other important disciplines in the forefront of higher education in agriculture in the twenty-first century. This book was written as a contribution towards that end.

Reference

Nakashima K, Ito Y, Yamaguchi-Shinozaki K (2009) Transcriptional regulatory networks in response to abiotic stresses in *Arabidopsis* and grasses. Plant Physiol 149:88–95

Index